LANCASTER

REAPING THE WHIRLWIND

MARTIN W. BOWMAN

LANCASTER
REAPING THE WHIRLWIND

The
History
Press

First published 2012

The History Press
The Mill, Brimscombe Port
Stroud, Gloucestershire, GL5 2QG
www.thehistorypress.co.uk

British Library Cataloguing in Publication Data.
A catalogue record for this book is available from the British
Library.

ISBN 978 0 7524 6467 1

Typesetting and origination by The History Press
Printed in India
Manufacturing managed by Jellyfish Print Solutions Ltd

CONTENTS

Acknowledgements   6

Milestones   7

1   Not a Cloud in the Sky   9

2   The Shining Sword   27

3   Old Man Luck   58

4   Chop City   81

5   Round-the-Clock   94

6   'This is war and somebody's got to die!'   121

7   The Means of Victory   154

8   Lancaster Legacy   182

Appendix 1   Bomber Command Lancaster Crew Victoria Cross Recipients   204

Appendix 2   Aircraft Sorties and Casualties 3 September 1939–7/8 May 1945   204

Appendix 3   Summary of Production   205

Appendix 4   Lancaster I Specifications   205

Appendix 5   Halifax/Lancaster Comparison at the End of 1943   206

Appendix 6   Comparative Lancaster and Halifax Squadrons   206

Appendix 7   Lancaster Squadrons at Peak Strength 1 August 1944   206

Appendix 8   Lancaster Squadrons formed late 1944–45   207

Notes   208

Index   212

ACKNOWLEDGEMENTS

I am particularly grateful to Dick Starkey; 'Johnny' Johnston and Eric Jones DFC; Derek Thomas, Secretary of 106 Squadron Association; Nigel McTeer; Alan Parr, Secretary of 49 Squadron Association; Philip Swan, who edited and annotated Campbell Muirhead's *Diary of a Bomb Aimer: Training in America and Flying with 12 Squadron in WWII*; and my friend and colleague Theo Boiten, with whom I have collaborated on several books, who provided all of the information on the *Nachtjagd* or German night-fighter forces contained herein. Aviation historians everywhere owe a deep sense of gratitude to his and all the other valuable sources of reference; in particular, those by the incomparable W.R. 'Bill' Chorley, Harry Holmes, Martin Middlebrook, Chris Everitt and Oliver Clutton-Brock.

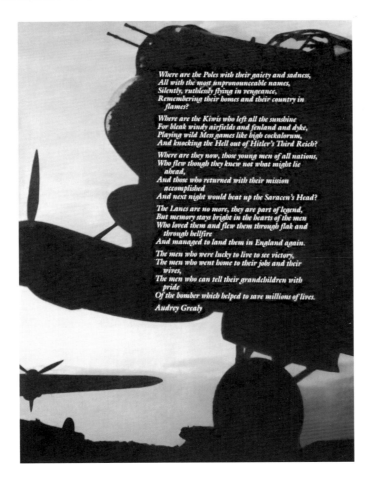

Where are the Poles with their gaiety and sadness,
All with the most unpronounceable names,
Silently, ruthlessly flying in vengeance,
Remembering their homes and their country in
    flames?

Where are the Kiwis who left all the sunshine
For bleak windy airfields and fenland and dyke,
Playing wild Mess games like high cockalorum,
And knocking the Hell out of Hitler's Third Reich?

Where are they now, those young men of all nations,
Who flew though they knew not what might lie
    ahead,
And those who returned with their mission
    accomplished
And next night would beat up the Saracen's Head?

The Lancs are no more, they are part of legend,
But memory stays bright in the hearts of the men
Who loved them and flew them through flak and
    through hellfire
And managed to land them in England again.

The men who were lucky to live to see victory,
The men who went home to their jobs and their
    wives,
The men who can tell their grandchildren with
    pride
Of the bomber which helped to save millions of lives.

Audrey Grealy

Lancaster and poem

MILESTONES

May 1936	Specification P.13/36 for a twin-engined bomber results in the Vulture-engined Avro Manchester designed by Roy Chadwick and his team.
July 1937	Manchester is ordered into production to specification.
July 1939	Manchester flies for the first time.
September 1939	Manchester III project with four Merlins first mooted.
Mid-1940	Avro is authorised to go ahead with the construction of a Manchester III prototype. The modified wing of increased span and the two extra engines results in a change of name from Manchester to Lancaster.
November 1940	Manchester enters squadron service with 5 Group, Bomber Command.

1941

9 January	First Avro 683 Lancaster (BT308) flies as the Manchester III and is originally fitted with triple fins.
13 May	Second prototype Lancaster (DG595) flies.
31 October	First true production Lancaster I (L7527) fitted with Merlin XX engines of 1,280hp flies. Lancaster Is continue in production until 1946.
21 December	Lancaster II prototype powered by American Packard-built Merlin engines flies.

1942

Early 1942	Deliveries of the first Lancaster Is to 44 Squadron at Waddington, followed by 97 Squadron at Woodhall Spa, begin.
3/4 March	Four aircraft of 44 Squadron fly the first Lancaster operation of the war with a mine-laying sortie in the Heligoland Bight.
10/11 March	Two aircraft of 44 Squadron make the first Lancaster night operation with a raid on Essen.
10/11 April	Lancasters drop the first 8,000lb bomb on Essen.
17 April	Twelve aircraft of 44 and 97 Squadrons, led by S/L J.D. Nettleton, carry out a low-level daylight attack on the MAN diesel engine plant at Augsburg. Nettleton is awarded the Victoria Cross for his part in the raid.
25/26 June	Last raid is flown operationally by Manchesters, after which the type is withdrawn from operations.
31 July/1 August	One hundred-plus Lancasters take part in a raid for the first time when Düsseldorf is bombed.
18/19 August	Lancasters of 83 Squadron take part in the first operation by the Pathfinder Force, with a raid on Flensburg.
17 October	Over 90 Lancaster Is of 5 Group bomb the Schneider Works at Le Creusot, a trip which involves up to ten hours' flying. Only one aircraft is lost.

1943

11/12 January	Two initial sorties by Mk.IIs against Essen prove abortive as they cannot reach the required operational ceiling, but five nights later the type goes into action successfully against Berlin.
4/5 February	Lancaster IIs are given an Italian target – Turin.
Spring	115 Squadron at East Wretham, Norfolk, in 3 Group and formerly flying Wellingtons, becomes first Lancaster II squadron.
March	Eighteen squadrons using Lancaster Is.
16/17 May	Eighteen Lancasters of 617 Squadron carry out a low-level attack on the Ruhr Dams. Eight aircraft are lost. W/C Gibson is awarded the VC.
23/24 May	Greatest non-1,000 raid of the war with 826 aircraft, including 343 Lancasters, on Dortmund.
20 June	Friedrichshafen bombed by 57 and 97 Squadrons, the Lancasters flying on to North Africa to refuel.
July	One hundred and thirty-two tons of bombs are dropped for every Lancaster lost on operations. This compares with only 56 tons for each Halifax lost and 41 for each Stirling.

| 15/16 September | Two Lancasters of 617 Squadron drop the first 12,000lb bombs during the disastrous low-level raid on the Dortmund-Ems Canal. |
| 11 November | First Lancaster to reach a civil register when G-AGJI is registered to BOAC to evaluate for airline use, which leads to the development of the Lancastrian, a 9–12 seat airliner. |

1944

15/16 February	More than 500 Lancasters take part in a raid for the first time when Berlin is bombed.
8 June	First deep penetration 12,000lb 'Tallboy' bombs are dropped by 19 Lancasters of 617 Squadron on the tunnel at Saumur.
June	Lancasters fly daylight tactical operations in support of the invasion forces around Caen in the days immediately following D-Day.
September	Last-ever batch of Lancasters (11 Lancaster IIIs) produced at Yeadon in West Yorkshire; the last delivered in October 1945. Most are used for ASR (Air/Sea Rescue) or for GR (General Reconnaissance).
12 November	Eighteen Lancasters of 617 Squadron and 13 of 9 Squadron drop 12,000lb 'Tallboy' bombs to capsize the 45,000-ton battleship *Tirpitz* in Tromsø fjord in Norway.

1945

17 January	Prototype Lancastrian (G-AGLF) flies and is delivered from Woodford to BOAC at Croydon on 18 February 1945. BOAC uses Lancastrians on long-range routes until September 1950.
March	No fewer than 56 squadrons of Lancasters operate on first-line duty (745 operational Lancasters and 296 more in OTUs).
14 March	First 22,000lb 'Grand Slam' bomb to be dropped, on the Bielefeld Viaduct, by a 617 Squadron Lancaster.
21 March	Fifty-six squadrons and 745 Lancasters on strength with 296 in OCUs.
25 April	Lancasters fly their last daylight operation of the war with a raid on Hitler's mountain retreat at Berchtesgaden.
25/26 April	Lancasters fly their final night raid of the war with an attack on an oil target at Vallø, Norway.
October	Statistics indicate that the total tonnage of high-explosive (HE) bombs dropped on primary targets by 6,500 of the 7,366 Lancasters built reached 608,612 in 156,192 sorties. Packed together, the bombs could fill a goods train 345 miles long. Lancasters accounted for two-thirds of the total tonnage dropped by the whole of Bomber Command from March 1942–May 1945. The average of 4 tons per bomber equalled 150,000 sorties using a total of 228 million gallons of fuel. Total incendiaries dropped by Lancasters: 51,513,106. Comparing the bomb tonnage dropped with aircraft lost, Stirlings amounted to 41 tons, Halifaxes to 51 tons and Lancasters to 132 tons. Between February 1942 and May 1945, 3,431 Lancasters and approximately 24,000 air crew failed to return from operations.

1945 and 1946

	Two Lancaster B.Is (PD328 Aries of the Empire Air Navigation School, Shawbury) and (PB873 Thor of the Empire Air Armaments School, Manby) make round-the-world and trans-polar flights.
2 February 1946	Last Lancaster, a Mk.I (FE), enters service. Some modified Lancaster B.Is continue serving on bomber squadrons until 1949–50 because of delays in Lincoln development and delivery.
9 July 1946	Lancaster B.I (FE)s of 35 Squadron set out on a goodwill tour of the USA.

1948

| | Two Lancasters operated by Flight Refuelling Ltd ferry petrol during the Berlin Airlift. Ten other Lancastrians are also used by Flight Refuelling Ltd as part of the air operation after the Soviets cut off land links with British- and American-occupied areas of West Germany. The Berlin Airlift lasts from 24 June 1948–11 May 1949. During the winter 6,000 tons of coal and petrol a day is delivered to the beleaguered city as well as 1,500 tons of food. Avro Lancasters and the Lancaster variants – Lancastrians, Yorks and Tudors – are among the British aircraft involved. |

1953 to 1956

December 1953	Last Lancaster (a PR.I) is retired from Bomber Command.
February 1954	Last Lancaster in RAF service, an MR.3, is retired after service in Malta.
15 October 1956	Farewell RAF ceremony at St Mawgan, Cornwall, as Lancaster RF325 at the School of Maritime Reconnaissance is retired and flown to Wroughton in Wiltshire to be scrapped.

NOT A CLOUD IN THE SKY

The Lancaster, coming into operation for the first time in March 1942, soon proved immensely superior to all other types in the Command. The advantages it enjoyed in speed, height and range enabled it to attack with success targets that other types could attempt only with serious risk or even certainty of heavy casualties. Suffice it to say that the Lancaster, in no matter what terms, was incomparably the most efficient of our bombers. In range, bomb-carrying capacity, ease of handling, freedom from accident and particularly in (low) casualty rate, it far surpassed the other heavy types.

Sir Arthur Harris

On Christmas Eve 1941 three ex-works Avro Lancasters arrived at Waddington – a magnificent Christmas present for 44 Squadron, which in September had become 'Rhodesia' Squadron because of the considerable number of Rhodesian volunteers now serving on it, many awaiting air-crew training. 'The Rhodesians – with their outlandish appearance and air of tough independence, skins bronzed by a warmer sun than England's and eyes used to wider distances – were so unmistakably not English,' recalled Pip Beck, a WAAF. 'I liked their easy, pleasant manner and lack of formality and soon came to know them well. The men came from places with strange musical names. Shangani, Umtali, Gwelo, Selukwe – Que-Que, Bulawayo – one could almost make a song from them, I thought.' It was with intense interest that everyone in Flying Control watched the Lancasters' approach and landing. As the first of the three taxied round the perimeter to the Watch Office, Pip Beck stared in astonishment at this 'formidable and beautiful' aircraft, cockpit as high as the balcony on which she stood, and the great spread of wings with four enormous engines. 'Its lines were sleek and graceful,' she purred:

yet there was an awesome feeling of power about it. It looked so right after the clumsiness of the Manchester, from which its design had evolved. Their arrival meant a new programme of training for the air and ground crews and no operations until the crews had done their share of circuits, bumps and cross-countries and thoroughly familiarised themselves with the Lancasters. There were one or two minor accidents at this time; changing from a twin-engined aircraft to a heavier one with four engines must have

presented *some* difficulties – but the crews took to them rapidly. I heard nothing but praise for the Lancs.

Snow blanketed the airfield and blew into great drifts. It was very beautiful to look out on from the Control room, wearing my loaned flying boots – but less enjoyable to walk through when the blizzard raged. When it stopped, all hands were called out to help clear the runways and it was an amazing sight to see hundreds of airmen, aircrew and some WAAFs shovelling away until well into the dusk to free the main runway. Between the efforts of the snowploughs and the toiling shovellers, piles of snow lay by the sides of the runway and the job was done. A little flying took place and two of the precious Lancs suffered a small amount of damage, though not serious, in spite of the dreadful weather conditions.

The weather eased up a little and one day the airfield seemed to be overrun with boys from the Lincoln Air Training Corps. I watched enviously as they had trips in a Lanc. If they could have this marvellous opportunity, why couldn't I? It was expressly forbidden for WAAF to go up in operational aircraft. The boys were wildly enthusiastic, of course ...[1]

The Rhodesian Squadron would suffer the heaviest overall losses in 5 Group and the heaviest Lancaster losses and highest percentage of Lancaster losses both in 5 Group and in Bomber Command. But they operated Lancasters longer than anyone else and they were the only squadron with continuous service in 5 Group.

Coinciding with the introduction of the Lancaster into squadron service was the arrival on 22 February 1942, at Bomber Command Headquarters at High Wycombe in Buckinghamshire, of a new AOC-in-C, Air Chief Marshal Arthur

Roy Chadwick, Chief Designer of the Lancaster, in his office at Avro. Born in Farnworth near Bolton in 1893, Chadwick began work in the drawing office in 1911, and in 1936 he designed the Manchester twin-engined bomber to specification P.13/36. When, by mid-1940, he knew that the bomber would not prove successful, he instructed his design staff to convert the Type 679 to a four-engined bomber using either Rolls-Royce Merlins or Bristol Hercules radials and the Lancaster was born. Unlike Reginald Mitchell, designer of the equally illustrious Spitfire, Chadwick witnessed the fruits of his endeavours during the war but he was killed on 23 August 1947 in the crash of the Avro Tudor II during a test flight.

Travers 'Bomber' Harris CB OBE, who was recalled from the USA where he was head of the permanent RAF delegation. Harris was directed by Marshal of the RAF Sir Charles Portal, Chief of the Air Staff, to break the German spirit by the use of night area rather than precision bombing and the targets would also be civilian, not just military. The famous 'area bombing' directive, which had gained support from the Air Ministry and Prime Minister Winston Churchill, had been sent to Bomber Command on St Valentine's Day, 14 February, eight days before Harris assumed command.

Harris did not possess the numbers of aircraft necessary for immediate mass raids. On taking up his position he found that only 380 aircraft were serviceable. Of these, only 68 were heavy bombers, while 257 were medium bombers. Another of the new generation RAF bombers, the Manchester, had been suffering from a plague of engine failures and was proving a big disappointment. During March, 97 'Straits Settlements' Squadron moved the short distance from Coningsby to Woodhall Spa, south-east of Lincoln, to become the second squadron to begin conversion from the Manchester to the Avro Lancaster. On 36 raids with Manchesters, 97 Squadron had lost eight aircraft from 151 sorties. Re-equipment would take time and early in 1942 deliveries began to trickle through. In January–February, 44 Squadron's Lancasters and their crews spent a frustrating time standing by to fly to Wick in Scotland to refuel and take off again to sow mines at the mouth of a

Norwegian fjord to prevent the *Tirpitz* from sailing, but the weather worsened and the Lancasters remained on the ground. On 23 February, the Lancasters were again loaded up with mines but the aircraft stood by all the next day and then were stood down until 1 March. Finally, on the evening of 3 March, with AVM John Slessor, the 5 Group commander, there to watch them, four aircraft led by S/L John Dering Nettleton, the South African CO of 44 Squadron, took off and flew the first Lancaster operation when they dropped mines in the Heligoland Bight. All the Lancasters returned safely. That same night the Main Force destroyed the Renault factory at Billancourt, near Paris. Just one aircraft (a Wellington) was lost. During March also, the first 'Gee' navigational and target-identification sets were installed in operational bombers; these greatly assisted aircraft in finding their targets on the nights of 8/9 and 9/10 March in attacks on Essen. Without 'Gee' these had been a difficult target to hit accurately. Just two Lancasters took part in the second of the Essen raids because on 8 March eight of 44 Squadron's Lancasters had been ordered to Lossiemouth for a possible strike on the *Tirpitz* near Trondheim. 'A Naval convoy escort fired at the Lancs as they flew over it,' recalls Pip Beck. 'A Spitfire next attacked the Lancs until frantic firing of the colours of the day convinced its pilot that they were not enemy aircraft.' Luckily no damage or casualties had resulted beyond irritated and indignant air crew. 'Oh well – the bloody Navy doesn't know its aircraft from its elbow! But they expected better from their own Service. If words could kill, the Spit would have gone down in flames.'

By command of the *Oberbefehlshaber der Luftwaffe*, the staff of the *Luftwaffenbefehlshaber Mitte* (*Luftwaffe Reich* or German Air Force Command Centre) had been set up in March 1941, and the following month the night-fighter division, under Generalmajor Josef Kammhuber, was placed in its command. In Holland and the Ruhr 6–10/10ths cloud was quite customary, so Kammhuber therefore concentrated his energies on the development of an efficient Ground Controlled Interception (GCI) technique – *Dunkle Nachtjagd* – later called Himmelbett (literally 'bed of heavenly bliss', or 'four-poster bed', because of the four night-fighter control zones). Kammhuber arranged his GCI positions in front of the searchlight zones and encouraged crews to attempt interception first under ground control; if that failed, the searchlights were then at their disposal. By the winter of 1941–42, the Kammhuber line was complete. The night-fighter formations, equipped mostly with Bf 110s, were stationed almost entirely in Holland, Belgium and north-western Germany. The completion of night-fighter bases for controlled Himmelbett fighting, with two giant Würzburgs and one Freya radar, progressed well and was planned to cover initially the north German coastal area and Holland, and later Belgium.

The first Lancaster night-bombing operation on a German target was inauspicious. Only 62 of the 126 Main Force crews dispatched claimed to have bombed Essen, which was obscured by unforecast cloud and industrial haze. Both of 44 Squadron's Lancasters returned safely, though one was hit by flak and the other had to land back at Docking in Norfolk. On the night of 13/14 March 1942, a single Lancaster piloted by Sgt George 'Dusty' Rhodes joined 61 Wellingtons, 13 Hampdens, ten Stirlings, ten Manchesters and nine Halifaxes in bombing Cologne. Rhodes got into difficulties on the return and landed without runway lights, overshooting the airfield. On 20 March, the second Lancaster squadron, No 97, flew their first operation with six aircraft out mining along the coast of Ameland and the Friesian Islands. One of the Lancasters machine-gunned a hotel and a party of soldiers for good measure and then climbed into cloud when a Bf 109 was spotted. All the Lancasters returned but one crashed near Boston and another landing at Abingdon crashed owing to the soft state of the ground. Two others landed safely at Upper Heyford and one at Bicester. On another mining operation, off Lorient on the night of 24/25 March, the first Lancaster casualties occurred when R5493 failed to return. A 420 Squadron RCAF pilot reported that he had seen a four-engined bomber heavily engaged with anti-aircraft (AA) fire over Lorient and the Royal Observer Corps reported flares out to sea. A search was made but no trace of South African F/Sgt Lyster Warren-Smith's crew was found.

On 25/26 March, seven Lancasters were included in the force of 254 aircraft, making it the largest force sent to one target so far, when Bomber Command attacked Essen again. Over 180 crews claimed to have bombed the city but the flare-dropping was too scattered and only nine HE bombs and 700 incendiaries fell on target. All the Lancasters returned safely but nine aircraft, including five Manchesters, were lost. Lancasters took no part in further operations until the night of 8/9 April, when the main Bomber Command operation was to Hamburg and 272 aircraft were dispatched, yet another record raid for aircraft numbers to one target. Over 170 Wellingtons and 41 Hampdens made up the bulk of the force, which included just seven Lancasters. That same night 24 Lancasters on 97 Squadron

Post-raid reconnaissance photo of the 17 April 1942 raid on Augsburg when 12 aircraft of 44 and 97 Squadrons, led by S/L J.D. Nettleton, carried out a low-level daylight attack on the MAN diesel engine plant. Nettleton was awarded the VC for his part in the raid.

carried out a mine-laying operation in the Heligoland Bight. Two nights later, eight Lancasters were included in the force of 254 aircraft that visited Essen again. Fourteen aircraft were lost but the Lancasters all returned safely.

Visiting a dispersal at Waddington, Pip Beck noticed on a neighbouring hard-standing a Lancaster with different engines.

'They,' I was told, 'are *radial* engines – it's a Mark II Lanc. Our Mark Is have *in-line* engines which, as you know, are Merlins. The Mark IIs are Bristol Hercules Is.' I was given some more technical detail, but that didn't stick. However, at least I could recognise a Lanc with radial engines and say, knowledgeably, 'Ah yes – a Mark II Bristol Hercules!' and feel rather smug.

During April, a suspicion grew in Control that something big was being laid on for sometime in the near future and 44 would not be the only squadron involved. There had been frequent visits from S/L Sherwood and F/L Penman on 97 Squadron at nearby Woodhall Spa. Low-level cross-country flight practices were taking place and we wondered what they presaged. On April 17th we found out.

On 11 April, 44 Squadron had been ordered to fly long-distance flights in formation to obtain endurance data on the Lancaster. At the same time 97 Squadron began flying low in groups of three in 'vee' formation to Selsey Bill, then up to Lanark, across to Falkirk and up to Inverness to a point just outside the town, where they feigned an attack, and then back to Woodhall Spa. Crews knew the real reason was that they were training for a special daylight operation, and speculation as to the target was rife. S/L John Nettleton was chosen to lead the operation. During the morning of 17 April he and S/L John Sherwood DFC*, the 97 Squadron CO, and selected officers were briefed that the objective was the diesel engine manufacturing workshop at the MAN (*Maschinenfabrik Augsburg-Nürnberg Aktiengesellschaft*) factory at Augsburg in southern Bavaria. P/O Patrick Dorehill, Nettleton's 20-year-old second pilot, wrote:

There was certainly some surprise on entering the briefing room to see the pink tape leading all the way into the heart of Germany. I can't say I felt anxious. I had an extraordinary faith in the power of the Lancaster to defend itself. And then flying at low level seemed to me to be the perfect way to outwit the enemy. I thought the only danger might be over the target and, even there, believed we would be in and away before there was much response.

In the late afternoon a force of 12 Lancasters, flying in two formations of six, set out to attack the MAN works. The Lancasters flew very low. The first flight, led by Nettleton, ran into German fighters when well into France. In the battle which ensued, four of the Lancasters were shot down. Nettleton flew on towards Augsburg. When they got near the target the light flak was terrific and another Lancaster was hit and crash-landed 2 miles west of Augsburg.

Patrick Dorehill continued:

It was only sheer bad luck that we flew past an enemy airfield to which their fighters were returning from the diversionary raids our fighters and Boston bombers had laid on to the North. Up they came and I shall never forget those terrible moments. I do not think there were as many fighters as our gunners reported; it was just that each made several attacks which made it seem like more. Being on the jump seat I stood up and saw quite a bit of the action. Maybe there were a dozen. At any rate I looked back through the astrodome to see Nick Sandford's plane in flames. He always wore his pyjamas on ops under his uniform. He thought it would bring him good luck.

This was followed by Dusty Rhodes' plane on our starboard catching fire. The rest went down except Garwell on our port side. There was nothing for it really but to press on. A passing thought was given to turning south and then out to the Bay of Biscay but we reckoned that as we had come so far we might as well see it through. By this time I can tell you I didn't give much for our chances. On we went and I marvelled at the peaceful countryside, sheep, cattle, fields of daisies or buttercups. Along came the Alps on our right, wonderful sight, Lake Constance looking peaceful. We had climbed up a bit by then, it being pretty hilly, and then down we came again getting close to the target. My recollection may be faulty but I thought we approached Augsburg from the south, following a canal or railway, factory chimneys appeared on the low horizon and then we came to the town. Large sheds were right in our path; Des Sands, the navigator, and McClure, the bomb aimer, had done a pretty good job of map reading.

Bombs away at about a hundred feet.

The flak zipped past and as we crossed the town to begin a left turn for home a small fire was apparent, gradually gaining strength, in Garwell's plane. Our gunners saw it make a crash landing, which seemed to go relatively well.

The trip home was uneventful, thank goodness … Nettleton did a brisk circuit and down we came to be almost out of fuel. Golly, I can tell you I was glad to feel those wheels touch the grass.[2]

The second formation of six, led by S/L Sherwood, encountered no fighters. F/L David Penman DFC, piloting *U-Uncle*, recalled:

Rising ground forced us a little higher and then we saw the final turning point, a small lake. At this stage, mindful of the 11-second delay fuses, I had dropped back a little from Sherwood's section and made one orbit before running in to attack. The river was a very good guide and it all showed up as predicted on the scale model. A column of smoke beyond the target came, presumably, from Garwell's aircraft and it was quickly joined by another as Sherwood received a shell through the port tank just behind the inboard engine. Escaping vapour caught fire and as he passed over the target he began to turn left. His port wing struck rising ground and the aircraft exploded in a ball of flame. I was convinced that no one could have survived and on my return reluctantly told Mrs Sherwood. She would not believe it and events proved her right. I met him again after the war; he had been thrown, complete with his seat, through the windscreen as the aircraft struck the ground – the only survivor.

As we ran in at 250ft tracer shells from light AA on the roofs of the buildings produced a hail of fire and all aircraft were hit. W/O Tommy Mycock on my left received a shell in the front turret which set fire to the hydraulic oil within seconds. The aircraft was a sheet of flame. It reared up and turned right, passing right over my head with its bomb doors fully open, before plunging into the ground, burning from end to end. A shell ripped the cowling from my port inner and F/O Deverill received a hit near the mid-upper turret at the same time which started a fire. Despite these distractions we held course, with my front gunner doing his best to reduce the opposition. Ifould, my navigator, was then passing instructions for the bomb run. As he finally called 'Bombs gone' we passed over the factory. I increased power and dived as Deverill passed me with one engine feathered and the remaining three flat out. I called him and he asked me to cover his rear as his turrets were out of action. Ours had been unserviceable since the Channel and as we had no wish to relinquish the navigation, I told him to remain in position.

Our attack had been close to the planned time of 20:20 hours and as darkness came over we climbed to 20,000ft for a direct run home over Germany. It says much for Deverill's skill that he remained in position until we reached the English coast and finally landed at Woodhall Spa. All surviving crews were grounded on return until after a press conference at the Ministry of Information in London.

Only five of the total force of 12 Lancasters returned, but eight of the 12 had bombed the target. Five of the delayed-action bombs had failed to explode. The others caused substantial damage, but the effect on production was slight, particularly since at least five of the MAN factory's licensees were building U-boat diesel engines at that time. In all, 85 air crew took part in the raid; 37 air crew, of whom 12 became POWs, failed to return. Nettleton, who landed his badly damaged Lancaster at Squires Gate, Blackpool, ten hours after leaving Waddington, was awarded the Victoria Cross. Sherwood, though recommended for a VC, was awarded the DSO. Early in 1943, Nettleton was promoted to wing commander and given command of a squadron, but on the night of 12/13 July he lost his life during a raid on Turin.

On 21/22 December 1942, P/O W.J. Dierkes RCAF and Lancaster B.I R5699 on 61 Squadron returned early from the raid on Düsseldorf due to a failure in the intercom system but got caught in a severe downdraft when the pilot tried to land at Syerston and crashed. There were no injuries to the crew.

In March and April the Ruhr suffered eight heavy raids in which 1,555 aircraft took part, beside three small attacks. In the same period Cologne was visited four times by a total of 559 aircraft. By the end of April about 75,000sq yd occupied by workshops in the Nippes industrial district had been damaged. Heavy bombs had completely destroyed buildings nearby covering an area of 6,000sq yd. The Franz Clouth rubber works, covering 168,000sq yd, had been rendered useless, much of them being razed to the ground. To the east of the Rhine a chemical factory and buildings beside it occupying 37,500sq yd was almost entirely destroyed. Severe damage had also been caused to the centre of the city. All this was confirmed by the evidence of photo reconnaissance. Twice Dortmund was heavily bombed – on 14/15 April and again on the next night – a group of factories in the Wiesenberger Strasse being extensively damaged. Hamburg endured five raids, those on 8/9 April and the 17th/18th being especially severe. The raid on Cologne on the 22nd/23rd by 64 Wellingtons and five Stirlings, all equipped with 'Gee' for blind-bombing, was largely experimental. Over 40 HE bombs and more than 1,200 incendiary bombs were dropped on the city – perhaps 12–15 aircraft loads – but others fell up to 10 miles away.

Ever since the successful raid on Lübeck in March, Harris had been keen to mount another series of fire raids against a vulnerable historic town where incendiaries could once again achieve the most damage. Rostock on the River Warnow,

Lancaster on a goodwill visit to the USA in 1942.

we turned to look, we saw debris flying high into the air. 'Look out, pilot,' shouted the navigator, as another stream of tracer shells shot up past the wing tips and we turned away to have a look at the target. All over the place, blocks of buildings were burning furiously, throwing up columns of smoke 3,000ft into the sky. We lost height to about 1,000ft and then flew across the southern part of the town, giving several good bursts of machine-gun fire. Sticks of bombs and incendiaries were crashing down everywhere and we certainly took our hats off to those anti-aircraft gunners. They continued firing even when their guns seemed to be completely surrounded by burning buildings. The last we saw of Rostock was from many miles away. We turned round and took a last look at the bright red glow on the horizon, then turned back towards England, very well satisfied with our first raid in our Lancaster.

The navigator of a Lancaster arriving towards the end of the raid told his captain that the fire he saw seemed 'too good to be true' and that it was probably a very large dummy. Closer investigation showed that it was in the midst of the Heinkel works and the Lancaster's heavy load of high explosives was dropped upon it from 3,500ft. Damage to the factory was considerable. The walls of the largest assembly shed fell in and destroyed all the partially finished aircraft within. Two engineering sheds were burnt out, and in the dock area five warehouses were destroyed by fire and seven cranes fell into the dock. Four bombers failed to return and three more crashed in England. Photographs taken in daylight after the second attack by just over 90 aircraft on Rostock the following night, when 125 aircraft were dispatched, show swarms of black dots near the main entrance to the station and thick upon two of its platforms. These were people seeking trains to take them away from the devastated city. Another 34 aircraft attempted the bombing of the Heinkel factory but it was not hit. A Wellington on 150 Squadron at Snaith crashed on take-off and three of the four-man crew on a 420 Squadron Hampden at Waddington died when the aircraft crashed near Sønderby Klint.

The third attack on Rostock the following night by 110 aircraft was met by strengthened flak defences yet no bombers were lost on the raid, nor on the one by 18 aircraft that attacked the Heinkel factory once more. Manchesters on 106 Squadron at Coningsby, commanded by 24-year-old W/C Guy Gibson DFC*, scored hits; the first time in the series of raids that the factory had been damaged. Rostock was

south of the Baltic, which like Lübeck had become a powerful member of the Hanseatic League in the fourteenth century and whose narrow streets characterised the *Altstadt* (old town) like those of Lübeck, offered the same possibilities for success. For four consecutive nights beginning on 23 April, Rostock was smothered with a carpet of incendiary bombs as had happened at Lübeck a month earlier, the only difference being that on the first three nights a small force of 18 bombers of 5 Group attempted a precision attack on the Heinkel aircraft factory on the southern outskirts of Rostock.

F/O John Wooldridge on 207 Squadron, heading for the Heinkel factory on his first trip in a Lancaster, could see the fire of the town glowing from almost 100 miles away.

It was an amazing sight. There hardly seemed to be any part of the town that was not burning. We dropped down to 5,000ft and skimmed across the factory. Lurid coloured shells seemed to be whizzing past in every direction. A building went past underneath, blazing violently. Then the nose of the Lancaster reared upwards as the heavy bomb load was released. Even from 5,000ft there was a clearly audible 'whoomph' as our heaviest bomb burst. As bombed again on 26/27 April, when just over 100 aircraft of seven different types split into two to attack the city and the Heinkel factory once more. The *Official History* describes the raid as 'a masterpiece with successful bombing by both parts of the force'. And all for just three aircraft lost. The RAF attacks on Rostock were followed by those on Stuttgart on 4 and 5 May, and on the night of 19/20 May, 197 aircraft – 13 of them Lancasters – attacked Mannheim. Most of the bombing was in open country. Eleven aircraft failed to return.

Harris had for some time nurtured the desire to send 1,000 bombers to a German city, and on the morning of 30 May he decided to send his bombing force to Cologne. On paper the actual number of serviceable aircraft totalled 1,047 bombers – mostly Wellingtons (602). The raid would also include the first Lancaster operations of 106 Squadron. All told, 5 Group detailed 73 Lancasters, 46 Manchesters and 34 Hampdens for the operation. At the end of the briefings at each station a message from Sir Arthur Harris was read out: 'Press home your attack to your precise objective with the utmost determination and resolution in the foreknowledge that, if you individually succeed, the most shattering and devastating blow will have been delivered against the very vitals of the enemy. Let him have it – right on the chin.'

Crews got their pinpoint south of the city and then turned north, with the Rhine to their right. Their aiming point was a mile due west of the Hindenburgbrücke bridge in the centre of the *Altstadt*. As the procession passed over the city, stick after stick of incendiaries rained down from their bomb bays, adding to the conflagration. The defences, because of the size of the attacking forces, were relatively ineffective and flak was described variously as 'sporadic' and 'spasmodic'.

David Walker, a deeply religious Scottish pilot on 44 Squadron, approached Cologne at 10,000ft and saw the city through the smoke haze:

We searched for an area that was not already burning for it seemed that Cologne then was ablaze from end to end. We had been briefed that the main post office was the aiming point. 'There are ammunition factories across the street,' we were told. Many of us, however, believed that we were bombing the civilian population because we knew that in most cities the main post office is not surrounded by factories.

The tension grew as the pilot opened the bomb-bay doors. The noise of the aircraft intensified. This was our most vulnerable moment. Our bomb, which seemed nearly as long as the four-engined aircraft itself, was now exposed. Coloured tracer bullets arched through the sky. If anything hit that bomb, we were finished!

The bomb aimer now took control of the aircraft. Pointing his sights towards the target area, he gave the pilot his instructions: 'Left … left … right … right … left a little … hold it … steady … on target. Bomb away!' The plane shuddered and I heard the 'whoosh' as the four-ton bomb fell away from the aircraft. An endless minute went by as we waited until the photoflash illuminated the area we had bombed. Once the damage had been photographed, we set off for home.

As we banked and turned steeply away, I could see the shocking sight of a city burning from end to end. Dense smoke could be seen drifting away leaving a brilliantly illuminated plan below. My immediate reaction was a mixture of sadness, fear and guilt. And into my mind flashed a comparison between the holocaust below and the preaching of the pastor at home. I thought about the men, women and children who had lost their lives. Why am I taking part in the slaughter of thousands of innocent people in this huge city?[3]

F/O Arthur 'Bull' Friend, a tall, 17-stone Rhodesian second pilot-navigator on a 97 Squadron Lancaster, recalled:

The dykes, the towns and sometimes even the farmhouses of Holland, we could see them all clearly as we flew towards Cologne soon after midnight. The moon was to our starboard bow and straight ahead there was a rose-coloured glow in the sky. We thought it was something to do with a searchlight belt, which runs for about 200 miles along the Dutch-German frontier. As we went through this belt we saw by the light of blue searchlights some friendly aircraft going the same way as ourselves and a few coming back. But the glow was still ahead. It crossed my mind then that it might be Cologne but we decided between us that it was too bright a light to be so far away. The navigator checked his course. It could only be Cologne.

It looked as though we would be on top of it in a minute or two and we opened our bomb doors. We flew on; the glow was as far away as ever, so we closed our bomb doors. The glare was still there like a huge cigarette-end in the German blackout. Then we flew into smoke; through it the Rhine appeared a dim silver ribbon below us. The smoke was drifting in the wind. We came in over the fires. Down in my bomb aimer's hatch I looked at the burning town below me. I remembered what had been said at the briefing, 'Don't drop your bombs on the buildings that are burning best. Go in and find another target for yourself.' Well at last I found one right in the most industrial part of the town. I let the bombs go. We had a heavy load, hundreds of incendiaries and big high explosive. The incendiaries going off were like sudden platinum-coloured flashes, which slowly turned to red. We saw many flashes going from white to red and then our great bomb burst in the centre of them.

As we crossed the town there were burning blocks to the right of us and to the left the fires were immense. They were really continuous. The flames were higher than I had ever seen before. Buildings were skeletons in the midst of fires. Sometimes you could see what appeared to be frameworks of white-hot joists. The blast of the bombs was hurling walls themselves across the flames. As we came away, we saw more and more of our aircraft below us silhouetted against the flames. I identified Wellingtons, Halifaxes, Manchesters and other Lancasters. Above us there were still more bombers lit by the light of the moon. They were doing exactly as we did: going according to plan and coming out according to plan and making their way home.

'Bull' Friend would return from Cologne but would die piloting a Lancaster on the Bremen operation on 27/28 June. Two of his crew were killed also; four men were taken into captivity.

Thirty of the 53 bombers that were lost were believed to have been shot down by night-fighters in the Himmelbett boxes between the coast and Cologne. It was estimated by Bomber Command that 16 of the 22 aircraft that were lost over or near Cologne were shot down by flak. In all, 898 crews claimed to have hit Cologne and almost all of them bombed their aiming point as briefed. Fifteen aircraft bombed other targets. The total tonnage of bombs was 1,455, two-thirds of this being incendiaries. Post-bombing reconnaissance certainly showed that more than 600 acres of Cologne had been razed to the ground. *The Daily Telegraph* the following day reported: 'At a Bomber Command Station, Sunday. On the 1001st day of the war more than 1,000 RAF bombers flew over Cologne and in 95 minutes delivered the heaviest attack ever launched in the history of aerial warfare.'

In England squadrons repaired and patched their damaged bombers – no fewer than 116 aircraft suffered damage, 12 so

badly that they were written off – and within 48 hours they were preparing for another 1,000-raid, against Essen. (The weather had proved unsuitable immediately after the raid on Cologne.) At nightfall on 1 June, 956 aircraft including 347 from the Operational Training Units (OTUs) took off and headed for Essen. Despite a reasonable weather forecast, crews experienced great difficulty in finding the target. Essen itself escaped lightly and Krupp's was once again left almost untouched. Although seemingly lacking the concentration of the earlier raid on Cologne, the bombing nevertheless was effective enough to saturate the defences. One skipper went as far as to say that the fires were more impressive than those of Cologne. A belt of fires extended across the city's entire length from the western edge to the eastern suburbs. Many fires were also spread over other parts of the Ruhr. Of the 37 bombers lost on the second 'Thousand-Bomber Raid' on Essen, 20 were claimed by night-fighters.

After Cologne and Essen, Harris could not immediately mount another 'Thousand-Bomber Raid' and had to be content with smaller formations. On 2/3 June just 195 aircraft, 97 of them Wellingtons, carried out a follow-up raid on Essen. On the night of 3/4 June, 170 bombers were dispatched on the first large raid to Bremen since October 1941. Crews reported only indifferent bombing results and 11 aircraft failed to return – eight of them shot down by *Nachtjäger*. On 5/6 June in the raid on Essen by 180 aircraft, 12 failed to return and the bombing was scattered over a wide area. The next night a force of over 230 aircraft was dispatched to Emden. It received another visit on 20/21 June, this time from 185 aircraft. However, only part of the bomber force identified the target and only about 100 houses were damaged. Eight aircraft, three of them Wellingtons, failed to return. On 22/23 June, 144 Wellingtons, 38 Stirlings, 26 Halifaxes, 11 Lancasters and eight Hampdens attacked Emden again for the third night in a row. 'Good' bombing results were claimed by 196 of the crews but decoy fires are believed to have diverted many bombs from the intended target. Six aircraft – four Wellingtons, one Lancaster and a Stirling – were lost. Emden reported that 50 houses were destroyed, 100 damaged and some damage caused to the harbour.

On the night of 25/26 June it was another 'Thousand-Bomber Raid', the third and final 'thousand' effort in the series of five major saturation attacks on German cities, when 1,067 aircraft – 96 of them Lancasters – attacked Bremen. The tactics were basically similar to the earlier 'thousand' raids except that the bombing period was now cut from 98 minutes, which was a feature of the Cologne raid, to 65 minutes. Attacks on Bremen were claimed by 696 Bomber Command aircraft. Generally the results were not as dramatic as at Cologne but much better than the second 'thousand' raid to Essen. Twenty-seven acres of the business and residential area were completely destroyed. The RAF plan to destroy the Focke-Wulf factory and the shipyards was not successful, although an assembly shop at the factory was destroyed by a 4,000lb bomb dropped by a 5 Group Lancaster. A further six buildings at this factory were seriously damaged and 11 buildings lightly so. The total of 48 aircraft lost was the highest casualty rate (5 per cent) so far. At Scampton two 83 Squadron Lancasters failed to return.

For three months – June, July and August – it was on only one night that 'Bomber' Harris was able to put into the air a force exceeding 500 aircraft. His wish to mount two or three raids each month of the order of 700 to 1,000 sorties was defeated by the weather, the rising casualties and the losses suffered by the OTUs. Despite the obvious harm that was being done to the training organisation, Harris used OTU aircraft and crews on four more operations to Germany. On three of these raids OTU aircraft and crews made up about a third of the force and on two of them the training units suffered a higher rate of losses than the squadrons. When OTU aircraft were included, raids were mounted by between 400 and 650 crews but in general Harris was compelled by the uncertain weather conditions to use only his operational squadrons. With these he achieved a high rate of effort, dispatching forces of the order of 200 aircraft ten times in June. A record 147 Bomber Command aircraft were destroyed by *Nachtjagd* (the German night-fighter arm) that month. From then until the introduction of 'Window' in July 1943, German night-fighters inflicted heavy losses on the bomber forces. ('Window' consisted of strips of silver paper, which when picked up by German radar gave a massive 'blip' covering the whole of their screens, preventing them from picking out individual aircraft.)

On the night of 27/28 June, when 144 aircraft visited Bremen again, 119 aircraft bombed blindly through cloud after obtaining 'Gee' fixes. It was believed that the raid was successful. Nine aircraft, four of them Wellingtons, two Halifaxes and two Lancasters and a Stirling, were lost.

Bomber Command dispatched 253 aircraft to Bremen on 29/30 June; 184 aircraft relying only on their 'Gee' fixes released their bombs within 23 minutes. This was the highest rate of concentration yet achieved. Fire caused extensive damage to five important war industries, including the Focke-Wulf factory and the AG Weser U-boat construction yard. Eleven aircraft failed to return.

At Scampton, 83 Squadron crews had assembled in the big station headquarters room used for briefings. 'Something' was in the wind. The squadron was stood down from operations for at least the next two weeks for a 'special operation in daylight'. On 11 July, the intelligence officer at Scampton drew back the curtain covering the large map on one wall and revealed what crews had waited a fortnight to find out. He must have savoured the gasp of astonishment when the unveiling revealed that the target was Danzig (now Gdańsk), a major U-boat repair and construction base far up on the Baltic, further even than Augsburg (from which, on 17 April, seven out of 12 Lancasters did not return). Forty-four Lancaster crews had been briefed for the 'special operation'; a round trip of 1,500 miles. The force flew at low level and in formation over the North Sea before splitting up and flying independently across the Jutland Peninsula, then south before swinging due east across southern Sweden to emerge over the Baltic heading south-east to Danzig. Crews had been promised blue skies all the way, which was an open invitation for fighters, especially when crossing Denmark, but none appeared. The target, as expected, was clear of cloud and the Lancasters bombed the U-boat yards from normal bombing heights just before dusk. Unlike the daylight mission to Augsburg, the plan worked well, although some Lancasters were late in identifying Danzig and had to bomb the general town area in darkness. Twenty-four Lancasters bombed at Danzig and returned. Two more were shot down at the target.

On 25/26 July it was Duisburg's turn, and 403 bombers, including 77 Lancasters, were dispatched. Twelve aircraft – two of them Lancasters – were lost. On the following night 403 aircraft – including 77 Lancasters – were dispatched to Hamburg which suffered its most severe air raid to date and widespread damage was caused, mostly in the residential and semi-commercial districts. Bomber Command was stood down on 27/28 July but this was the full-moon period and, on the night following, a return to Hamburg was announced at briefings. Crews were told that the raid would be on a far bigger scale than two nights before, though no Lancasters took part because of bad weather at 5 Group bases. The month of July and the 'moon period' ended with a raid on Düsseldorf on 31 July/1 August when 630 aircraft,[4] including 100-plus Lancasters for the first time, were dispatched. More than 900 tons of bombs were dropped and some extensive

A factory worker cleans the Perspex bomb aimer's panel on the Avro production line.

On 15 August the first step in the creation of the Pathfinder Force (PFF), to precede the Main Force and drop brilliant flares and incendiary bombs over the target, both to 'mark' the target and provide a beacon for the Main Force of bombers, was taken when Bomber Command issued an instruction that a Pathfinder Force was to be formed at Wyton. The new force, whose motto was 'We light the way', comprised 7 Squadron flying Stirlings at Oakington, 35 Squadron flying Halifaxes at Graveley, 83 Squadron flying Lancasters at Wyton, 156 Squadron flying Wellingtons at Warboys nearby, and 109 Squadron flying Mosquitoes at Wyton. The force was commanded by Australian Group Captain (later AVM) Donald Bennett CBE DSO. After the PFF began operations, the rate of bombing rose from 17 tons a minute to more than 50 tons a minute and the loss of personnel per ton of bombs was halved. The first operation in which the PFF technique was used was the attack on Flensburg on the night of 18/19 August 1942, in which 31 heavy bombers took part. Gradually refinements were made in PFF technique, particularly as the result of the improvement in radar methods.[6] Thirty-one bombers, 21 of them Wellingtons, were lost and a Lancaster crashed at Waddington on return.

Twelve bombers failed to return from the 4/5 September raid on Bremen by 251 Wellingtons, Lancasters, Halifaxes and Stirlings. For the first time the Pathfinders split their aircraft into three forces. 'Illuminators' lit up the area with white flares, 'visual markers' dropped coloured flares if they had identified the aiming point and then 'backers-up' dropped all-incendiary bomb loads on the flares. The weather was clear and the PFF plan worked well. Bremen reported heavy bombing and the Weser aircraft works and the Atlas shipyards were among the industrial buildings that were seriously hit.

Cloud and haze on the following night prevented concentrated bombing by the 207 bombers sent to hit Duisburg. Eight aircraft were lost. On 8/9 September, 249 aircraft were dispatched to Frankfurt-am-Main. Five Wellingtons and two Halifaxes were lost. Two nights later, training aircraft of 91, 92 and 93 Groups swelled the numbers in a 479-bomber raid on Düsseldorf. The Pathfinders successfully marked the target using 'Pink Pansies' in converted 4,000lb bomb casings and containing a red pyrotechnic, benzol, rubber and phosphorus, for the first time. All parts of Düsseldorf except the north of the city were hit, as well as the neighbouring town of Neuss. As a result of the raid, 19,427 people were bombed out. Thirty-three aircraft failed to return and the OTUs were hard

damage was inflicted but 29 aircraft, two of them Lancasters, failed to return. At Waddington there was no word from *L-London* flown by F/Sgt Norman Tetley, a South African, which crashed near Mönchengladbach. All seven crew died. Among them was Sgt Peter Rix, the Rhodesian mid-upper gunner who two months earlier had married Jean, a WAAF at Waddington. Her favourite song-of-the-moment was *Not a Cloud in the Sky* which everyone loved to hear her sing. After Peter went missing no one saw her cry. Neither did they hear her sing again.[5]

hit. Many training aircraft from various OTUs and conversion units were included in the force of 446 bombers which took off for Bremen on the night of 13/14 September. Almost 850 houses were destroyed and considerable damage was caused to industry, with the Lloyd Dynamo works being put out of action for two weeks and various parts of the Focke-Wulf factory for from two to eight days. Twenty-one aircraft failed to return.[7]

When, on the next night, 202 aircraft attacked Wilhelmshaven, the Pathfinder marking was accurate and the city suffered its worst raid of the war to date. Two nights later, 369 aircraft, including for the last time Wellington and Whitley aircraft on OTUs, carried out a strong Bomber Command raid on the Krupp works at Essen. Much of the bombing was scattered but effective, with 33 large fires and 80 'medium' fires being started. Much damage was caused to housing, and eight industrial and six transport premises were struck by bombs. The Krupp works were hit by 15 high-explosive bombs and by a crashing bomber loaded with incendiaries. Forty-one bombers were lost, including nine Lancasters, two of which were on 9 Squadron at Waddington, the first of this type's losses for the unit since converting from Wellingtons.

On the night of 19/20 September, 118 aircraft were dispatched to Saarbrücken and 68 Lancasters and 21 Stirlings went to Munich. At Saarbrücken the Pathfinders had to mark two targets but ground haze caused difficulties and the bombing was scattered to the west of the target. At Munich about 40 per cent of crews dropped bombs within 3 miles of the city centre, the remainder dropping them in the western, eastern and southern suburbs. Four bombers went missing in action. Four nights later, Lancasters of 5 Group bombed the Baltic coastal town of Wismar and the Dornier aircraft factory nearby. Many of the crews went down to under 2,000ft but just four aircraft were lost. On 6/7 October the Pathfinders succeeded in illuminating the Dummer See, a large lake north-east of Osnabrück, which was used as a run-in point, to which 237 aircraft were dispatched. The bombing was well concentrated. Six aircraft were lost. No further Main Force ops were flown until the 13th/14th, when the target for 288 aircraft – 82 of them Lancasters – was Kiel. One of the eight aircraft lost was a Lancaster. On 15/16 October, 289 aircraft including 62 Lancasters went to Cologne. This time 18 aircraft – five of them Lancasters – failed to return.

On the afternoon of 17 October, 94 Lancasters of 5 Group took off for the large Schneider factory at Le Creusot on the eastern side of the Massif Central. The factory was 200 miles south-east of Paris and more than 300 miles inside France so only Lancasters would have the necessary performance. Le Creusot was regarded as the French equivalent to Krupp's and produced heavy guns, railway engines and, it was believed, tanks and armoured cars. A large workers' housing estate was situated at one end of the factory. Bomber Command had been given this as the highest priority target in France for a night attack but only in the most favourable of conditions. AM Sir Arthur Harris decided to attack by day, despite the failure of the Augsburg raid exactly six months earlier. Eighty-eight aircraft led by W/C Leonard 'Slosher' Slee DFC, the 49 Squadron CO at Scampton, were to bomb the Schneider factory and the other six were to attack a nearby transformer station which supplied the factory with electricity. Crews claimed a successful attack on the factory but photographs taken later showed that much of the bombing had fallen short and had struck the housing estate nearby. Some bombs had fallen into the factory area but damage there was not extensive. The *Daily Express* claimed that the Lancasters had done a 'Grand National over hedges to blast French Krupps'. One Lancaster was lost, when it crashed into the transformer station during its low-level bombing run. Forty-one aircraft landed at their 'home' base in Lincolnshire but 45 'dropped in' at 23 airfields south and south-west of Lincolnshire, one as far west as Exeter.

On 22/23 October, 112 Lancasters of 5 Group and the Pathfinders set out for the historic port city of Genoa to recommence the campaign against Hitler's Axis ally to coincide with the opening of the Eighth Army offensive at El Alamein in Egypt. It was a perfectly clear moonlit night and the Pathfinder marking was described as 'prompt and accurate'. This was the first heavy and really concentrated attack made on Genoa, 180 tons of bombs being dropped in the ideal conditions. No Lancasters were lost, although one from 97 Squadron at Woodhall Spa crash-landed at North Luffenham after running low on fuel. The bombers returned to Genoa on the night of 23/24 October. The next day, 88 Lancasters of 5 Group were sent to attack Milan in daylight. This time a fighter escort of Spitfires accompanied the force across the Channel to the Normandy coast. The aircraft flew on independently over France close together and very low, hedge-hopping in the manner in which the Augsburg and Le Creusot raids had been made possible and using partial cloud cover, to a rendezvous at Lake Annecy. The Alps were then crossed and at four minutes after five o'clock in the afternoon, the first Lancaster nosed down through the heavy clouds and unloaded. They came in rapidly, pinpointing their targets and wheeling round. Mixed in the general delivery of HEs and incendiaries were a goodly proportion of 4,000-pounders. Some of the Lancasters went down to 50ft to bomb their targets. The raiders caught Milan by surprise and the bombing was accurate, 135 tons of bombs being dropped in 18 minutes. Only when they were well on their homeward run did darkness come down and afford them any protection. Three Lancasters that failed to return were all presumed lost over the sea. A fourth Lancaster, piloted by F/L Dorian Dick Bonnett DFC on 49 Squadron at Scampton, crashed on approach to Ford airfield in Sussex. Bonnett and five crew members died and two men were injured, one of whom died six days later.

Lancasters returned to Italy on the night of 6/7 November, when a large number of flares had to be dropped by the Pathfinders to light up the port of Genoa for the 72 Lancasters of 5 Group. Flak was heavier than before, especially from the dock area, and most bombs fell in residential areas. Two Lancasters were lost. The next night the force of 175 aircraft, including 85 Lancasters, bombed Genoa again. 'It was as if we were flying in the slipstream of another aircraft', was the description given by a pilot later. One front gunner reported that as his plane penetrated cloud, ice came crackling through the turret ventilators in handfuls, while up-currents threw the bombers about. The night was bitterly cold and a thick hoar frost coated the windscreens, each pilot watching the luminous altimeter needle as he climbed above the clouds. Over the Alps snow was falling steadily. Ice flew in large chunks from the whirling airscrews, rattling against the windscreen like hail. A Lancaster on 207 Squadron at Langar and a Halifax went missing over France. All members of the Lancaster crew died. Returning to Waddington, two Lancasters on 9 Squadron were destroyed and all 14 crew members killed when one collided with the other in the circuit.

On the night of 9/10 November, 213 aircraft – 72 of them Lancasters – returned to bombing targets in Germany again with a raid on Hamburg. The bombers encountered cloud and icing and winds which had not been forecast. Fifteen aircraft, five of them Lancasters, were lost over Hamburg. Genoa was bombed again on the night of 13/14 November by 67 Lancasters of 5 Group and nine Stirlings of the Pathfinder Force. Among the individual targets attacked was the Ansaldo works, to the west of the port, which constructed naval armaments and marine engines. There was a bright half-moon and visibility was good. 'The burned and destroyed areas in Genoa are so large and many that it is useless to attempt to give details of the damage,' wrote the eyewitness. 'Seen from a height, one has the impression that half of the town has been destroyed and in the port almost everything seems to be destroyed, burned, or damaged. Among the few exceptions in the port area are the silos where the grain for Switzerland was stored. Lack of small craft in the port, as a result of a large number having been sunk or severely damaged, is causing difficulties to the port authorities.'

On the night of 18/19 November, 77 aircraft attacked Turin and all aircraft returned safely. Bomber Command returned to bombing Turin on the night of 20/21 November with a force of 232 aircraft – the largest raid on Italy during this period. German night-fighters were posted far to the south inland from the French coast to attack a force of Bomber Command invaders should it return. At Holme-on-Spalding Moor 101 Squadron was flying Lancasters for the first time since converting from the Wellington and they had a number of encounters with German pilots. Their wing commander's aircraft was attacked three times. First W/C D.A. Reddick had to fight it out with a Ju 88 on the French side of the Alps. By diving, the wing commander got below the Junkers and the mid-upper gunner got in a burst that sent the German airman winging away. It was shortly afterwards that a twin-engined Me 110 rushed in, with cannon blazing. The Lancaster replied with fire too hot for the Messerschmitt pilot. He also disappeared. Later, on the way back, the Lancaster was approached by another Ju 88, but by rolling off into a patch of cloud an encounter was avoided. This was another successful attack. Three aircraft, none of them Lancasters, were lost.

Bomber Command returned to the bombing of a German target 48 hours later, on Sunday 22 November, a clear, crisp day; cold but invigorating. The destination for 222 aircraft was Stuttgart but the route was planned to make the Germans think that Italy was again the target; the flight would go down through France until south of Paris, where the force would swing east and, hopefully, 'catch Jerry with his pants down'. It was, however, the full-moon period and normally fighter attacks higher up in full moonlight could be expected. A thin layer of cloud and some ground haze concealed the target area so the Pathfinders were unable to identify the city centre. Heavy bombing developed to the south-west and south of Stuttgart, and the outlying residential districts of Vaihingen, Rohr, Mohringen and Pliengen, all about 5 miles from the centre, were hit. Two of the bombers attacked the centre of Stuttgart at low level and dropped bombs on to the main railway station, which caused severe damage to the wooden platforms and some trains in the station. Ten aircraft, five of them Lancasters, failed to return.

On the night of 28/29 November, 106 Squadron set off for the Fiat works at Turin with 117 Lancasters, 45 Halifaxes and 19 Wellingtons. 'Turin received a packet last night,' W/C Guy Gibson DFC* gave as his opinion when he returned. Gibson, who was to earn the VC months later for his brilliant attack on the Möhne dam, and F/L 'Bill' Whamond dropped the first 8,000lb bombs on Italy. P/O F.G. Healy had also set out carrying an 8,000lb bomb but as the Lancaster's undercarriage would not retract, its bomb had to be jettisoned into the sea. A great many high explosives and 100,000 incendiaries were dropped on the city's targets by 228 Lancasters, Stirlings, Halifaxes and Wellingtons. The 'packet' made up three wide areas blazing from end to end.

On 29/30 November, 29 Stirlings and seven Lancasters of 3 Group and the Pathfinder Force were dispatched to Turin with the Fiat works again the target. Weather conditions were poor and only 18 aircraft – 14 Pathfinders and four Stirlings of 3 Group – are known to have reached the target and bombed. On the night of 2/3 December, when 112 aircraft – 27 of them Lancasters – were dispatched to Frankfurt, Master Bomber W/C Sidney Patrick 'Pat' Daniels DSO DFC*, commanding 35 PFF Squadron at Graveley, was used for the first time. Six aircraft – one of them a Lancaster – were lost. On the night of 8/9 December, when the RAF paid the first of three more visits that month to Turin, fires were raised by incendiaries and more 8,000-pounders dropped by the force of 133 aircraft of 5 Group. The target was marked well by the Pathfinder Force and bombing was very accurate. Residential and industrial areas were both extensively damaged. One Lancaster was lost. When 227 aircraft of Bomber Command returned the following night the inferno had not completely subsided. The raid was described as 'disappointing' as the Pathfinders were unable to mark as well as on the night previously and the smoke from the old fires partially obscured the target area. A Wellington and a Lancaster failed to return and two Lancasters crashed in England.

On 11/12 December, the 82 Turin-bound bombers – 20 of them Lancasters – flew through a thick curtain of snow. One Lancaster bomb aimer reported: 'Only a small part of the glass of my compartment was clear of ice and through it I searched

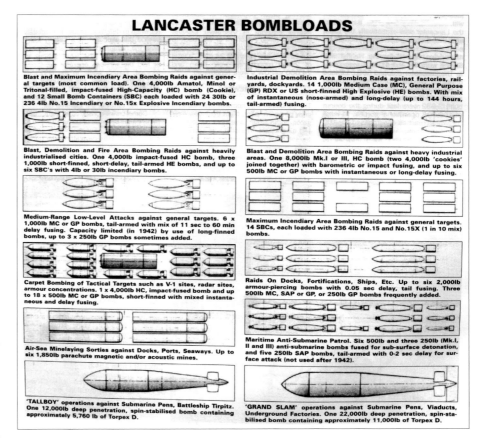

Lancaster bomb-load chart.

for a gap in the clouds. Ice had formed all over the aircraft and snow fell as we approached the Alps. When I went into the front turret after bombing I found several inches of snow behind the gunner's seat.' Another Lancaster captain recounted:

Shortly after crossing the Channel we ran into thick cloud. The further we went the worse it got. Then we struck an electrical storm. Vivid streaks of bluish light darted along the glass in front of my cabin and there were bright sparks between the barrels of the guns. At first we thought they were flashes of flak. The windows began to ice up and then we ran into driving snow. Very quickly the fronts of my windows and of the gun-turrets were covered with ice and snow, which also interfered with the smooth running of the engines. We began to lose height. Three of my engines cut out for short periods and for about ten seconds only one engine was working. The flight engineer got the engines running again but I had to decide not to risk crossing the Alps. The weather was equally bad coming back.

More than half of the force turned back before attempting to cross the Alps and only 28 crews claimed to have bombed Turin. Three Halifaxes and a Stirling failed to return.

Due to bad weather no major night raids to Italy or the greater Reich were possible so minor operations to small German towns and mine-laying trips were flown. On 17/18 December, in bright moonlight and without cloud cover, 27 Lancaster crews of 5 Group were en route to raid eight small German towns, and 16 Stirlings and six Wellingtons were detailed to attack the Opel works at Fallersleben. Another 50 aircraft were to lay mines from Denmark to southern Biscay and five OTU crews were being dispatched on sorties to France to find their feet.[8] At Waddington three Lancasters on 44 Squadron were shot down on the operation to Nienburg, *D-Dog* on 9 Squadron went missing in action on the raid on Diepholz and *S-Sugar* piloted by Sgt J. Wilson RCAF went down at Berkhausen between Varel and Rastede on the operation to Cloppenburg. The Canadian pilot and five of his crew survived and were taken into captivity but Sgt William Harvey Penn RCAF, an American mid-upper gunner from Tulsa, Oklahoma, was killed. Two Lancasters on 97 Squadron at Woodhall Spa failed to return from the raid on Neustadt, and another two on 50 Squadron at Skellingthorpe were lost on the operation to Soltau.

On the night of 20/21 December, over 230 aircraft, including 111 Lancasters, attacked the docks at Duisburg-Ruhrort. Two Lancasters collided shortly after take-off from Waddington with the loss of all 14 crew members. Twenty miles from Duisburg crews could see the Rhine gleaming in the frosty moonlight. The customary haze lay in patches along the valleys but they could pick out quite easily the course of the Ruhr itself and found that the target area was clear. One Lancaster pilot observed far below him the vapour trails spread by other aircraft going low to make their bombing runs. It was a concentrated attack in the new style that had developed from the thousand-bomber experiments. Huge explosions were seen and large fires spread, and much damage was claimed by the returning bomber crews. Twelve bombers that had flown to the target were missing, mostly shot down by night-fighters or flak. There were no survivors from the six Lancasters that were shot down.

The final Main Force operation in 1942 was the attack by over 130 aircraft of 1 and 5 Groups and the Pathfinders on Munich, which took place on the night of 21/22 December. The force was made up mainly of 119 Lancasters, of which one, a 61 Squadron aircraft, crashed shortly after take-off from Syerston. Most of the bombs fell in open country, although 110 aircraft claimed to have bombed the city. Three Stirlings, a Wellington and eight Lancasters failed to return.

F/L W.E. Vaughan RCAF, a navigator on Pathfinder operations, recalls:

There are three things that immediately come to mind. First is the crew. We became closer than family as we shared the hazards of operations. Though three of them were killed on our final trip and though I have not seen the others since the war ended, I remember them as if I saw them yesterday. And of course, there was our aircraft – the Lancaster. As far as we were concerned there was no other aircraft that could touch it as a heavy bomber. Also, always in the back of my mind was the bombing of civilians. Sometimes we hit strategic targets and other times there were near misses. But civilians were always killed. Although I was a navigator, there were times when I had to release the bombs over the target. More than once I wondered, 'How many people will those bombs kill?' However, you couldn't dwell on it. That's the way war is.[9]

By this time, Josef Kammhuber's Himmelbett defences had been completed and the Lichtenstein AI-equipped[10] *Nachtjagd* aircraft were now capable of exacting a toll of up to 6 per cent bomber casualties on any deep penetration raid into the Reich. Thus, Bomber Command losses rose rapidly during the first few months of 1943, although in early January operations were of necessity on a limited scale, normally to the U-boat bases on the Atlantic coast. On the night of 3/4 January, three Pathfinder Mosquitoes carried out a 5 Group Oboe-marking experimental raid on Essen by 19 Lancasters. Three of the Lancasters failed to return. Further raids on Essen using Oboe took place on 9/10 January when three Lancasters were lost, and on three successive nights – 11–14 January – when 66 Lancasters and three Oboe Mosquitoes took part. This time four Lancasters were lost to add to the two lost on the two previous nights. Two of the Oboe Mosquitoes had to return without marking, and the sky-markers of the third aircraft failed to ignite above the cloud. German aircraft also appeared to have dropped decoy flares to distract the Lancasters.

Oboe, the most accurate form of high-level blind bombing used in the Second World War, took its name from radar pulses which sounded like a musical instrument. The radar pulses operated on the range from two ground stations ('Cat' and 'Mouse') in England to the target. The signal was heard by both pilot and navigator, and was used to track the aircraft over the target. If the aircraft was inside the correct line, dots were heard; if outside the line, dashes; a steady note indicated that the target was on track. Only the navigator heard this signal. When the release signal – consisting of five dots and a two-second dash – was heard, the navigator released the markers or bombs.

On the night of 14/15 January, Bomber Command carried out the first of eight area attacks on Lorient on the French Atlantic coast. Some 122 aircraft, including six Lancasters, were dispatched. The Pathfinder marking of the U-boat base was accurate but later bombing by the Main Force was described as 'wild'. Over 150 bombers returned to Lorient the following night when bombing was more accurate, with at least 800 buildings destroyed.

On the night of 16/17 January, Sir Arthur Harris sent 190 Lancasters and 11 Halifaxes[11] on their way to Berlin – the first attack on the German capital in 14 months – with the words, 'Tonight you are going to the Big City. You will have the opportunity to light a fire in the belly of the enemy that will burn his black heart out.' This raid saw the first use of vari-coloured markers, or Target Indicators (TIs), dropped on the aiming point by selected crews. The operation, however, was a disappointment. Thick cloud en route and haze over the target caused problems and the bombing was scattered. The Berlin flak had proved light and ineffective, and it was assumed that the greater altitude of the attacking force had surprised the German gunners. Only one Lancaster was lost.

Harris repeated the raid on the 'Big City' the following night when the weather was better. The routes taken by the 170 Lancasters and 17 Halifaxes to and from Berlin were the same as those followed on the previous night and German night-fighters were able to find the bomber stream. Nineteen Lancasters and three Halifaxes were lost. On both raids the Pathfinders were unable to mark the centre of Berlin and bombing was inaccurate. (The experiments with the Lancaster-Halifax force using TIs against the 'Big City' now ceased until the H_2S navigational radar system became available.) Broadcaster Richard Dimbleby reported on the raid for the BBC, flying with W/C Guy Gibson DSO* DFC*, the CO of 106 Squadron at Syerston, in a 9-hour 15-minute round trip.

It was Gibson's 67th op. Next morning British listeners tuning in to the BBC Home Service heard Dimbleby's broadcast on their wireless sets. Nineteen Lancasters and three Halifaxes did not return.

On 21/22 January, 79 Lancasters and three Mosquitoes went to Essen, where they encountered 10/10ths cloud and bombs were dropped on estimated positions. Four Lancasters failed to return. The next Main Force raids were on Lorient and Düsseldorf on 23/24 January. Two Lancasters failed to return from the Düsseldorf operation, where some bombs fell in the south of the city. Nearly 160 aircraft returned to Lorient on 26/27 January when two Wellingtons and a Lancaster were lost. The following night Düsseldorf was attacked again when Oboe Mosquitoes carried out 'ground-marking' for the first time. Bombing was well concentrated on the southern part of the city. Three Lancasters and three Halifaxes were lost. On the night of 30/31 January, 148 heavies, 135 of them Lancasters, went to Hamburg and carried out the first H_2S attack of the war with Pathfinder Stirlings and Halifaxes using the new device to mark the target. It was not successful and five Lancasters were lost.

In February attacks were made on German cities, on French seaports on the Atlantic coast, and raids were resumed against Italy. The night of 2/3 February, when 161 aircraft, including 116 Lancasters, went to Cologne, was cloudy and a further experiment was made using a four-engined bombing force with various forms of Pathfinder techniques. Markers were dropped by two Oboe Mosquitoes and H_2S heavy-marker aircraft. Damage was caused right across the city. Five aircraft – three of them Lancasters – failed to return. The night following it was Hamburg, when 263 aircraft – 62 of them Lancasters – braved icing conditions over the North Sea. The Pathfinders failed to mark the target effectively and bombing was scattered. Sixteen aircraft, including a Lancaster and eight Stirlings, were lost. On 4/5 February, when 77 Lancasters and 111 other aircraft were sent to Turin, the 156 aircraft that reached the target caused serious and widespread damage. Three Lancasters, including one that crashed in Morocco with the loss of all seven crew, failed to return. That same night four Pathfinder Lancasters were sent to attack the port of La Spezia with new 'proximity fused' 4,000lb bombs, which exploded between 200ft and 600ft above the ground to widen the effects of the blast. Three of the Lancasters dropped their bombs successfully and all returned safely.

Lorient was attacked again on 7/8 February, when just over 320 aircraft – 80 of them Lancasters – were dispatched.

A typical bomb load for 20 Lancasters. It includes 20 4,000lb bombs and mixed incendiaries; typical blast, demolition and fire area-bombing loads, one impact fused 4,000lb bomb, four 1,000lb GP HE bombs and SBCs (Small Bomb Containers) of 4lb incendiaries.

The Pathfinder marking plan worked well and the two Main Force waves carried out a devastating attack. Seven aircraft, three of them Lancasters, were lost. The next big raid was on Wilhelmshaven on the 11th/12th, when 177 aircraft – 129 of them Lancasters – were dispatched but cloud completely covered the city. The Pathfinders were forced to carry out sky-marking by parachute flares using H_2S, but it was accomplished with great accuracy and bombing was remarkably successful. Three Lancasters failed to return.

The raid by 466 aircraft – including 164 Lancasters – on Lorient on 13/14 February was the heaviest attack on this port by Bomber Command during the whole war; more than 1,000 tons of bombs were dropped for the first time. Of the seven aircraft lost, three were Lancasters. When, on 14/15 February, just over 140 Lancasters of 1, 5 and 8 Groups visited Milan, fires could be seen 100 miles away on the return flight. Two Lancasters were lost. Two nights later, 377 aircraft, 131 of them Lancasters, returned to Lorient and 363 aircraft dropped mainly incendiary loads in clear visibility. Over 190 aircraft, including 127 Lancasters, went to Wilhelmshaven on 18/19 February. The Pathfinders claimed accurate marking in clear visibility but most of the bombs were dropped in open country west of the target. Four Lancasters were lost. Over 330 bombers – 52 of them Lancasters – returned to Wilhelmshaven the following night but the raid failed again and 12 aircraft, four of which were Lancasters, were shot down. A raid on Bremen took place on 21/22 February, and the night following 115 aircraft of 6 and 8 Groups returned to

Wilhelmshaven. Bomber Command rounded off the month with raids on Nuremberg and Cologne, which resulted in the loss of 19 aircraft, nine of them Lancasters, and St-Nazaire on 28 February/1 March. Just over 150 Lancasters were included in the force of 437 aircraft and two Lancasters were among the five aircraft that failed to return.

On the first night in March just over 300 crews – 156 of them Lancasters – were briefed for Berlin. AM Sir Arthur Harris was to say later, 'We can wreck Berlin from end to end if the USAAF will come in on it.' However, the Pathfinders experienced difficulty in producing concentrated marking because individual parts of the extensive built-up area of the capital could not be distinguished on the H_2S screens. Though the attack was spread over 100 square miles, because larger numbers of aircraft were now being employed and because those aircraft were now carrying a greater average bomb load, much damage was caused to the south and west of Berlin. Seven Lancasters were among 19 aircraft that failed to return. Main Force crews were then stood down for one night and the only activity was by 60 aircraft on mine-laying operations. One Lancaster and two Wellingtons failed to return.

Main Force operation resumed on the night of 3/4 March, when 417 aircraft including 149 Lancasters and 14 H_2S-equipped Pathfinders were dispatched to Hamburg, but of the 344 crews who had confidently reported bombing, only 17 had actually hit the city. Six Pathfinders suffered radar failures, including one PFF aircraft whose crew decided to 'press on regardless'. Nonetheless, the radar operator still managed to pick out distinctive features such as the Alster Lake in the centre of the city and the point where the Elbe narrowed at Hamburg. However, the tide was out and revealed extensive mudbanks, which on H_2S screens looked like a river narrowing several miles downstream of the city. It was not the Alster Lake that the H_2S operator had seen; it was the Wedel Lake, 13 miles downstream. The marker flares went down over the small town of Wedel where most of the bombing was concentrated, although in Hamburg a proportion of the bombing force did hit the city and the fire brigade had to extinguish 100 fires. Ten aircraft – four of them Lancasters – failed to return.

The night of 5/6 March has gone into history as the starting point of the Battle of the Ruhr. Essen, the most important and most difficult target in the Ruhr-Westphalia region to find, was the destination for 442 aircraft whose targets included the Krupp works. Two-thirds of the bomb tonnage was incendiary. One-third of the HE bombs, which included

150 4,000-pounders, were fused for long delay. For most of the way out the route was cloudy, and 53 bombers turned back early because of technical problems and other causes. Fifteen miles from the target the weather cleared, although pilots reported that natural valley mists were still seeping in from the river. These and industrial haze did not affect the outcome of the raid. Five Oboe Mosquitoes marked the centre of Essen perfectly and the Pathfinder 'backers-up' dropped their TIs blind on the target. Some 367 aircraft were thought to have reached the target and 345 crews claimed to have hit it. The Main Force bombed in three waves, which took 40 minutes to complete. In the third and final wave were 157 Lancasters, four of which failed to return. A week later photo reconnaissance revealed that an area larger than 160 acres was laid to waste, as well as 450 acres extensively damaged by fire.

The next Main Force raid was on 8/9 March, when 335 bombers, including 170 Lancasters, went to Nuremberg, but haze prevented accurate visual identification of the target area and more than half of the bombs fell outside the city boundaries. Eight aircraft, including two Lancasters, were lost. On the following night 264 aircraft – 142 of them Lancasters – visited Munich. Eight aircraft – of which five were Lancasters – failed to return. On 11/12 March, the fourth night of attacks in the Battle of the Ruhr, Stuttgart was the target for 314 Lancasters, 109 Halifaxes and 53 Stirlings. The attack was not successful as the marking was inaccurate. Eleven aircraft failed to return. 'Happy Valley was brighter and merrier than ever,' one pilot proclaimed when, on 12/13 March, 457 aircraft – 156 of them Lancasters – raided Essen again. 'They seemed determined that we shouldn't get through,' another said. 'As we approached the town I saw three huge cones with at least fifty searchlights in each.' Twenty-three aircraft – eight of them Lancasters – were lost. The next major raids would not take place until late in the month, when almost 360 aircraft set off for St-Nazaire; four nights later over 450 aircraft attacked Duisburg. In the meantime, small numbers of Wellington and Lancaster crews dropped their 'vegetables' on 'Gardening' sorties to 'regions' that stretched from the French Atlantic coast to as far as the Kattegat.

J.W. Henderson, a gunner on 50 Squadron, describing an operation on Duisburg, recalled:

On the run to the target you could feel the Lancaster lifting up and down from the blast of the anti-aircraft guns. Searchlights absolutely filled the sky over all the Ruhr. It

was like daylight inside the aircraft, with everything visible. The target was covered in smoke and fire-tracks. Our bomb aimer, once he had seen the barrage of flak that the Germans were sending up, cried out to the pilot, 'How the hell do you go through this?' But undaunted, the skipper flew on. It was nerve-wracking, having to fly straight and level to bomb the target precisely. You could hear the guns above the throb of your engines. You imagined that if you got home your aircraft would be filled with holes.

On the night of 27/28 March, 396 aircraft – 191 of them Lancasters – raided Berlin again. Nine aircraft, including three Lancasters, were lost. Two nights later, 329 aircraft, including 162 Lancasters, returned to the capital. Eleven Lancasters failed to make it back.

Essen was the target for 348 aircraft on 3/4 April, when the Pathfinders prepared a plan both for sky-marking and ground-marking the target because the weather forecast looked ominous for the operation. When the bombers arrived at Essen they found no cloud over the target, and the two types of marking confused the Main Force aircraft. Bombing, however, was accurately delivered and there was widespread damage in the city centre and in the western half of Essen. Fourteen Halifaxes and nine Lancasters failed to return. Twenty-four hours later, 577 bombers were dispatched to Kiel, the largest 'non-1,000' bombing force of the war so far. The Pathfinders encountered thick cloud and strong winds over the target so that accurate marking became very difficult. Decoy fire sites may also have drawn bombing off from the target. Twelve aircraft failed to return. Nineteen bombers – six of them Lancasters – were lost on the Duisburg raid of 8/9 April from a force of 392 aircraft. Another eight Lancasters failed to return from a force of 104 Lancasters and five Mosquitoes that went to Duisburg again the following night. Thick cloud again caused a scattered attack and bombs fell over a wide area of the Ruhr. A force of 502 aircraft, including 136 Lancasters, raided Frankfurt on the night of the 10th/11th. Five Lancasters were among the 20 aircraft that were lost.

On 13/14 April, 208 Lancasters and three Halifaxes attacked the San Vim arsenal, the shipyards and the submarine base at La Spezia in Italy, which was an outward journey of 700 miles. This operation was regarded as a maximum-range target and therefore the balance between fuel and bomb load was critical if the crews were to return safely to England. Four Lancasters were lost and three more, either damaged or in mechanical difficulties, flew on to land at Allied airfields in North Africa. The following night Stuttgart was the destination for 462 aircraft – 98 of them Lancasters. The Pathfinders claimed to have marked the target accurately but only a few bombs fell in the centre of the city. The main bombing area developed to the north-east, along the line of approach of the bombing force. This was caused by 'creep-back', a feature of large raids, which occurred when Main Force crews and some Pathfinder 'backers-up' failed to press through to the centre of the marking area but bombed or re-marked the earliest markers that were visible. Twenty-four bombers failed to return.

On 16/17 April, 327 aircraft – 197 of them Lancasters – were given the Skoda armaments factory at Pilzen in Czechoslovakia, but the Pathfinders mistook a large lunatic asylum 7 miles away. Though 249 crews claimed to have attacked the target, only six had got their bombs within 3 miles of it. Enemy fighters took advantage of the bright moonlight and got into the bomber stream early. Thirty-seven aircraft – 18 of them Lancasters – failed to return. Three of the Lancasters that were lost were on 460 Squadron at Breighton, which took the Australian unit's losses to 11 crews in just 16 days. Three more Lancasters were lost on 467 Squadron at Bottesford.

When 175 Lancasters and five Halifaxes returned to La Spezia on 18/19 April, the main railway station and many public buildings were hit. One Lancaster was lost. Eight other Lancasters laid mines off La Spezia harbour. On 20/21 April, 339 bombers – 194 of them Lancasters – visited Stettin, more than 600 miles from England. This proved to be the most successful attack beyond the range of Oboe during the Battle of the Ruhr. Visibility was good and the Pathfinder marking was carried out perfectly. Approximately 100 acres in the centre of the town were claimed as devastated. Twenty-one aircraft, including 13 Lancasters, failed to return.

On 26/27 April, Duisburg was raided by 561 aircraft, including 215 Lancasters. Seventeen aircraft – three of them Lancasters – were lost. The Pathfinders claimed to have marked the target accurately but most of the bombing fell in the north-east of Duisburg. Two massive mine-laying operations took place on the nights of 27/28 and 28/29 April, which cost a total of 23 aircraft, including eight Lancasters. It was the heaviest loss of aircraft while mine-laying in the war. At Wickenby four Lancasters on 12 Squadron, known as 'The Shiny Dozen', failed to return. Essen was again the target on 30 April/1 May. Just over 300 bombers – 190 of them Lancasters – were dispatched. Six Lancasters and six Halifaxes were lost.

May opened on the 4th/5th with the first major attack on Dortmund, by nearly 600 aircraft – becoming the largest 'non-1,000' raid of the war to date. Included in the force were 255 Lancasters. The initial Pathfinder marking was accurate but some of the backing-up marking fell short. Half of the large force did bomb within 3 miles of the aiming point and severe damage was caused in central and northern parts of Dortmund. The operation cost 42 aircraft, 11 of which either crashed or were abandoned in bad weather over England. It was one of the worst nights of the war for 101 Squadron at Holme-on-Spalding Moor, which lost six Lancasters. On 12/13 May, there was a trip to Duisburg-Ruhrort for 572 bomber crews, including 238 Lancasters. Thirty-five aircraft, including ten Lancasters, were lost and three bombers crashed in England. The following night 442 aircraft – of which 98 were Lancasters – were detailed to bomb Bochum, and another 156 Lancasters and 12 Halifaxes were to bomb the Skoda armaments factory at Pilzen again. The raid on Bochum began well but after 15 minutes what were believed to be German decoy markers drew much of the bombing away from the target. Twenty-four bombers were lost, one of them a Lancaster. The returning crews reported a fairly concentrated attack, with lots of fires and smoke. Eight Lancasters failed to return from the raid on Pilzen and a ninth overshot Fiskerton on return.

Three nights later, the switchboard at Scampton was swamped with calls. 'On the night of 16 May,' recalls Gwen Thompson, a WAAF telephone operator:

many senior figures sporting 'scrambled egg' (gold braid) on their hats had arrived at the base. We knew then that something significant was happening. The Dam Busters had bombed the Möhne, Eder and Sorpe dams in Germany, causing widespread havoc and great loss of lives on both sides. We had to send out 56 telegrams to the next of kin, which had been received from the local GPO at Skegness. Those young crews gave so much. For their families, it was a pyrrhic victory.[12]

2

THE SHINING SWORD

The Lancaster surpassed all other types of heavy bomber. Not only could it take heavier bomb loads, not only was it easier to handle, not only were there fewer accidents with this than any other type throughout the war, the casualty rate was also considerably below other types. I used the Lancaster alone for those attacks which involved the deepest penetration into Germany and were, consequently, the most dangerous. I would say this to those that placed that shining sword in our hands – without your genius and efforts we could not have prevailed, for I believe that the Lancaster was the greatest single factor in winning the war.

Air Chief Marshal Sir Arthur Harris

After a nine-day break in Main Force operations the target for the night of 23/24 May 1943 was Dortmund in the Ruhr Valley area of Germany. Bomber Command detailed 829 aircraft to go to the region sarcastically nicknamed 'Happy Valley' for what would be, except for the 'Thousand-Bomber Raids' of 1942, the largest raid of the war. No fewer than 343 of the aircraft involved were Lancasters, the rest comprising 199 Halifaxes, 151 Wellingtons, 120 Stirlings and 13 Mosquitoes. Zero hour had been timed for 0100hrs with the attack to take place between 0058hrs and 0200hrs. As an aid to navigation, 11 Oboe Mosquitoes dropped yellow TIs at a key turning point. They then went on to drop three red TIs in salvo on the aiming point, with 33 'backers-up' dropping green TIs on the centre of the reds. The first wave of 250 aircraft comprised the best crews from all Groups, with all remaining non-Lancasters in the second wave and the Lancasters in the final wave. PFF aircraft were at the head of each wave. The Pathfinders marked the target exceptionally accurately in cloudless conditions and the ensuing attack went according to plan. Seven hundred and fifty-four bombers dropped 2,042 tons of bombs – the biggest bomb load ever dropped anywhere in a single night – and large areas in the centre, north and east of Dortmund were completely devastated. Almost 2,000 buildings were destroyed and many industrial premises were hit, in particular the large Hoeschstahlwerke (Hoesch steelworks), an important centre for the German war industry, which ceased production. German records agree that the attack was accurate; in addition to extensive damage to houses, a large number of industrial premises were destroyed or damaged.

Ground defences in the target area and night-fighters on the return route put up a strong opposition and 38 aircraft, 18 of them Halifaxes, failed to return. Eight Lancasters, six Stirlings and six Wellingtons were also lost. On the morning after the raid there came more congratulations and another promise from the commander-in-chief, addressed to all the crews in Bomber Command. 'In 1939, Goering promised that not a single enemy bomber would reach the Ruhr,' Sir Arthur reminded them. 'Congratulations on having delivered the first 100,000 tons on Germany to refute him. The next 1,000,000, if he waits for them, will be even bigger and better bombs, delivered more accurately and in much shorter time.'

On the night of 25/26 May, 759 bombers were dispatched to bomb Düsseldorf. Twenty-seven bombers were lost – 21 of which were due to night-fighters. The raid was a failure owing to the difficulty of marking in bad weather. When Essen was subjected to a Main Force raid by 518 aircraft on 27/28 May, 23 bombers failed to return. Minor operations were flown on the following night when 34 crews, some of them undoubtedly flying their first op, sowed mines off the Friesians. On the moonless night of 29/30 May, Bomber Command flew one of the most significant raids of the Battle of the Ruhr. The area bombing of Barmen at the north-eastern end of the long and narrow town of Wuppertal involved 719 aircraft – 292 of them Lancasters – in five successive waves. The town was within easy Oboe range and 11 of the Mosquito markers were dispatched. Sixty-two of the aircraft turned back early with technical problems. Six hundred bombers dropped 1,822 tons of bombs on the Barmen district and a large fire started in the narrow streets of the old town centre. About 4,000 houses were burned to the ground and 2,450 to 3,400 people died as a minor firestorm raged. Around 118,000 people were made homeless. It took almost eight hours to put out the fires. The high-explosive bombs and fires devastated more than three-fifths of Barmen, the area of severe damage covering almost 700 acres. Thirty-three bombers were lost, three more crashed on landing and 60 bombers returned with serious damage caused by flak; two after they were hit by night-fighters. Six of the aircraft returned with damage caused by incendiary bombs falling from above.

In June 1944, 'Rhodesia' Squadron departed Waddington. Pip Beck watched disconsolately from control as they winged away, one by one, to Dunholme Lodge:

> The last two were airborne – but something was happening … they circled the airfield and then, on an obviously pre-arranged plan, formated and flew directly at the Watch Office at 'nought' feet, only climbing away at the very last moment, one to port, the other almost vertically. We had watched, mesmerised, and a gasp of indrawn breath came from all the observers of this final, brilliant display. It was a magnificent beat-up and a great flourish of farewell. And the two pilots responsible were F/L Pilgrim – and Cliff Shnier. Some weeks later, news found its way back that Cliff had been commissioned and transferred to 97 Squadron, Pathfinders, at Bourn. He went missing on a Hamburg raid at the end of July. I hadn't believed it could happen to him – not Cliff, with his superb skill, confidence and daring …[1]

In June German night-fighters claimed a record 223 victories, one of the high loss rates of the month being achieved on 11/12 June, a very bright moonlit night, when 76 aircraft set out for Münster and 860 heavies, 326 of them Lancasters, took

ED382 *J-Joe* on 101 Squadron at Holme-on-Spalding Moor shortly before leaving for Ludford Magna in June 1943.

off to bomb Düsseldorf. The Münster operation was really a mass H_2S trial; the raid, which lasted less than ten minutes, was accurate and much damage was caused to railway installations and residential areas of the city. Just five aircraft were lost but 44 aircraft failed to return from Düsseldorf, 29 of which were shot down by German night-fighters. Five of the 14 Lancasters lost were from 12 Squadron at Wickenby, which reported the worst loss rate suffered by the squadron in a single night for the whole of 1943. Only three men survived from the 35 missing air crew. Haze made visibility difficult at Düsseldorf but 130 acres were claimed destroyed. It was estimated that over 1,200 people were killed, 140,000 were bombed out and scores of other buildings and 42 war industries were destroyed or seriously damaged.

On 12/13 June, 24 of the 503 aircraft raiding Bochum were destroyed by flak and fighters. The raid took place over a completely cloud-covered target, but accurate Oboe sky-marking enabled all 323 Lancasters and 167 Halifaxes to cause severe damage to the centre of the city and 130 acres were destroyed. On the night of 14/15 June, Oberhausen was cloud-covered but once again the Oboe Mosquito sky-markers were accurate. Seventeen of the 197 Lancasters dispatched failed to return. Fourteen Lancasters were lost from a force of 202 Lancasters and ten Halifaxes of 1, 5 and 8 Groups that raided Cologne on 16/17 June. The marking for this raid was not by Oboe, but by 16 heavy bombers of the Pathfinders using H_2S. The target area was cloud-covered and some of the Pathfinder aircraft had trouble with their H_2S sets. The sky-marking was late and sparse, and the bombing of the all-Lancaster Main Force was thus scattered. Several hundred aircraft approached Cologne but because of the bad weather, only the first 100 bombed, the remainder turning back. Most of the damage in Cologne was to residential areas, with 401 houses destroyed and nearly 13,000 suffering varying degrees of damage.

Three nights later, 290 aircraft of 3, 4, 6 and 8 Groups took off to bomb the Schneider armaments factory and the Breuil steelworks at Le Creusot; 26 of the H_2S-equipped Pathfinders who released flares at Le Creusot were to fly on to drop flares over the electrical-transformer station at Montchanin. By the light of these flares a further 26 Lancasters of 8 Group were to bomb the target. The Main Force crews were detailed to make two runs from between 5,000ft and 10,000ft, dropping a short stick of bombs each time, but the smoke from the many flares that were dropped obscured the targets and results were poor.

All crews bombed within 3 miles of the centre of the target but only about one-fifth managed to hit the factories. Many bombs fell in residential areas. At Montchanin most of the attacking crews mistook a small metals factory for the transformer station and bombed this instead. A few crews did identify the correct target but no bombs scored hits on it. Two Halifaxes failed to return from the raid on Le Creusot.

The following night, in Operation Bellicose, 56 Lancasters of 5 Group and four Pathfinder crews on 97 Squadron in 8 Group in the first shuttle raid to North Africa carried out an attack on the Zeppelin works at Friedrichshafen on the north shore of the Bodensee (Lake Constance), the south shore of which is in Switzerland. The Master Bomber was G/C 'Slosher' Slee, who had led the Le Creusot operation, his deputy being the Australian, W/C Cosme Gomm. In addition there were two controllers, either of whom could take charge of the proceedings in the event of an emergency. The force took off late in the evening of 20 June and over France they progressively lost height down to 10,000ft as they passed Orléans and then lower still to between 2,500 and 3,000ft. After crossing the Rhine they began to climb to their attack altitude, which was 5,000ft for the Pathfinders and 10,000ft for the 5 Group aircraft. Cosme Gomm took over the control of the operation after Slee's Lancaster lost an engine. At the target flak was heavier than anticipated and Gomm ordered the Main Force up to 10,000ft. The TIs were dropped along the shoreline around the two pre-designated points so that the rest of the Lancasters could bomb indirectly. After the attack all except one of the Lancasters, which had been driven off track by thunderstorms, orbited the target to form up before heading south for airfields at Maison Blanche and Blida in Algeria. Six of the Lancasters were damaged by flak, one beyond repair, but considerable destruction was caused to the factory. The new bombing procedures tried during Operation Bellicose were deemed successful: 10 per cent of the bombs hit the target factory and many of the near misses destroyed other industrial premises. By flying on to North Africa after the raid the bomber force confused the German night-fighters that were waiting for them to return directly to England. Eight of the aircraft that bombed Friedrichshafen remained in North Africa for maintenance. On 23 June, 52 of the Lancasters attacked the oil depot at La Spezia and flew on to England.

On 21/22 June, before the moon period was over, in good visibility 705 aircraft, including 262 Lancasters, attacked Krefeld near the German border with Holland. In order to

Crews on 50 Squadron back from Milan in 1943.

saturate the German fighter defences along the route of the attack, the bomber crews kept together in a concentrated 'stream'. A total of 44 aircraft – nine of them Lancasters – were lost, 38 of them to night-fighters. The Pathfinder was almost perfect and 619 bombers dropped 2,306 tons of bombs on the markers. The whole centre of Krefeld – approximately 47 per cent of the built-up area – was burnt out; over 5,500 houses were destroyed and 72,000 people lost their homes. On 22/23 June, 565 bombers raided Mülheim and destroyed 64 per cent of the city. Thirty-five bombers, eight of them Lancasters, were lost. Wuppertal was the destination for 630 aircraft on 24/25 June. The Pathfinder marking was accurate and the Main Force bombing started well, but the 'creep-back' became more pronounced than usual and 30 aircraft bombed targets in other parts of the Ruhr. Altogether, 34 aircraft including eight Lancasters were lost.

Just 24 hours later, 473 bombers took off to attack Gelsenkirchen and Bochum. The target was obscured by cloud and the raid was a failure, with bombs falling on many other Ruhr towns. Thirty-one aircraft, including 13 Lancasters, failed to return. Sergeant J.S. 'Johnny' Johnston, a Melbourne Scot, was the flight engineer on *Z-Zebra* on 103 Squadron at Elsham Wolds, flown by F/Sgt Alan E. Egan RAAF, which was shot down at Bechtrup, north of Lüdinghausen. Johnston had begun his RAF service in 1936 as an apprentice at Halton

Bardney, 27 July 1943.
Sgts R.W. 'Reg' Moseley
(flight engineer),
J.W. 'Clem' Culley
(wireless operator) and
William 'Bill' Siddle
(pilot) on 9 Squadron in
line (left).

at which the flak guns reached for the first time. This type of night-fighting was hastily introduced under the command of Ritterkreuzträger Oberst Hans-Joachim 'Hajo' Herrmann, a bomber pilot and one of the foremost blind-flying experts in the Luftwaffe, who had been agitating for a long time without success for permission to practise freelance single-engined night-fighting. Herrmann had reasoned that by the light of the massed searchlights, Pathfinder flares and the flames of the burning target below the freelance single-engined *Wilde Sau* (Wild Boar) Fw 190 and Bf 109 pilots could easily identify enemy bombers over a German city. By putting a mass concentration of mainly single-engined fighters over the target, his pilots could, without need of ground control but assisted by a running commentary from a *Zentraler Gefechtsstand* (Central Operational Headquarters), visually identify the bombers and shoot them down.

At Waltham, John Rolland's crew on 100 Squadron had begun their operations at the height of the Battle of the Ruhr and had flown their first trip on Krefeld (21/22 June), followed by Mülheim (22/23 June), Cologne twice (3/4 and 8/9 July) and then Essen. They were to go on to complete their tour, one of only two crews at the time at Waltham to achieve this. Rolland's bomb aimer was Bill Couzins, a Plymouth lad who forsook his family's naval traditions by joining the RAF. Bill Couzins recalled:

One of the most memorable raids was the first of the big attacks on Hamburg when, from the nose of Lancaster *P-Peter* I saw the dramatic effect of the dropping of 'Window'. I think we were about eight or 10 miles from the target with all the guns and searchlights going when suddenly they stopped. The lights, which normally looked so orderly, went straight up in the air and the gunfire almost stopped. What had happened was that the window had confused the radar predictors on the lights and guns and, instead of seeing a steady stream of aircraft, they were confronted by the images of what must have looked like a hundred thousand. Because of that we just sailed through. After that the anti-aircraft guns resorted to box barrage firing which, while unpleasant, was nowhere nearly as dangerous as the radar-controlled flak we had previously had to face. 100 Squadron crews were lectured by a senior officer from 1 Group HQ at Bawtry who told us that in future we should stop their jinking runs over the target and instead go in at 21,000ft and go straight through. Some of the lads started booing and telling him to get stuffed. But

aged 15 years and 179 days. Four years later he was in the Battle of France, a member of the permanent ground staff on 226 Squadron, which was equipped with Fairey Battles. He was lucky to get away by sea from Brest on 19 June 1940, after France surrendered.

Johnston was captured and told to climb into the back of a truck. It was dark and he could just see some boxes. He put his hand out and felt a flying boot. He thought, 'Gee, that's all right, I want a pair of flying boots.' Johnston put it on and felt around for the other one. Suddenly he realised that they were soaking wet and he thought it was blood. He was sitting on top of coffins. Afterwards he found out that four of them were members of his crew. Johnston spent the next 20 months in *Stalag Luft VI* in East Prussia before escaping.

On 28/29 June, Cologne was attacked by 608 heavies – 267 of them Lancasters. Twenty-five bombers, including eight Lancasters, failed to return. On 3/4 July, 653 bombers returned to the city. The aiming point was that part of Cologne situated on the east bank of the Rhine where much industry was located. Pathfinder ground-marking was accurately maintained by both the Mosquito Oboe aircraft and the 'backers-up', and the Main Force were able to carry out another heavy attack. Thirty-one bombers, including another eight Lancasters, failed to return, a dozen of them being claimed by the *Nachtjagd Versuchs Kommando* (Night-fighter Experimental Command), which was allowed to operate over a burning city above the height he explained that with barrage fire it was much safer to go through the target area as quickly as possible and I think he got his message through.

Height was always a problem. The crews believed, with some justification, that the higher they were, the safer they were. At this stage of the war, most of the squadrons from Lincolnshire and Yorkshire climbed to 18,000ft before the bomber stream assembled, often over Mablethorpe, depending on where the target was. Then they headed across the North Sea. On some raids Couzins saw crews deliberately dropping part of their bomb load in the North Sea to lighten their aircraft and enable them to gain height, principally because they tended to carry heavier bomb loads than squadrons in other groups and consequently could not achieve the same altitude. However, Couzins believes those mainly responsible were the Halifaxes from the Yorkshire squadrons because later he would see some of those aircraft operating at up to 22,000ft or even 23,000ft, where a fully laden Halifax had no right to be. He recalled Stirlings, Wellingtons and some of the earlier Halifaxes being silhouetted against the clouds below by fighter flares:

just like flies on a table cloth. I still marvel at how we survived that tour. We were attacked a couple of times by night-fighters but each time we got away with it. Perhaps the hairiest moment we had was one very foggy night when four Waltham Lancasters crashed when we were coming in. We picked up the funnel but the clouds must have been down to about 200ft because as we came in over the wireless station at New Waltham the warning lights on top of the masts were either side of us. Then our wheels just brushed across the roof at Waltham Toll Bar School. At that point an enthusiastic controller in the caravan at the end of the runway realised *P-Peter* was having problems and fired a Very flare. This emerged right in front of the Lancaster and momentarily blinded the pilot. We ended up missing the runway completely and coming down on the grass. As we were bouncing along I was just hoping no one had left a vehicle out. *P-Peter* was a well-used Lancaster which had been rebuilt after an earlier crash but her four Merlins ran very sweetly. Often on their way home they would call Waltham while over the Wash and report they were in the circuit. This meant they were quickly allocated a landing slot and would not have to wait their turn in the circuit. Then it was a case of putting the nose down, giving the

Sgt Geoff Maddern, a West Australian pilot on 103 Squadron, with WAAF Intelligence Officer Lucette Edwards. In a bomber air crew the pilot was the crew captain; whether an NCO or not, he was superior to any officer in the crew. He made the decisions, although, if sensible and there was time, he would consult the crew first.

engines full boost and flying flat out across Lincolnshire for home. *P-Peter* went on to fly 72 ops before being lost later in the war. The ground crew were always there to meet them. They were great lads and they looked after us well.

On 8/9 July, it was Cologne again. Over 280 Lancasters of 1 and 5 Groups devastated the north-western and south-western sections of the city and a further 48,000 people were bombed out, making a total of 350,000 people who lost their homes during the series of three raids in a week. Seven Lancasters were lost. It was on 9/10 July that the next Main Force raid was directed to Gelsenkirchen. Thirteen 'Musical Mosquitoes' again marked with Oboe sky-marking, but the equipment failed to operate in five of the aircraft and a sixth Mosquito dropped sky-markers in error 10 miles north of the target so the raid was not successful. Seven Halifaxes and five Lancasters went missing from a force of 418 aircraft. On the night of 12/13 July, 295 Lancasters of 1, 5 and 8 Groups attacked Turin. Thirteen Lancasters, including *Z-Zebra* flown by W/C John Dering Nettleton VC commanding 44 Squadron at Dunholme Lodge, were lost. Nettleton's Lancaster was shot down by a German night-fighter over the English Channel on his return. All seven on the crew were lost without trace.

At the start of the Battle of the Ruhr 'Bomber' Harris had been able to call upon almost 600 heavies for Main Force operations

packets of metal strips, which when dropped in bundles of a thousand at a time at one-minute intervals produce almost the same reactions on RDF equipment as do aircraft and you should stand a good chance of getting through unscathed.

Strips of black paper with aluminium foil stuck to one side and cut to a length (30cm by 1.5cm) were equivalent to half the wavelength of the Würzburg ground and Lichtenstein airborne interception, radar. Bomber Command planned to use 'Window' in the first 'Thousand-Bomber Raid' on Cologne on 30/31 May 1942, but at the last minute William Sholto Douglas, head of Fighter Command, had halted deployment of 'Window' because he was afraid the Germans might learn to use it against his own night-fighters. At last, at a staff conference on 15 July 1943, Churchill stated that he would accept responsibility for initiating 'Window' operations.[2] 'Window' was carried on the 791 aircraft – 347 of them Lancasters – which set out for Hamburg. During the Battle of Hamburg, 24 July–3 August 1943, 'Window' prevented about 100-130 potential Bomber Command losses.

In his message of good luck to his crews, Harris said:

> The Battle of Hamburg cannot be won in a single night. It is estimated that at least 10,000 tons of bombs will have to be dropped to complete the process of elimination. To achieve the maximum effect of air bombardment this city should be subjected to sustained attack. On the first attack a large number of incendiaries are to be carried in order to saturate the fire services.

and at the pinnacle of the battle, near the end of May, more than 800 aircraft took part. Innovations such as Pathfinders to find and mark targets with their TIs and wizardry such as Oboe, which enabled crews to find them, were instrumental in the mounting levels of death and destruction. Little, it seemed, could be done to assuage the bomber losses, which by the end of the campaign had reached high proportions. There was, however, a simple but brilliant device which at a stroke could render German radar defences almost ineffective. On the night of Saturday 24 July 1943 when Harris launched the first of four raids, code-named Gomorrah, on the port of Hamburg, each station commander was authorised to tell the crews:

> Tonight you are going to use a new and simple counter-measure called 'Window' to protect yourselves against the German defence system. 'Window' consists of

Led by H_2S PFF aircraft, 740 out of the 791 bombers dispatched rained down 2,284 tons of HE and incendiary bombs in two and a half hours upon the suburb of Barmbek, on both banks of the Alster, on the suburbs of Hoheluft, Eimsbüttel and Altona and on the inner city. The advantages enjoyed by Kommodore Josef Kammhuber's Himmelbett night-fighter defence system, dependent as it was on radar, had been removed at a stroke by the use of 'Window'. The German fighter pilots and their *Bordfunkers* (radio operators) were blind but 12 bombers, including four Lancasters, were lost in action. To many it appeared that every section of this huge city was on fire. An ugly pall of smoke was blowing to the south-west. It looked the way that one might imagine hell to be.

One of the Lancaster pilots who returned safely was Sgt William 'Bill' Siddle on 9 Squadron at Bardney, who had flown as a 'second dickey' on the trip with another crew. This was standard practice in Bomber Command, the aim being to give a new pilot first-hand experience of an actual attack on a target before asking him to take his own crew to war. 'Although a "sprog" pilot was usually teamed up with an experienced crew for his initiation,' recalled Sgt Clayton C. Moore, Siddle's Canadian rear gunner:

it was not unknown for a new crew to lose its skipper on one of these trips, so we were greatly relieved when old Bill presented himself in the Mess the following day and gave us a vivid account of the ride.

Our skipper described the trip as having been mainly uneventful. There had been a marked absence of fighters and the ground based defences had appeared to be in serious disarray, with the searchlights waving aimlessly around the sky and with the flak bursting in confused patterns at a height well below that of the main concentration of bombers. He had seen two aircraft being shot down in the target area, but both of these had been from the lower echelon, which consisted mainly of Stirlings and Halifaxes and it had been obvious that the searchlight and flak battery crews were being forced to rely entirely on visual sighting because of the effect that 'Window' was having on their equipment.

On the night of 25/26 July, 'Window' was still effective but bad weather over north Germany prevented a return to Hamburg by the Main Force, so 705 heavies were dispatched to Essen. Brigadier General Fred L. Anderson, commanding VIII Bomber Command, flew as an observer on the raid; Canadian F/L 'Rick' Garvey's crew on 83 Squadron had the honour of flying the American general on *Q-Queenie*. Twenty-six bombers, including five Lancasters, failed to return, of which 19 were destroyed by night-fighters.[3] In all, 627 aircraft out of 705 dispatched dropped 2,032 tons of bombs upon Essen. Harris was not exaggerating when he said that 'they inflicted as much damage in the Krupps works as in all previous attacks put together'.

Harris was intent on sending his bombers back to Hamburg for another major strike. He said, 'I feel sure that a further two or three raids on Hamburg, then probably a further six raids on Berlin and the war will finish.' On the night of Tuesday 27 July, 787 aircraft followed a longer route out to Hamburg and back to include a longer flight over the North Sea with the intention of confusing the *Jägerleitoffiziers* (JLO, or GCI-controllers)

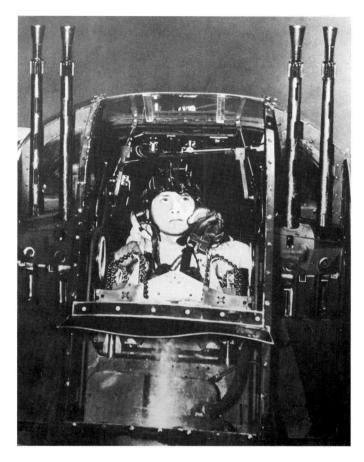

The cramped, Frazer Nash FN 20 turret with four 0.303s remained standard on Lancasters. The two gunners had lonely jobs and could not relax. The lives of the rest of the crew rested upon their vigilance. The tail gunner never saw another member of his crew throughout the flight. They discouraged any attack and were rarely given second chances by attacking fighters.

as to the intended target. This meant that each aircraft had to carry a smaller bomb load than normal, so it was decided to include a higher proportion of incendiaries.

Clayton Moore on Bill Siddle's crew on 9 Squadron wrote:

At 22.00 hours we took off in Lancaster DV198 (WS-U) for what was to be our first operational flight as a complete crew. It was with a mixture of excitement, fear and foreboding that I prepared myself for whatever the next five hours in the air might bring. Darkness was beginning to creep steadily westward and the sky above Lincolnshire was filled with the sight and the sound of the big bombers as they wheeled and struggled to lift themselves and their heavy loads to their operational height. My location in the aircraft left me with the impression that we were almost on the point of stalling as, with tail down and engines roaring

FN 82 rear turret armed with .50-calibre machine guns, which appeared on late-production aircraft at the end of the war.

steadily, WS-U continued to claw for height. As we reached 10,000ft the order was given to don oxygen masks and I adjusted the level of illumination on my reflector sight and set my four Brownings to 'fire'. At last we turned eastwards on a course that would take us out over the coast in the direction of Germany.

As we made our way out over the North Sea, still climbing, the other Lancasters around us gradually faded from my view as the darkness closed in. Now the only evidence of their presence was the turbulence of a slipstream, or the occasional glimpse of streaking tracer bullets as some over-zealous gunner tested his weapons. A routine which was generally accepted as standard, this was one of which I did not approve. Apart from the obvious danger of scoring a hit on an unseen friendly aircraft in the vicinity, I was of the belief that a correctly serviced and adjusted Browning could always be relied on to function when required.

The butterflies had been giving me trouble ever since I had learned that we were 'on' that night, but their effect was beginning to lessen now that we were on our way to the target. This was it. A year of intensive training was behind me and I knew that if I was to see a fighter out there in the darkness tonight, it wouldn't be a friendly Beaufighter or Hurricane, armed with nothing more lethal than a camera gun. There wouldn't be the comforting presence of an instructor to tell me that I wasn't getting it quite right. This time it was for real and my reward for dropping a clanger would be a lot more severe that a low assessment or a telling off. The defence of this big bomber and its crew and cargo was the responsibility of Dick Jones [the mid-upper gunner] and me and we couldn't afford to give or expect any mercy if attacked. In accordance with the established crew policy, it was our job to spot the fighter first and dodge a confrontation. But would we see him in time? After all, I hadn't seen that Beaufighter back at the OTU until it was too late.

As we droned out over the water, I began the now familiar search pattern and congratulated myself on having sighted a couple of Lancasters forging through the murky darkness that surrounded us. I was later to learn that this was unusual, it being possible to complete a trip along with several hundred other aircraft without seeing a single one, except in the vicinity of the target.

As we forged steadily onward, with only the pin-points of light from the stars to relieve the darkness, I drew comfort from the business-like way in which we were functioning as a crew. Conversation was being kept to the minimum, with only regular crew checks by Bill and the occasional course correction from Dick Lodge [navigator] breaking into the familiar background noise of the engines. At one point, the aircraft began to buck mildly as it ran into the turbulence being kicked up by another one somewhere ahead of us and our navigator remarked smugly that somebody else must be right on track. Such mild humour, although unnecessary, was nonetheless appreciated because it served to relieve the tension that affected all of us.

At last, after more than two hours in the air, we turned to starboard on the heading that would take us across Hamburg. Bill announced over the intercom that the target was dead ahead and that he could see a lot of fires still burning from the previous raid. The sky around us was beginning to grow much lighter now and I sensed a feeling of nakedness as the protective curtain of darkness was gradually drawn aside. Glancing over my left shoulder, I could see the big black starboard rudder and one distant wing tip held in stark profile against the bowl of flickering red and yellow light that lay ahead of and beneath us.

As we drew nearer to the city, I could see frequent flashes of blinding white light as the photoflashes exploded near the ground, interspersed with the dull red flashes of the 4,000-pounders and other high explosive bombs as they slammed into the streets and buildings far below, each load accompanied by a long, narrow swath of twinkling light thrown up by the incendiary bombs igniting on impact.

As we neared the action that lay ahead of us, things began to happen with increasing frequency. We were into the bombing run. Ken, lying flat in his compartment in the nose, had ordered the bomb doors to be opened and I could feel the aircraft making gentle course corrections as he calmly issued commands to the skipper. Beneath us, long strings of light ack-ack shells climbed lazily upwards, looking like strips of multi-coloured fairy lights, only to burn out before reaching us. The heavy batteries appeared to be aimed at about our height, each shell burst giving off a dull red flash which was quickly consumed by a cloud of thick black smoke left hanging in the air and looking remarkably like a barrage balloon as it drifted past. The heavy ack-ack shells gave no visual indication of their approach, so that the first warning that you got of their coming was the 'whump' of the explosion, a frightening sound that was frequently audible

The Lancaster did not carry a second pilot but instead a flight engineer monitored the instruments constantly, in particular checking fuel consumption, and aided the pilot generally. He also helped the bomb aimer by 'Windowing' to confuse enemy radar. He kept a log of all his activities which would be checked at debriefing with the squadron engineering officer.

above the noise of the engines, depending on the nearness of the blast.

Early in my career of operational flying, I was to learn that the run up to the target could be the most traumatic part of a raid. This was because, at a time when you were openly exposed to the full destructive potential of a well defended target and when the survival instinct was screaming at you to get the job finished and get the hell out of danger, the taking of evasive action was forbidden, except in cases of extreme necessity. There were two reasons for this: first, the bomb aimer required a steady run up in order that he could take accurate aim on the target and I found this acceptable. It was the wisdom of the remaining directive which I considered to be iniquitous. This was concerned with the action to be taken after the bomb load had been released. Each aircraft carried a fixed on-board camera

which was automatically set to take a picture of the area beneath the aircraft at the exact time that the bombs would be detonated. Also involved was the photoflash, the function of which was to provide illumination of the target, thereby enabling the camera to record the result of the drop. In order for this to happen, it was imperative that the aircraft should be flying straight and level, so that the camera lens would be aimed correctly at the time when the exposure was to be made. Depending on the height at which the bombs were released from the aircraft, it could take in excess of one full minute (in addition to the run-up) for them to reach the target and this was one hell of a long time to spend in morbid contemplation of a possible outcome.

As I saw it, there were two distinct reasons for taking a photograph, the first being that it would be of value in determining the degree of accuracy obtained by the bomb aimer and would thus provide a good indication of the damage inflicted on the target. Again, I concurred with the need for this, except in circumstances when the target was either covered by cloud or obscured by smoke, in which case the recording of ground detail would be impossible and only parachute flares (if in use) would appear on the film. The second reason for exposing a valuable aircraft and its crew to possible destruction and death was to prove that they had actually carried out the attack. A small number of crews were not keen to attack their objective and some ingenuity had gone into the perfection of ploys which were designed to cheat the system. However, their existence did not warrant the exposure of honest men to unnecessary danger.

Now that we were almost directly above the target, the excitement (and the fear) was intense. The light from the fires, photoflashes and searchlights; the smoke drifting up, the thundering flak bursts and the flash of the bombs as they exploded far beneath, had turned the darkness into near daylight. Other Lancasters could be seen all around us, each with its bomb doors gaping open and with the bombs spinning and tumbling as they began their unstoppable descent. Amidst the turmoil and confusion of the battle that was taking place all around, I got the feeling that all eyes must be on us and I noticed for the first time the perspiration that fear could generate as I frantically searched the sky above us for the fighter attack that I felt must surely come.

'Left …' came Ken Hill's voice over the intercom.

'Steady … steady … bombs going.' While this was going on, I was searching the starboard quarter, high.

'Tally Ho!' warned Dick from the mid-upper turret. 'Fighter astern, diving port to starboard.'

I quickly swung my turret around, saw and recognised our enemy at once.

'It's a Focke Wulf 190, skipper, but he's after somebody else. Watch him, Dick. I'll cover the tail.'

There then followed the long, nail-biting ordeal as we waited for the camera to function. Meanwhile, all hell was breaking loose around us. The gunners down below were sending up a box barrage and I reckoned that they had the range just about right. Flak shells were bursting all over the place, causing the Lancaster to wallow drunkenly each time one got a bit near. Down on the starboard quarter, somebody copped one and I watched for the first time the spinning, flaming hulk of what had been a Lanc as it fell into the inferno below, comet-like, with orange and white flames streaming to a point behind it.

A wandering searchlight beam flicked across us, paused and then returned. Some eagle-eyed operator had spotted us and he soon had us held captive in his all-revealing beam of light. A couple of others locked onto us, followed by several more and Bill began to throw us around the sky in a vain attempt to shake them off, while I cursed our luck at having been coned by searchlights on our first trip.

As we careered around the sky in a corkscrew that threatened to tear the wings off, I felt totally disorientated by the violence of the manoeuvre and the effect that the blinding light was having on my sense of direction and surroundings.

'For God's sake everybody, keep your eyes open for fighters,' Bill shouted down the intercom at us as, with the Merlins screaming in torment, the Lanc continued to dive, bank and climb, subjecting me to stresses that I had never before experienced in an aircraft. Now our predicament was being compounded by the action of the ack-ack gunners who had started to fire visually up the beams at us. Despite the confusion and noise that was all around us, I could distinctly hear the shrapnel rattling against the fuselage each time a shell burst got near.

During our participation in subsequent raids, I was to become sufficiently callous as to rejoice at the sight of someone else being coned as we were at this moment, because we could be assured that most eyes (both friendly and unfriendly) were indeed on the unfortunate sods in the cone. This meant that, for the time being at least, the heat of

battle was a little less intense for the rest of us. I was also to learn that the chances of surviving such an experience were about even.

'Some bastard down there doesn't like us!' Ken remarked in broad cockney.

'Shut up, bomb aimer,' Bill admonished and then added, 'There's no future in this. I'm going to do a power dive. See if we can shake them off that way.'

Almost at once, there was a falling off in the accuracy of the flak and the searchlights began to thin out until only a couple still held us. Bill pulled out of the dive and went into another corkscrew and we were soon surrounded once more by the comforting darkness. We had been held in the cone for a total of six minutes.

The emergency over, Bill began calling up each member of the crew for damage reports. It was then that we were told

The navigator was the hardest worker in the crew – pilot included. A flight could last from four to nine hours and he would be working all the time. The pilot could take a rest because most of the time the aircraft would be on 'George', the automatic pilot. Timing was the navigator's responsibility, but even at best he was only as good as his pilot flying the correct course, height and speed advised by the navigator. This was duly noted in his log and woe betide any pilot who did not conform because the navigator's log was always checked at debriefing after an operation. It was a very skilled trade carried out under trying, cramped conditions with an oxygen mask on most of the time, which did not help concentration. Navigating involved measuring angles and distances on the Mercator Projection Chart in poor light, taking a radar fix every six minutes, fixing wind speed and direction, and estimating time of arrival. Winds could change very quickly and could be 100mph at 20,000ft.

The wireless operator, or WOp as he was known, was responsible for operating the radio, the radar-jamming devices and worked 'Fishpond' which was radar that picked up other aircraft close by. He gave the gunners warnings of fighters about.

that Dick Jones had suffered an injury. A piece of shrapnel had penetrated the Perspex dome of his mid-upper turret, striking him in the left shoulder. The wireless operator was immediately detailed to take over the position while the flight engineer attended to the wound. Fortunately, this was only superficial and Reg Moseley soon had Dick patched up and back in position. As for damage, the aircraft had collected a few small flak holes in the central fuselage, but nothing of vital importance was involved and all engines and instruments were functioning normally.

A course for base having been set, I continued to search the sky behind us, this being made easier by the light from the inferno that was Hamburg. The city was well alight, with flames and smoke rising several hundred feet. A large ball of flame, probably from an exploding gasometer rose steadily upwards amid the waving searchlight beams and pinpoints of flak.

As we drew further away from the target area, I observed high above the holocaust a lone bomber with its black surfaces looking surprisingly silver as it tumbled frantically this way and that, seemingly supported by a giant tripod of white light that stood among the burning streets below. Watching the drama I fervently hoped that our fellow combatants would escape as we had. But soon, the moth-like silver speck suddenly glowed orange and began to tumble slowly downwards, breaking up as it fell.

'Aircraft going down in flames over the target,' I reported to the skipper.

Although I was not to know it then, we were taking part in Operation *Gomorrah* and this was the night of the firestorm, a man-made horror that was to be extensively chronicled and questioned by historians of both sides in later years. I was as yet very much inexperienced in such matters and cared little about the political reasoning behind the conflict in which I was involved. Nevertheless, I was capable of imagining something of the terrible degree of death and destruction that was being inflicted on the City of Hamburg and its unfortunate inhabitants even as I watched. No doubt a lot of people, some too young to hold political beliefs, were dying at that minute.

Soon we were out over the North Sea and Hamburg was nothing more than a deep red glow in the dark sky behind us. We touched down at Bardney at 3.20 a.m. to be met at dispersal by 'Chiefy', the F/Sgt in charge of our ground crew. During all of my term of service with Bomber Command, I never ceased to marvel at the keenness and devotion to duty displayed by these much underpaid and undervalued servants, whose expert skills were so essential to us. No matter what the hour of the night might be and regardless of weather conditions, there was always at least one of them waiting to welcome us back, sometimes several. In the main, these men spent much of their tour of duty on the remote and isolated dispersal, with nothing more than a makeshift hut of their own construction in which to shelter when the weather turned nasty.

We had hardly enough time to give Chiefy a run down on the extent of the damage before the transport arrived to take us over to the flight offices for debriefing. Dick Jones had been picked up soon after the landing and was already on his way to the station hospital for treatment. The debriefing room was already thronged with weary yet spirited young men like ourselves and we were pleased to be told that all of the squadron aircraft had returned safely. There were four long tables in the room, at each of which was seated a WAAF and a bomber crew. While waiting for a table to become vacant, we each helped ourselves to steaming hot cups of tea which we eagerly drew from a tall, chromed urn in a corner of the room.

After the debriefing, we all trooped over to the Mess for breakfast. During the meal, we discussed the trip and agreed on the need for some modification to our existing crew tactics, these to be based on our experiences during the flight. In general, Bill was satisfied with our performance, but ruled that in future, only crew positions would be used when calling someone over the intercom, there being a danger of confusion arising because of the duplication of Christian names within the crew. It was also ruled that, in similar circumstances, we would adopt the shallow dive tactic on being coned by searchlights over the target. Using the corkscrew manoeuvre would only prolong the duration of our presence in the danger area, thereby increasing the chance of us copping a packet from the ack-ack gunners, as had happened that night. The diving technique of evasion would not only shorten our stay within the area covered by the guns, but would also make more difficult the chance of the defences to hang onto us, because of our increased speed and rapid change of altitude.

On arising from a well-earned sleep in the early afternoon, I was pleased to learn that Dick had been passed as fit for flying. I was also told that the squadron was 'dicing' that night and that we were again on the battle order. It certainly looked as if we were going to complete our tour in record time at the rate we were going.

A firestorm was started by a combination of the high temperature prevailing (about 30°C at six o'clock in the evening) and a total of 2,417 tons of bombs was dropped by 739 bombers. The firestorm raged for about three hours and only died down when there was nothing left to burn. About 16,000 multi-storey apartments were destroyed and 40,000 people died; most of them by carbon monoxide poisoning. Seventeen aircraft failed to return.

After a late lunch Clayton Moore and Dick Jones on 9 Squadron caught a transport over to the dispersals at Bardney to check out their turrets and guns for the coming operation. 'Because we had been flying the previous night,' recalls Moore:

another crew had done the NFT on our aircraft during the morning and the armourers were already busy winching the bombs on board when we arrived. On asking, we were told that the bomb load was almost identical to that carried by the squadron aircraft on the previous night's trip, but that the flight duration could not be calculated until orders concerning the fuel gallonage were announced. We were soon to realise that, once these details were known, the ground crews were quite capable of predicting the approximate location of the target with remarkable accuracy. For the pending trip [29/30 July] we had been allocated a Mk.III Lancaster, ED666 (WS-G), which had arrived new on the squadron in May. Quite by chance, WS-J, another 'A' Flight Lanc, was parked in the next dispersal to ours. Johnnie Walker/Still Going Strong! (as she had been named) was the pride and joy of fellow Canadian 'Mitch' Mitchell and his crew, one of which was none other than Eric Plunkett, the one acquaintance left over from my months spent in training.

At the briefing, it was revealed that the target was again to be Hamburg, making a total of three attacks on the city by Bomber Command in just four nights. Like some of the others present, I was concerned by this display of repetition in target selection and I wondered at the wisdom of sending a large force to the same target at such close intervals. I feared that, having been subjected to a pounding on two out of the three previous nights, the German defences would by now have realised that we were intent on the total destruction of Hamburg and would be ready and waiting for us to put in an appearance. My belief was further strengthened by a warning from the squadron Gunnery Leader that we could expect to find an increase in fighter activity in the vicinity of the target.

The Met Officer, standing beside his blackboard and easel with pointer in hand, explained the meaning of the multi-coloured chalk sketches he had prepared for us. Cloud over the target was expected to be sparse and it had been planned that our return would beat the arrival at our bases of a mass of low cloud that was moving in from the North West. We would be flying at 20,000ft and the winds were expected to be moderate. [The objectives for 777 aircraft that were detailed were the northern and north-eastern districts of Hamburg, which had so far escaped the bombing.] A somewhat elderly intelligence officer then rose to give an outline of the known defences we could

H$_2$S was a 10cm experimental airborne radar navigational and target location aid developed by Dr Lovell at the TRE (Telecommunications Research Establishment) at Malvern, Worcestershire, which produced a 'map' on a CRT display of a 360° arc of the ground below the equipped aircraft. It was first used in January 1943.

expect to encounter in the area of the target and warned that there was photographic evidence of mobile flak and searchlight batteries being moved into position. However, window was again in use and this could be expected to limit the effectiveness of the German defences. He then went on to report on the detail contained in the reconnaissance photographs taken of the target earlier in the day. 'There are a lot of fires burning out of control in many parts of the city, but there are still a few buildings with their roofs still intact,' he said. Then, with obvious distress, he announced that he had just attended the funeral of some close relatives who had been killed in a raid on the East End of London. In the respectful and awkward silence that followed, he tearfully implored us to 'go over there and give them hell,' then added, 'I won't be happy while there's still a roof left in Hamburg.'

After the briefing, we next visited the aircrew Mess for the traditional pre-flight meal of bacon and two eggs, followed by the equally traditional bowl of prunes and custard. It was widely suspected that the dessert course was the brain child of some high-up Air Ministry medic who, being aware of its propensity for keeping one 'functional' had prescribed the treatment. On discussing the value of the prunes as a

laxative, Ken put forward his own evaluation which, roughly translated, was to the effect that they fell far short of the effect that could be produced by a German flak barrage.

Take off was scheduled for 22.50 hours, so we had about ninety minutes in which to prepare ourselves for the coming trip. Our next call was at the crew locker rooms, where we all got dressed for the part. The locker rooms resounded to the boisterous babble of anxious young men intent on dispelling the apprehension that they undoubtedly felt. Only a few of the jokes to be heard were new and the laughter was too loud and spontaneous. After a time, I learned the art of separating the new crews from those which had completed more than the first half-dozen flights. While the new boys were noticeably noisy at this stage of their preparation, the 'veterans' went about the chore of donning their flying kit in a spirit which bordered on quiet resignation.

The air gunners had to wear more clothing than the other crew members because of the lack of heating in the turrets. In fact, the rear turret was open to the slipstream because of the cut-out clear vision panel in front of the unfortunate occupant. This admitted a fierce gale, particularly when the turret was trained to one beam or the other. As a consequence, the 'clobber' consisted of: battledress trousers and tunic; one-piece electrically-heated suit; zip-up canvas flying suit (with huge imitation fur collar and spacious knee pockets); two pairs of thick woollen socks and leather flying boots. On top of this livery went the Mae West and the parachute harness and the ensemble was completed by a fleece-lined leather helmet, complete with goggles, oxygen mask and intercom (microphone and earpieces). Finally, leather flying gauntlets with electric inner gloves were worn and most gunners included a thick roll-neck sweater worn beneath the battle dress tunic.

On leaving the crew room, fully clad, the next task was the collection of the various items issued for use during the flight. These amounted to a small sealed emergency escape pack containing a map of the territory over which we would be flying; a supply of the local currency; forged identity documents; a small compass; a selection of concentrated food capsules (including Horlick tablets), chewing gum and a few pieces of chocolate. This pack was only to be opened in the event of us being shot down and was otherwise surrendered on our return. The pack, together with one bar each of Fry's Sandwich and Fry's Chocolate Cream was presented to us by an equally apprehensive-looking young

WAAF who stood just inside the exit. She also had for disposal a supply of boiled sweets, plus a limitless supply of 'wakey-wakey' pills. These were always in great demand, not because of their faculty for preventing one from dozing off during the trip, but because it was popularly believed that they were an effective aphrodisiac. Resulting from this, the pills were jealously hoarded in readiness for the next leave period, during which they would be hastily consumed.

There now remained only a visit to the parachute section in order to collect and sign for a pack before boarding the trucks or buses that would take us around the airfield perimeter tracks to our waiting aircraft. Having been dropped off at WS-G's dispersal, we at once busied ourselves in stowing our gear and running a final check on our equipment. This done, we then sat on the grass at the edge of the pan, discussing the coming flight and making last-minute decisions on tactics.

As I climbed into my rear turret, I was somewhat perturbed to find it piled high with packets of window on either side of my seat. I at once lodged a complaint with a nearby armourer, who informed me that this incursion into my already cramped domain had been decided on by 'the powers that be' and that the rear gunner would in future be expected to eject a bundle by means of his clear vision panel at regular intervals during the time that the aircraft was over enemy territory. I could see that the idea was fraught with problems and decided to challenge the order at the first opportunity.

Although darkness had fallen, the weather looked promising. Soon it was time to take up our positions prior to the engine run-up, after which the wheel chocks were dragged away by the ground crew and we taxied out to join the long line of Lancasters making their way to the end of the runway for take-off. As we rumbled slowly along, I carried out one last check on my turret, tested the intercom and fastened my seat belt. Came our turn for take-off and Bill lined us up with the runway. The intercom was buzzing with the voices of Bill and Reg as they made ready for receipt of the green light from the nearby control caravan, at one end of which stood the familiar group of well-wishers (the 'Press-on Gang', as it was known). As the Lancaster ahead of us lifted its heavy load from the runway and clawed for height, leaving four heavy black trails of exhaust streaming in its wake, we got the green light and Bill opened the throttles wide. My turret began to bounce as the wheel

brakes were held firmly against the wayward pull of the Merlins. Then, satisfied that the engines were set at the correct condition for take-off, the brakes were released and we began to roll down the runway, steadily gathering speed as we went.

Our Lancaster began to pound down the runway as the navigator began calling out the ever increasing airspeed. The tail wheel lifted and I could see the big port rudder working as Bill countered the tendency of all Lancasters to pull over to the left on take-off. As Dick Lodge called 'One-twenty', the tail dipped low and the ride suddenly became smooth as we left the runway and soared out over the boundary fence.

Once airborne, we joined in with the rest of the heavily-laden bombers as they circled for height above the darkened Lincolnshire landscape. At 10,000ft we got the order to don oxygen masks and then turned onto the course that would take us out over The Wash on the first leg of the flight. We were still striving for altitude and most of our comrades had faded from sight by the time we finally reached our operational height of 20,000ft, somewhere out over the North Sea.

Two hours into the flight, Bill announced from the cockpit that the fires of Hamburg could be seen in the distance.

Dropping bombs in the right place, especially at night, required precision and courage. It was no use flying all the way to the target, putting every member of the crew at risk and then missing the target. The bomb aimer alone would be looking straight down at the flak. It was an awesome sight to see the flak coming straight at you and bursting all around; to see a Lancaster, silhouetted against the burning target, being hit right in front of you; to see one of your own bombers dropping a stick of bombs on to another Lancaster; or to watch the searchlights groping around ever closer, knowing what was in store for you if they caught you in their beams. Throughout an attack the bomb aimer would calmly give instructions to the pilot and keep his nerve until the target was in his bombsight. Then he would go through the bombing sequence with only quarter-inch Perspex shielding him from the flak.

A bomb aimer sighting through the revised, enlarged arming panel introduced on the B.III Lancaster. Once the bombs were dropped, the bomb aimer called 'Bombs gone' and the bomb doors remained open so that the aircraft flew level to enable accurate photographs to be taken. The heavy-breathing silence continued until the bomb aimer called 'Bomb doors close' to the pilot and everybody heaved a sigh of relief.

'Seems a waste of time sending the Pathfinder boys out to mark this one,' remarked Ken from the nose. One hour later, we were lining up for the dreaded bomb run. During much of the approach, I had been busy throwing out bundles of 'Window'. Soon the heavy flak began to burst near us, so I dispensed a few more, just for good measure. 'Bomb doors open. Markers in sight.' The big stuff was bursting all over the place and at about our height.

'Two going down in flames to starboard' (this from the mid-upper).

'Another one astern,' I put in. 'Steady ... right a little ... steady ..., bombs going ..., Bombs gone!' Now came the long wait for the photoflash to explode.

'Flash gone!' came the welcome retort from Ken. Bill at once threw us into a tight diving turn to port. 'Let's get the hell out of here,' he yelled, just as a searchlight beam flashed across us, followed by two others.

The diving turn had the effect of shaking off the searchlights before they could catch us in a concentrated cone. It also did strange things to me and my turret: My stomach did a loop and the sudden application of zero gravity sent the remaining bundles of window cascading in all directions around the turret, where they became lodged in various nooks and crannies – ammunition belts, gun

mounts, on the floor, even down the back of my neck. As I retrieved them and hurled them out, I vowed that I would raise a strong objection to their presence in my turret on our return to base.

The remainder of the flight proved uneventful and we touched down at Bardney just before 4.30 a.m., tired, but pleased at having got one more trip 'in the book'.

Compared to our first op, this one had been a piece of cake. Not so for Hamburg. That unfortunate city had been well alight by the time we dropped the bombs and – even from a height of almost four miles – we could see whole streets plainly outlined by the burning houses on both sides. [The Main Force had been detailed to approach Hamburg from almost due north but the Pathfinders arrived 2 miles too far to the east and marked an area south of the devastated firestorm area. 'Creep-back' stretched about 4 miles along the devastated area and heavy bombing was reported in the residential districts of Wandsbek and Barmbek and parts of Uhlenhorst and Winterhude. In all, 726 aircraft dropped 2,382 tons on the city and caused a widespread fire area but there was no firestorm. Twenty-eight aircraft were shot down.] It was clear that vast areas of Hamburg were totally destroyed and, try as I might; I found difficulty in imagining the hell that the people down there must be experiencing. The treatment of the injured and the disposal of bodies would be a severe strain on the public utilities and it was likely that loss of power, fractured water mains and streets littered with the debris of fallen and burning buildings must be making the task of the unfortunate fire fighters and ambulance crews quite impossible.

After the debriefing, at which I complained about the stowing of window in the rear turret, we enquired about the fortunes of the other squadron crews involved and were told that they had all returned to base safely. Not so on the following night however (July 30th). We were not down to fly that night, when 9 Squadron had been sent to bomb Remscheid. During this attack JA692 (WS-D) went down, taking with it the highly experienced crew of F/L Charles Fox, plus an unfortunate 'second dickey'. The aircraft had only been with the squadron for a month and had completed less than one hundred flying hours since its arrival from the factory.[4]

On 2 August, after a day of heavy thunderstorms, 740 bomber crews were briefed for the fourth raid on Hamburg.

They were told that the weather was extremely bad and that cumulonimbus clouds covered the route up to 20,000ft. Thirty aircraft failed to return as a result of enemy action and icing, taking total losses on the four raids to 79. More than 6,000 acres of Hamburg smouldered in ruins. Over four nights, 3,095 bombers were dispatched, 2,500 attacked and 8,621 tons of bombs were dropped, 4,309 tons of them being incendiaries. Half the city had been totally devastated. Paralysed by 'Window', *Nachtjagd* and the *Flakwaffe* were unable to offer any significant resistance. On average, British losses during the Hamburg raids were no more than 2.8 per cent, whereas in the previous 12 months losses had risen from 3.7 to 4.3 per cent. 'Window' neutralised the Würzburg GCI and GL radars and short-range AI and completely destroyed the basis of GCI interception. Controlled anti-aircraft fire was almost completely disrupted at night and fixed box barrages only remained possible. The new British tactics also combined the use of PFF, the massed bomber stream and new target-finding equipment (H_2S). This combination resulted in total chaos to the German night-fighter defence system, which was unable to obtain a true picture of the air situation or control the night-fighters in the air. Until they received better equipment, the only method German fighter pilots could use to overcome the crisis caused by 'Window' was the Wild Boar tactic.

On the night of 30/31 July, nine of 105 and 109 Squadrons' Mosquitoes, each carrying four red TIs, marked Remscheid, a centre of the German machine-tool industry on the southern edge of the Ruhr, which was the target for over 270 aircraft. This was the first raid on the town. Weather over the target was clear and the raid was well concentrated on a town with a population of 107,000 that manufactured machine precision tools from high-grade steel. The largest factory, Alexanderwerk AG, employed nearly 3,000 workers and covered 75 acres. It produced special machinery for the chemical industry, motor components and small arms. There were 25 smaller factories in the town, as well as railway workshops that specialised in the repair of goods wagons. The Oboe ground-marking and the bombing by the comparatively small Main Force were exceptionally accurate and 83 per cent of the town was destroyed. Remscheid was still blazing on Saturday at midday with smoke to a height of 14,000ft over the town. This brought the Battle of the Ruhr to an end after 18,506 sorties in which 57,034 tons of bombs had been dropped for a loss of 1,038 aircraft.

At Bardney following the crew's second operation on Hamburg on 29/30 July, F/L Biddle's crew were not called upon to fly again until the morning of Monday 2 August, as Clayton Moore recalled:

We were detailed to take ED975 (WS-Y) for an hour of local flying, just to keep us on the ball. At a later date, this aircraft was to figure prominently in our fortunes (and otherwise). On return to base, we were told to get ready for ops that night. For this one, we were allocated ED654 (WS-W). At briefing, we were surprised to learn that the target was again to be Hamburg and one comedian was heard to remark on the danger of pilots becoming redundant because the squadron aircraft knew their own way to Hamburg and back. Our route was to be longer this time, with the main force jinking around northern Germany in an attempt to keep the Jerry defences guessing about the identity of our intended objective until almost the last minute.

Take-off was later than usual – ten minutes before midnight – so it was to be yet another sleepless night for us and the survivors of Hamburg. Nevertheless, I reckoned that the rear turret of a Lancaster would provide greater comfort and safety than a Hamburg air raid shelter under the circumstances and I much preferred my role in the proceedings to that of the unfortunate people down there beneath us. 'Window' was again in use, but I was pleased to find that, as a result of my complaint (and that of others), the responsibility for its disposal had been passed to another member of the crew. For us at least, the trip was again fairly straight forward and uneventful, although bad weather proved troublesome and was probably the cause of the high loss rate (30 aircraft in all, representing almost 7 per cent of those sent). The defences in the target area were less effective than before and the smoke rising from the many fires which seemed to be raging totally out of control made bombing tricky. Nevertheless, I saw a number of aircraft going down, including one Lancaster which was under attack by a Focke-Wulf 190. On arrival back at base after almost six hours in the air, we learned that one of the squadron aircraft had failed to return. This was ED493 (WS-A). It had been on its 37th operation and had been piloted by Sgt D. Mackenzie.

On 7/8 August, 197 Lancasters of 1, 5 and 8 Groups attacked targets in Genoa, Milan and Turin. One of the Pathfinders who led the way for the Main Force was W/C John Searby of 83 Squadron, who was the Master Bomber or the 'Master of

Ceremonies' at Turin. The Master Bomber arrived over Turin first to take charge of the dropping of marker flares by the Pathfinders and then continued to circle over the target area to orchestrate the bombing more effectively. When he noticed that 'creep-back' was taking place he instructed the later bombers to aim forward of the fires so that bombing was more concentrated in the target area. This achieved only limited success but this was a trial in preparation for the role Searby would fulfil in the successful raid on Peenemünde later that same month.

At Bardney, F/L Siddle's crew returned from a week's leave and apart from one bombing and air-to-sea firing exercise, most of their time was taken up by routine training and 'lounging around' the airfield. 'One hour before midnight on Monday, August the 9th,' recalls Clayton Moore:

we lifted off in DV198 (WS-U) for a raid on Mannheim. 'U' was a recent arrival on the squadron and was in spanking condition. We had a reasonably easy trip in her, but the short sea crossing and the much extended duration of flight over enemy territory called for a prolonged period of alertness on the part of us all. The target defences were on form and we were treated to the now familiar assortment of punishment we had come to expect. I saw a few aircraft being shot down and in the light of the target I saw an unfortunate Lancaster suddenly blow up for no apparent reason. It was on the bombing run (as we were at the time) and there were no flak bursts in the immediate vicinity. I could only presume that the Lanc had copped a load of bombs from above. One of our duties was to report and log all sightings of aircraft going down, and it fell to the navigator to note details of the time and approximate bearing in relation to our aircraft so that the point of impact could be plotted. No doubt such information was of use to the intelligence bods, but it could be time consuming and I tended to dislike any duty which detracted from my main responsibility of searching the sky around us for enemy fighters.[5]

On the night following our Mannheim trip, the squadron attacked Nürnberg and ED654 (WS-W), piloted by P/O Newton, returned with the rear gunner (Sgt McFerran) wounded and the mid-upper gunner [Sgt Percy Lynam, 24] dead. WS-W, the aircraft we had taken to Hamburg eight nights previously, had been attacked by a German fighter. Our crew had been rested that night, so did not take part in the Nürnberg raid.

Eighteen aircraft did not return. One was ditched in the English Channel and the crew taken into captivity. Five more crashed in England.

The night of Thursday/Friday 12/13 August was a long one for 321 Lancasters and 183 Halifaxes targeting Milan, while 152 aircraft of 3 and 8 Groups were detailed to bomb Turin. Turin was hit by 112 Stirlings, 34 Halifaxes and six Lancasters, and two Stirlings failed to return. On 14/15 August one Lancaster was shot down on the raid on Milan and seven were lost on the raid on the same city the night following, mostly to German fighters, which were waiting for the bombers' return over France.

By early August the Allies had acquired evidence that Peenemünde was the site of advanced weapon research, especially rocket technology, and the War Cabinet had scheduled the site as a high-priority target for Bomber Command. Operation Hydra – a force of 596 heavies, including, for the first time, Lancasters of 6 (Canadian) Group – was dispatched on 17/18 August with G/C John Searby commanding 83 Pathfinder Squadron, as Master Bomber. The attack was on three main target areas. The first and second waves would attack on Pathfinder TIs but 5 Group squadrons were to bomb last in the third wave and would use the new time and distance bombing method as the target would probably be obscured by smoke from the first and second wave attacks. The objective for the third wave aircraft was the accommodation block and the experimental works, which consisted of over 70 small buildings containing vital development data and equipment. The German ground controllers were fooled into thinking the bombers were headed for Stettin, but in the wild mêlée over the target *Nachtjagd* claimed 33 aircraft, with total claims from the Peenemünde force amounting to 38 victories, some by crews flying Bf 110s fitted with *Schräge Musik* ('Oblique Music'). This device, invented by an armourer, Paul Mahle of II./NJG5, comprised two 20mm MG FF cannon mounted behind the rear cockpit bulkhead of the Bf 110 and Ju 88 night-fighters, and was arranged to fire forwards and upwards at an angle of between 70° and 80°.

The daylight reconnaissance 12 hours after the attack revealed 27 buildings in the northern manufacturing area destroyed and 40 huts in the living and sleeping quarters completely flattened. A number of important members of the technical team were killed. The attack of 1 Group on the assembly buildings was hampered by a strong crosswind but substantial damage was inflicted and this left only

5 and 6 Groups to complete the operation by bombing the experimental site.

August–September was a grim time for 12 Squadron at Wickenby. Twenty-five aircraft went to Peenemünde and one, piloted by S/L Fraser Burstock Slade DSO, crashed. The 'A' Flight commander and his crew were killed. Five more aircraft were lost before the end of the month. In September four Lancasters on 12 Squadron were lost. One of them was flown by an Australian, P/O Jack Pierce Hutchinson RAAF, who lost three engines after being attacked on his way back from a raid on Munich. His wireless operator managed to get a distress call off and later five of the crew, including the pilot, were picked up 60 miles off the Lincolnshire coast. Learning from this experience, the whole squadron was encouraged to practise dead-engine flying and dinghy drill, and visits were arranged to the air-sea rescue unit at Grimsby, Hutchinson's rescuers. Some crews appeared to be jinxed. F/L Rowland was attacked by night-fighters in successive raids in October on Hanover. His rear gunner was killed in the first and three members of his crew injured in the second. Hutchinson, the man rescued from the dinghy, brought a badly damaged Lancaster back from Frankfurt. In the same raid P/O Jack Currie, who lost an aileron over Hamburg, had his aircraft hit by bombs from another Lancaster and had to crash-land at West Malling. The following month Jack Hutchinson, who by now had transferred to 626 Squadron, crashed at the same airfield after being shot up by a night-fighter over Berlin. He was later promoted to flying officer and was awarded the DFC but was killed on the last op of his tour on Schweinfurt on 24/25 February 1944. P/O Currie's crew, whose aircraft had been hit by lightning and by falling bombs, was one of three 626 Squadron crews who beat the odds and completed their tour in yet another Berlin raid on 28/29 January. Currie's rear gunner, Sgt Brettell, completed his second tour at the same time. There was quite a party at Wickenby that night.

At Fiskerton on 22 August, 20-year-old Sgt Eric Jones on 49 Squadron waited to fly his first op. The former builder's clerk from Newent in Gloucestershire and four of his crew had come together at 29 OTU at Bruntingthorpe near Leicester, where they trained on the Wellington before transitioning on to the Lancaster at 1661 Lancaster Conversion Unit at Winthorpe near Nottingham:

Some crew members had met on previous courses; some had met in the few days they had been on the station. I knew some

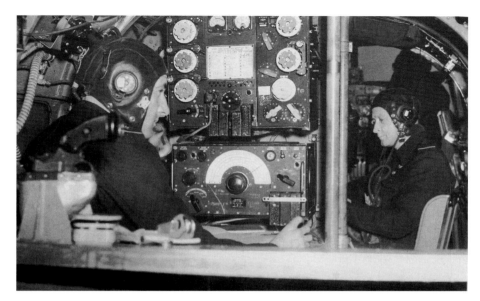

Wireless operator at his position, the TR 1154/55 in front of him with the W/T transmitter on top and the receiver below. To his right is the flight engineer.

of the pilots, we had trained together, but I didn't know any of the aircrew. Very cleverly the precise number of aircrew for the pilots available had been mustered into the room. I stood back, not knowing the abilities of any of the aircrew and also, I suspect, realising that friendship did not necessarily account for ability. I just let things happen; any positive choice I might have made might have been the wrong one. I was tall, very tall – 6ft 5ins – and also quite thin. My weight must have been around 11 stones and perhaps I did not look strong enough to handle a four-engined aircraft. On occasions like this I did not exactly ooze with confidence so I suppose I was somewhat surprised that, before the room thinned out too much, I had started to collect my crew. Except for Steve, a bomb aimer, they all looked so very old. Ken Blackham, an ex-London policeman and my navigator-to-be was six years older. Herbert (call me 'Peto') Whiteley who would look after the rear turret was ten years older, red faced with a balding head. Clarence (call me that and I'll punch you on the nose, you call me Pat) Peacock was a tough, square, smallish Canadian wireless operator, obviously older than myself. 'Steve' Stevenson who would look after the front turret and drop the bombs was a Londoner and had been in the Blitz and couldn't wait to get at the Germans. Steve was my age. I could not have wished for a better crew. Knowing the future that lay ahead of us and the absolute necessity of pulling together we all quickly became firm friends.

The hydraulically powered Frazer Nash FN 50 mid-upper turret which was developed from the FN 5 front turret and was also armed with two 0.303in Brownings. In March 1942, this turret was given a fairing which incorporated a cam track fitted to avoid areas of the aircraft being hit by its own fire. A modified FN 50, the FN 150, was designed in late 1944; this incorporated a better gun sight and improved controls. A Martin-type gun turret fitted with twin 0.5in Browning machine guns equipped the Austin-built Mk.VII after trials by the A&AEE. This turret was positioned further forward than normal and took up too much space in the aircraft, to the irritation of the crews. Trials were undertaken with remote-controlled barbettes and also the Bristol B.17 dorsal turret with two 20mm Hispano cannon on JB456 in May 1944, which was later fitted to Lincolns.

Conversion onto the Lancaster was a comparatively short course lasting only approximately four weeks and in that time I had to learn everything about the plane because my life and the lives of my crew depended on it. When I first saw a Lancaster bomber I thought it massive and wondered with trepidation how I would ever manage to fly it. The blow was somewhat softened by first being introduced to the Manchester, a twin-engined aircraft which, with two huge and unreliable Vulture engines, was not an operational success and was given a training role. We did a few hours on the Manchester and I was fascinated by the propellers on those huge Vulture engines. They seemed to rotate so slowly I could see the individual blades. Fortunately they kept turning for me until we were transferred onto the 'Lanc'. Big she was, but she handled beautifully and although we were not flying with a full load I could sense the power of those four Merlins and felt confident that they would handle all that was demanded of them. As on the Wellington, asymmetric flying was practised again and again, feathering first one engine and then, on the Lancaster, two engines. The 'Lanc' held height quite satisfactorily on two engines when unloaded and many a pilot having lost an engine on the way into the target was able to press on and deliver his bombs before turning for home. Either the pilot's seat or the rudder pedals adjusted to accommodate my long legs and I did fit quite comfortably into the 'Lanc' which was just as well as our longest trips were to take over nine hours. There was a tube suitably placed between my legs which did cater

for half of nature's requirements. If a direct result of using this device was to create a block of ice which fell on some German's head – who cares?

My crew of five was now inadequate to man the Lancaster so 19-year-old Ron Harris, the engineer, and a wee Scot, 'Jock' Brown joined the crew as the mid-upper gunner. Ron's task was to care of and nurse the engines, monitor the petrol supply from the various petrol tanks, take stock of all the pressures etc and would be constantly at my right hand side at all times in the air. He was the youngest member of the crew and I was second youngest and at first it sounded a little odd when they referred to me as 'Skipper'.

After completing 34 hours on the Lancaster the 'powers that be' must have considered the crew ready for operations so we were duly posted to an operational squadron. So it was to Fiskerton, home of 49 Squadron with the motto 'Cave Canem' (Beware the dog) that our course from Winthorpe journeyed. We now knew that we had finally arrived at the cutting edge of the war with Germany and that every time an operation was 'laid on' we would crew one of a maximum of eighteen participating Lancasters. We also knew that on most operations perhaps one, two or even more aircraft would fail to return. We would have to complete 30 operations before our tour was completed. From the moment of arrival at Fiskerton and for some weeks to come we thought our chances of survival rather slim, but it was a fact that the odds against us were accepted as a matter of course and that somehow these would be overcome and our crew would always return. Living was very much on a day to day basis and our main objective was to survive each 'op' and catch that crew bus into Lincoln on the next non-operational night. But we were not pitched into battle straight away – it was standard routine for each 'sprog' crew to do a 'Bullseye'. This exercise entailed stooging over a number of English cities at night to give their searchlight crews a bit of practice and also to get us used to being 'coned' in a pyramid of searchlights. We did not do too much evasive action so we did spend a considerable amount of time in the dazzling glare of these lights.

With that trip behind us the next step was the 'second dickey' operation. This involved the sprog pilot, in this case me and not his crew, flying with an experienced crew on a real live operation. On August 22nd I went to Leverkusen in the Ruhr with F/Sgt B.W. Kirton who was on his 14th trip and very little happened. There was complete cloud cover, very little flak and no fighters.[6] S/L G.A. Day our Flight Commander must have considered this operation a poor example of the real thing so the next night I was detailed for another 'second dickey'. In the meantime the crew were left kicking their heels on the base awaiting their first baptism of fire. Sometimes the aircraft carrying the 'second dickey' would be shot down and then the waiting crew would sometimes wait around for weeks for a spare pilot or fill in their time making up other crews to full strength. On entering the Briefing Room the first thing I noticed was the large map at the far end of the room and on it the route markers pointed straight to Berlin. I was in the very first Lancaster to get airborne in the Battle of Berlin. On only my second op the weather forecast was a clear sky over the target with no cloud! The weather men turned out to be correct. My skipper for this second 'second dickey' trip was F/Lt Munro, a New Zealander with the DFC. Although I felt a little minnow in the presence of such austere company I also thought, as I looked at him, 'This is experience and I am going to come back from this trip.' The trip I had made the previous night bore no comparison to this one. The 'flak', the marker flares on the ground, the fires, the bombs bursting, the searchlights and the fighter flares all contributing to a scene I had never thought possible. As we left the target area I reflected with amazement on our survival and thought of the remainder of the tour which I had to complete with my crew.

I returned to a barrage of questions from the crew. 'Well, what was it like?' 'Was the flak heavy?' 'Did you see any fighters?' etc, etc. How could I adequately describe the scene over the target? They would have to wait and see for themselves.[7]

Berlin had been the target for 335 Lancasters, 251 Halifaxes and 124 Stirlings. Seventeen Mosquitoes were used to mark various points on the route to the city in order to help keep the Main Force on the correct track and W/C 'Johnny' Fauquier DSO** DFC, the Canadian CO of 405 'Vancouver' Squadron RCAF, was the Master Bomber. Later, he would step down from the prized rank of air commodore to take command of 617 Squadron from W/C James 'Willie' Tait DSO* DFC.[8]

The Pathfinders were unable to identify the centre of the city by H$_2$S and had marked an area on its southern outskirts. The Main Force arrived late and many bombers cut a corner and approached from the south-west instead of using the

planned south-south-east approach. This resulted in more bombs falling in the sparsely populated southern suburbs of Berlin and in open country than would otherwise have been the case, 25 villages reporting bombs. Even so, it was the capital's most serious bombing raid of the war so far, with a wide range of industrial, housing and public properties being hit and over 2,600 individual buildings destroyed or seriously damaged. The flak and night-fighter defences were extremely fierce and 63 aircraft were lost or written off; 17 of them Lancasters.

At Fiskerton on 27 August, Sgt Eric Jones and his crew were on the battle order for Nuremberg, which was the target for 674 aircraft, 349 of them Lancasters:

Lancaster cockpit with the pilot in his seat and the flight engineer monitoring the gauges. Woe betide any pilot who flogged his engines. They were the pride and joy of the engineering officer and would be needed on the next operation. He had usually done at least 30 operations himself and knew a lot about engines.

Lancaster II powered by Hercules engines. The first unit to operate the 1,650hp Bristol Hercules VI-engined Lancaster II was 61 Squadron, which had one flight equipped with this variant from October 1942 to March 1943. The first fully operational squadron was 115, who flew the Mk.II from March 1943 to April 1944.

The large map at the end of the Briefing Room told us that our target was to be a city deep into the heart of Germany – Nuremberg. The proceedings commenced with the arrival of the Station Commander, G/C Windell. W/C Alexander Annan Adams, Squadron Commander, called all those present to attention and the CO took his seat. We had already heard rumours that the CO had been known to sit in his quarters and shoot out his windows with his service revolver. A story confirmed later to be absolutely true. Maybe he was one of those commanders who felt some stress at watching us young chaps go out to do battle or maybe he was just drunk. Then we had all the specialist officers. They told us of the weather en route (to the accompaniment of low hisses), the petrol load and bomb load we were carrying, which wave of aircraft we would be in (there were usually three) and the time on target. Also the defences we were likely to encounter and all the other information deemed to be necessary for the trip. The emergency runways available at Manston and Marham – they were 3,000yd long instead of the usual 2,000 – and the single searchlight on the coast at Mablethorpe maintained under all conditions as a guiding beacon for those in distress.

Then a good luck message from the CO and a message sent in from our Commander-in-Chief, Bomber Harris. There was not a single aircrew member who would not have followed Bomber Harris into hell and back had the occasion arisen, such was his immense popularity. He seemed to make all the hazards we were facing worth all the risks and he always believed and made us believe that what we were doing could end the war without the necessity of invading Europe.

We left the Briefing Room to while away the hours until take-off time. During this time no one was allowed off the station. With everyone on the station keyed up for the off a 'scrub' or cancellation went down like a lead balloon. It meant that the whole procedure would have to be gone through the next night and maybe the next night. And for us, no trips being deducted from tour requirements. Operations were never laid on during the full moon period. Too much light meant too many losses. But we weren't idle during these moon lay-offs – the time was always filled in with training flights.

Little Audrey was our crew bus driver throughout our tour. Audrey would not have won any beauty contests but she had a heart of gold and we all loved her. After the ground crew that flagged us into dispersal after an op, she was the first person to greet us. As time went by she was to remain our driver and was to wait patiently for our return at all hours of the night and in all kinds of weather. We would not have swapped her for any other driver on the station. Audrey transported us out to *G-George* for our first operation. I walked round the aircraft carrying out the laid down external check. Ron, our engineer, chatted to the ground crew to ensure everything was OK from his angle and then the crew prepared to board. We had heard from some of the old hands that it was good luck to pee on the tail wheel before boarding the aircraft. It was a good idea anyway thinking of those long hours ahead. So we all queued up to carry out this ritual having previously ensured that there was sufficient content to carry out this task. A ritual to be repeated on all our trips and I suspect, by most other crews. Entering the aircraft, Peto turned left towards the rear gun turret (housing four Browning machine guns) and Jock, the mid-upper, only had a few steps to reach his two gun turret but the rest of crew had to clamber over the notorious centre spar.

As take-off time approached engines were started up, intercom checked with each member of the crew and all the necessary cockpit checks made; many of these with Ron who would be at my right-hand side at all times. Finally, a signal to the waiting ground crew to pull away the wheel chocks; a last thumbs up to them and we start to roll. The ground crew would remain on duty until 'their' aircraft returned, hopefully with a minimum of damage or none at all. It was always 'their' aircraft and we were only its custodians for a few hours, but of course we thought differently. So we took our place in the queue of eighteen or so Lancasters making their way slowly to the take-off point; slowly, because the perimeter tracks were narrow and our undercarriage wide. One Lanc dropping a wheel into the soft earth at the side of the track could abort the trip for everyone behind. At this time, loaded with maximum petrol and bomb load, the aircraft weighed over 30 tons.

Our turn for take-off eventually arrived by way of a green Aldis lamp flashed from the small black and white chequered hut at the end of the runway and we slowly moved out onto the 2,000-yard runway lowering 20° of flap to give added lift to the wings for take-off. With the pitch of the propellers in fully fine I lined up the Lanc on the runway, applied the brakes, came to a stop and tested all the engines at 0lb boost. I mustn't hesitate too long, the aircraft behind me would all be anxiously awaiting their turn so, to vigorous waves from the small knot of station personnel who always turned up for ops take-offs. I slowly eased the four throttles forward. The Lanc started to swing – it always did – and I checked it by correction on the throttles. As the tail came off the ground and I had rudder control I moved all the throttles to fully open. I then left the throttles to Ron whose hand has been following closely behind mine and I concentrated on the take-off. The throttles fully open gave 18lb boost which was in excess of the recommended setting for the Merlin engines. We could only hold this throttle setting for three minutes. Ron shouted out the airspeed. I had no time to look at the instruments, 60, 70, 80 and at about 110mph I felt the aircraft wanting to leave the 'deck', the end of the runway looming ever nearer until suddenly we were airborne. I reached down and selected the undercarriage lever to the 'up' position and immediately I gained a few more miles per hour. We were still only a few feet into the air as we roared over the edge of the airfield and if an engine failed now we had had it. The speed was now 145mph and if an engine failed now we might just make it. The three minutes were up and I throttled back to a powered climb at 9lb boost. At

about 300ft I started to ease off that 20° of flap. If I did it too quickly I would sink back into the ground. So we slowly, very slowly, climbed away until I reached a safe height where I could throttle back to a normal climb at 7lb boost and 2,650 revs per minute.

It was customary to obtain at least 10,000ft (oxygen height) over England before setting course for the continent and one of the more usual points from which the Group set course was Fakenham in Norfolk. It could still be half light at this time and we would be able to see Lancasters amassing all around us. Halifaxes from 4 Group and other bombers from other Groups would be carrying out a similar exercise in other parts of the country. As soon as darkness fell one felt all alone, suspended in the night sky with only the occasional glimpse of another Lanc to reassure that there were others heading towards the enemy coast. Normally, when flying on operations at around 22,000ft we would be above the weather but in unstable air conditions the cloud tops could go up to 30,000ft plus and inside these clouds flying conditions could be very unpleasant with the aircraft being buffeted around like a cork in an angry sea.

From the outset of the operation strict radio silence was maintained, hence the use of the Aldis lamp for permission to take off, and although Pat, our wireless operator, kept a

PM-H on 103 Squadron at Elsham Wolds, which was badly damaged on the 23/24 August raid on Berlin and was later repaired and transferred to 576 Squadron, completing more than 50 trips before the end of the war. Those in the photo include Doug Finlay, Jim Wivers (killed in action in September 1943), Sandy Rowe, Sgt Harry Wheeler (who was killed on 23 August when Lancaster W4323 PM-C was destroyed by fire at Elsham), Bill Gillespie, Sgt Steel, Ian Fletcher, John McFarlane and the MT driver, Peggy Forster.

L-London on
103 Squadron at **Elsham**
Wolds in the summer
of **1943**. Note the wolf
emblem beneath the
cockpit.

continuous watch for a possible recall etc, he was unable to send any messages. As our tour progressed, Ken, our navigator, might be asked to send back new wind speeds and directions he may have assessed. These were given to Pat, he would send them back to base and they would be redirected back to main force aircraft. This may have happened once or twice at the end of our tour when Ken's reliability would have been proven. Pat also had the capability of transmitting the roar from our engines into the frequencies used between German night-fighters. This was code-named 'Tinsel'.

As our Lanc was heading out over the North Sea, gradually gaining height, on the intercom could be heard: 'Rear gunner to Skipper, permission to test guns.'

'OK rear gunner; go ahead but take a good look round first. This goes for you mid-upper and you bomb aimer.'

Guns were always tested at height over the North Sea and as one bullet in every five was a tracer shell it was possible to see the line of fire as the bullets streamed away from the aircraft.

We knew that the fighters were there and if we didn't keep a constant look out we could be shot from the sky without knowing what had hit us. I would be gently weaving our Lanc either side of course to give the crew, particularly the gunners, a greater field of vision. It was almost a hypnotic situation, staring fixedly at the instruments with the

occasional glance ahead when suddenly I would be jolted from my reverie by a cry from Peto in the rear turret or Jock in the mid-upper. 'Fighter on the starboard quarter,' and instantly the whole crew was 100 per cent wide awake and a little shiver would run down my spine. We would wait to see if he had spotted us and if we were his next intended victim. Sometimes it was another Lancaster – we could not be too careful – but sometimes it was a German fighter. Peto would say, 'Standby by to corkscrew,' and as the fighter closed in on us it would be 'Corkscrew port – go,' and I would immediately jab the control column forward and the same time pulling the starboard wing up and over and in no time at all we would be entering a powered turning dive. As the speed built up I would wing over in the opposite direction expecting all the time to see the fighter's tracer bullets streaking past. As the airspeed built up to around the 260mph mark I would start to ease the 'stick' back easing the Lancaster over into an opposite turn and up we would go turning and twisting in this fixed pattern until that blessed shout from Peto, 'OK Skipper we have lost him, back onto course.'

If we were attacked on the outward journey we would still have our full bomb load and most of our petrol. The Lanc would be heavy and it was desperately hard work and if the fighter was still there the whole procedure had to be repeated. The 'corkscrew' followed a fixed pattern – practised in training it enabled the gunners to sight their guns with reasonable accuracy. The exertion required in flying the Lancaster through this manoeuvre was intense and I always finished up bathed in sweat.

Unlike my 'second dickey' to Berlin which was a successful raid with much damage being caused (but with the highest losses Bomber Command had experienced to date) Nuremberg was not a success.[9] Although we must have finally completed our bombing run I do not recall it but we might have been forgiven if our bombing effort was one of those which 'crept back'. We operated between 20,000 and 24,000ft except for the climb out of the target area. The height was always dependent on how cold the night was. The colder the night the higher we could fly, the engines developing more power from cold dense air. Sometimes the outside air temperatures could be as low as -500 Centigrade. One cold night I reached 29,000ft climbing out of the target area. I distinctly remember the manner in which the plane wallowed in the thin rarefied air, way beyond the design ceiling of the Lancaster.

On the return journey I contacted Ken on the intercom and told him we were over the North Sea and that I would start to descend. It was a very dark night and inexperienced me, after hours of flying on instruments, thought we must have reached the North Sea. Ken replied, 'Hold your height Skipper; we are still over enemy territory.' An argument ensued, the only one we ever had in the air, but I did hold my height and he was correct and we were still over Holland. To maintain height over enemy territory was of paramount importance, to lose height too soon brought the aircraft into the range of even more weaponry. In the future it was to be either consultation with or taking instructions from Ken when it came to the navigation of the aircraft. My job, so far as he was concerned, was to steer an accurate course no matter what (even though we weaved from side to side) and to maintain a consistent altitude.

The 'rookie crew' survived their first sortie over Germany and confidence had probably increased by .01 per cent, and the pattern of visits into Lincoln, operating whenever the moon and weather and sleeping permitted, began to take shape.

On 30/31 August, 660 heavies targeted the twin towns of Mönchengladbach and Rheydt – the first major attack on these cities since 11/12 August 1941. Unlike the two-phase operations of 1944–45, which would allow a two- or three-hour gap between waves, this was a two-minute pause while the Pathfinders transferred the marking from the former to the latter. The Main Force crews exploited accurate marking in what was the first major raid on these targets, and over 2,300 buildings were destroyed. About half of the built-up area in each town was devastated. Twenty-five aircraft failed to return, 22 of which were shot down by *Zahme Sau* (Tame Boars).

On the last night of August, the Main Force assembled in a giant stream and headed for the 'Big City' once more. Of the 622 bombers detailed for the raid, over 100 aircraft again were Stirlings. Nine Mosquitoes of 105 and 109 Squadrons route-marked for the heavies by dropping red TIs near Damvillers in north-east France and Green TIs near Luxembourg. The enemy used 'fighter flares' to decoy the bombers away from the target and there was some cloud in the target area. This, together with difficulties with H_2S equipment and enemy action, all combined to cause the Pathfinder markers to be dropped well south of the centre of the target area and the Main Force bombing to be even

W/C John Searby of 83 Squadron, who was the Master Bomber or the 'Master of Ceremonies' on the Peenemünde raid on 17/18 August 1943.

further back, bombs falling up to 30 miles back along the line of approach. The intensely bright white flares dropped in clusters of a dozen or more from about 20,000ft at the corners of the target area and a double strip apparently dropped by rapidly moving aircraft around the perimeter of the area and igniting at about 17,000ft lasted for several minutes and served to illuminate the bomber stream. About two-thirds of the 50 aircraft that were lost were shot down over Berlin by night-fighters.

In the daylight reconnaissance 12 hours after the Peenemünde attack on 17/18 August 1943, photographs revealed 27 buildings in the northern manufacturing area destroyed and 40 huts in the living and sleeping quarters completely flattened. The foreign labour camp to the south suffered worst of all and 500–600 mostly Polish workers were killed. The whole target area was covered in craters. The raid is adjudged to have set back the V-2 experimental programme by at least two months and to have reduced the scale of the eventual rocket attack on Britain. (*Australian National Archives*)

Forty-seven aircraft – ten of them Lancasters – were missing. A Lancaster on 106 Squadron piloted by F/O H.D. Ham crashed at Romney Marshes in Kent on return, as Sgt N.D. Higman, the 18-year-old tail gunner, recalled:

The outward trip had gone well, with no opposition until the target approach leg. Then flak came up in a terrifying hail of shells and tracers, exploding all around the wallowing bomber. For nearly ten minutes the Lancaster was buffeted and bounced around the sky, but it released its bomb load over the target and then swung out of the flak zone in a fast climbing curve, heading for safer air. In the rear turret the eighteen-year-old gunner was 'paralysed with fear and feeling distinctly sick in the guts'. Nothing in his training had prepared him mentally for such an ordeal; while the sight

of another Lancaster exploding in mid-air in a huge gout of flame merely added to his terror. Fighting against an almost overwhelming desire to faint, the gunner remembered his skipper's last words before take-off; 'It may become a bit rough over there, but for Christ's sake keep watching for fighters, Higman. I'll take care of anything else.'

The Lancaster survived the holocaust of flak and was drumming steadily homewards when Higman saw a shadow in the port quarter behind. It was a fighter! Yelling a warning into his microphone, Higman let loose with all four Brownings as the silhouette grew larger, with strings of tracer leaping directly at the rear turret, then rushing past Higman on each side. The ensuing few minutes were a kaleidoscope of lights, wheeling clouds and moon, cordite stink and rattling; with Higman valiantly trying to keep his seat and still bring his guns to bear. Unbeknown to the gunner, cannon shells had ripped out most of the pilot's instrument panel, smashed the wireless set, wounded the navigator and knocked him unconscious, and slashed chunks out of the forward length of fuselage. The port inner engine had been put out of action. Ham eventually evaded any further fighter onslaught and then took stock. He had no way to obtain a position fix, his navigator was in no state to help and the Lancaster was sloppy on the controls. The port inner engine began pluming smoke ominously – he could expect a fire any second. A rough check on the time told him he should be somewhere near the coast, but where? The intercom was obviously 'out' and to add to his problems he was heading into a sky of dense black cloud which eliminated any hope of guidance from the stars.

Determined to try for home, Ham set a guessed course north-westwards and struggled with reluctant controls for as long as he could. Then, on the point of physical exhaustion, he realised that the fuel state was almost zero – the fighter must have hit the tanks or the supply lines. The port outer engine grunted to a stop, starved of fuel and then the starboard inner began to slow. Yelling to the bomb aimer to tell the others, he gave the order to bale out. Each man made his way to the escape hatch. The navigator, now conscious though still dazed, was helped into his parachute and through the hatch by the wireless operator. The mid-upper gunner climbed down from his turret, made his way back to Higman's rear turret and banged on the doors. Certain his knocking had been acknowledged, he went forward and baled out. Higman stayed in his seat, still searching for fighters and he only relaxed when they were well out of Luftwaffe range. Receiving nothing on the intercom, Higman flipped open his rear doors and looked down the fuselage. To his horror he realised he was on his own. Panic set in. Closing the door, he rotated the turret to the beam, having grabbed his parachute pack from its stowed position and then back-flipped out of the turret.

Coming to earth at the edge of a small copse, Higman made for cover and got rid of his parachute and harness, burying them as best he could. Then, in inky darkness, he set off. For the next 18 hours he lay low, avoiding all paths and roads and snatching sleep in a thick patch of gorse. As darkness approached again he set out in a vaguely northern direction. Near-exhausted, hungry and dying for a drink of water, he came out of a small clump of trees – and nearly tripped over an RAF corporal making love to a blonde WAAF! The astonishment was mutual. Higman then learned that for 18 hours he had successfully 'evaded capture' within 2 miles of a tiny RAF signals unit near the coastline of Kent.

On 3/4 September, because of the high casualty rates among Halifax and Stirling aircraft in recent raids on the 'Big City', the raid on Berlin consisted entirely of 316 Lancasters. Wynford Vaughan-Thomas, a BBC Home Service commentator, and Reginald Pidsley, a sound engineer, flew with F/L Ken Letford and his crew on *F-Freddie* on 207 Squadron at Langar, and recorded his impressions on a one-sided wax '78' disc 4 miles high over Berlin:

Now and again, as we watch, we see a burst of flak, a bright light winking among the concentrated beams. They have got every single searchlight you could imagine out there to catch us. We are coming up to them all the time, waiting for it. In a moment it will be our turn to pass through them. A dark shape is going out ahead of us, another Lancaster, to lead in. There goes the flak again, a winking burst up among the searchlights. They must be having a go at us all right. Away to port another constellation is coming up. They work in great groups, trying to stop and grapple you as you come in over the coast. All the time, they are moving in. It is disconcerting to see that welcome waiting for us. I am counting the time, watching the hand on my watch creeping round. I know that it will be our turn in exactly three minutes' time. [Pause] In the cone of searchlights, they caught one of our aircraft. Up goes the flak around him, bursting in vivid flashes. Now there are winks from the ground below us. They may be after us, because the

searchlights are starting to move away. They have left that other bomber and they are moving now slowly towards us, feeling for us all the time. They are pumping up the flak in a steady stream.

You suddenly see a white flash on the ground, then just seconds later, there is a vivid burst among the searchlight cones. There goes the flak, bursting in that cone of searchlights, darting from vivid white pinpoints, moving all the time, trying to follow that bomber. Again they come bending, the whole lot of them. They seem to bend towards us, following a master beam. We are moving away to starboard and it looks as if, this time, we've slipped through.

Cloud up to 21,000ft provided uninterrupted cover and hampered the flak and the fighters. On the outskirts of Berlin the cloud broke up, leaving a clear space over the aiming point, and after the attack the cloud drifted back over the city. Four Mosquitoes dropped 'spoof' flares well away from the route to Berlin to decoy night-fighters away from the bomber stream, which approached the city from the north-east. The marking was mostly short of the target and the bombing that did reach the built up area fell in residential districts.

'It is pretty obvious now, as we are coming in through the searchlight cones,' continued Vaughan-Thomas:

it is going to be hell over the city itself. There is one comfort. It is going to be soundless because the roar of our engines is

U-Uncle on 103 Squadron at Elsham Wolds with W/O R.G. Cant's crew, who were shot down on Mannheim on 5/6 September 1943 on *S-Sugar*. L–R: W/O R.B. Cant (pilot); Sgt G.F. Thomas (navigator); F/Sgt G. Dickson (engineer); Sgt S. Horton (WOp); Sgt W.R. Milburn (mid-upper gunner); Sgt T.D.G. Teare (bomb aimer); Sgt D.R. Parkinson (rear gunner). All the crew survived and most escaped, either back to the England or to Switzerland. Only Thomas was taken prisoner.

Lancaster III DV238 on 49 Squadron being towed by a tractor driven by WAAF driver LACW Lillian Yule at Fiskerton in the summer of 1943. DV238, which formerly served on 619 Squadron, was later transferred to 44 Squadron at Dunholme Lodge where it became KM-M. DV238 and P/O Donald Albert Rollin DFC's crew were lost on the Berlin raid on 16/17 December 1943 when the Lancaster crashed at Diepholz. Rollin and six of his crew were killed. One man survived.

drowning every other sound. We are running straight into the most gigantic display of soundless fireworks in the world. We are due over our target in two minutes' time. Bill our bomb aimer is forward, he is lying prone over the bombsight and the searchlights are coming nearer all the time. There is one cone, split again and then it comes together. They seem to splay out, like the tentacles of an octopus waiting to catch you. Then suddenly they come together.

'Half a minute to go. OK, boys,' said the skipper in a deadpan tone.

'Left … left,' said the bomb aimer, or bombardier as he was called on this aircraft.

'Hello, bombardier, ready when you are, bomb doors open.'

'Steady … steady … bit longer yet. Steady … level a bit. Steady, bombs going. One, two, three, bombs still going.'

F-Freddie was attacked by a night-fighter but Letford's gunners, W/O Fieldhouse and Sgt Devenish, shot it down. Needless to say, Vaughan-Thomas and the others were relieved to see England again:

There it goes, our first sight of England, just a little light from a beacon, flashing at us from the darkness below. After the giant glare of lights we left behind in Berlin, it seems small and frail. But everyone on board *F for Freddie* is mighty glad to see it. Con our navigator has got out the goodies that we were not able to tackle during the hectic hours over

Germany. We are cracking open our fruit juice. Someone has got his mouthful of chocolate and Scottie the engineer, who is just ahead of me, is pouring out from his thermos a cup of hot tea. We have got our oxygen masks off. There is a sense of freedom throughout the whole of *F-Freddie*. Ken our skipper has just said over the intercom: 'Boys, we won't be the first home, but we're damned glad to be home at all.' (Skipper continues after pause) 'Hello, bombardier, English coast should be coming up now. Will you tell me when you cross it?'

'OK, I'll let you know … I can see it coming up ahead.'

'Thank you.'

'NAV lights on, skipper.'

'OK, navigator, nav lights on.'

Bomb Aimer: 'We should be over it in a few seconds. There it is. Good to see old England again.'

Skipper: 'Yes, after that trip, boy, that's a sight for sore eyes, that is.'

Navigator: 'ETA at base: 15 minutes.'

'OK, navigator, thanks very much … Hello, engineer, skipper here. How's the petrol going? Are we doing all right?'

'Oh yes.'

'Everything running OK?'

'Just lovely.'

'Good show, boy.'

The wax disc made of the operation was rushed to London and broadcast on the ten o'clock news. In all it was broadcast nine times in English and on numerous European and foreign programmes as well as the American network. It was described as 'the outstanding broadcast of the war'. Twenty-three Lancasters failed to return.

On 5/6 September, just over 600 crews were detailed to make a double attack on Mannheim and Ludwigshafen. At Bardney, P/O Bill Siddle's[10] crew on 9 Squadron were on the battle order that night for their second raid on Mannheim in less than a month, as Clayton Moore recalled:

We found the target to be unsparing in the kind of activity we had come to expect of it: searchlights, flak bursts, flares, a lot of smoke and a lot of aircraft – including several that were seen burning fiercely as they fell from the sky. Window was again in use, but it was becoming obvious that our worthy opponents were fast adapting themselves to the situation. There was a reduction in the amount of flak coming up and the searchlights seemed to be concentrating on lighting up as much of the sky above the target as possible, thus increasing the likelihood of our bombers being sighted. Because Window had for the present neutralized the German radar system, thus making the directing of their fighters onto us almost impossible, large numbers of single-seater aircraft were being deployed as freelance hunters. The pilots of these relied entirely on interception by sight and were ably assisted by their practical liaison with the ground defences. Of course, this system of co-operation was only effective in the vicinity of the target and I envisaged its advantage over us if the target was covered in cloud as often it was. In such circumstances, we would appear to a high-flying fighter as so many flies crawling across an illuminated opaque screen and would have difficulty detecting the approach of our attacker as he dived on us from the darkness above. We were to witness this phenomenon later, during the Battle of Berlin, when the target was often found to be covered by unbroken cloud. As a crew, we were showing promise. Siddle had a flair for leadership which produced a degree of co-operation and discipline that was not to be found in some of the other crews. (The 'chop rate' amongst new and inexperienced crews appeared to be highest during the first six trips and we had just passed the magic figure.) We were now into the second stage, during which complacency could produce the 'it'll never happen to us' syndrome which was so often accompanied by a slackening of concentration. This was probably the most dangerous of the three segments of a tour and it was a pity that so many otherwise good crews succumbed so needlessly to careless overconfidence at this stage. The third and final segment of the tour (the remaining ten trips) affected the outlook of a crew in an entirely different way. During the months that you had spent with the squadron so far, you had come to estimate your chances of survival and had rated them as very slim. Then, having entered number twenty in the log book, you suddenly realised that maybe you might just make it after all and the resulting degree of caution could in itself be a danger. The raid was successful but 34 aircraft – 13 of them Lancasters – were missing.

The following night found us again listed on the battle order. This time the target was Munich – again deep inside Germany – and we still had ED975 (WS-Y) as our mount.

We witnessed a few casualties during the outward journey, but had come to expect this. The victims were

Air and ground crew of *X-X-Ray* on 103 Squadron, which was flown by F/L (Later S/L) Van Rolleghem, a Belgian pilot who became one of the squadron's most decorated pilots. John Lamming (third left, front) had just painted the Belgian and British flags on the Lancaster and the insignia marking *X-X-Ray*'s 21st operation.

usually some unfortunates who had wandered off track and strayed away from the collective safety of the main stream and this usually resulted in them being caught up in the concentrated firepower of a built-up area. As a general rule, however, the casualty rate was highest in the area of the target and on the homeward leg of the flight. This was because, until then, the fighter force was uncertain of the intended target, thus precluding an organized concentration of resources. Having attacked the target, our next objective (home) was plain and the Jerry fighters were always quick to seize on the advantage. Once again I began to experience the now familiar and unpleasant feeling in the pit of the stomach and the dryness of the throat as we began our run to the target. Although the actual assault on our objective was probably no more hazardous than the rest, it was always a time of intense vigilance and tension, caused largely by the feeling of naked exposure one got when having to traverse an illuminated area which contained so much activity.

Ken was busy giving directions as we lined up on the target with the bomb doors open. There seemed to be just the usual enemy activity: plenty of searchlights combing the sky for customers; one poor sod coned over on the port beam; the expected haphazard mix of light and heavy flak coming up; there were no fighters to be seen so far, but I wasn't complaining about that.

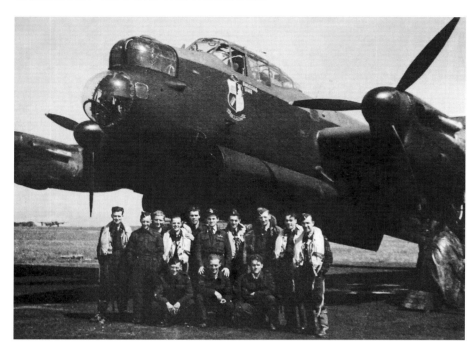

The air and ground crew on *R-Roger* on 576 Squadron at Elsham Wolds.

'Bombs gone!' It was all routine so far. Just the wait for the photo-flash and then nose down and head for home, weaving like hell.

I was traversing my turret to the starboard beam in search of trouble when we got it. Above and just to starboard and directly in my line of sight, I beheld a blinding, dull red flash and heard the deafening 'whump!' as a heavy ack-ack shell burst a few yards away. Our wounded Lancaster at once dropped her right wing to the vertical as if recoiling from the searing heat of the explosion and it looked like we were going to turn turtle completely. But Bill must have been fighting furiously with the controls, for she slowly levelled out and resumed her normal attitude. The engine and flight controls were responding normally, so it appeared that we had been extremely lucky to have escaped with so little damage. Once clear of the immediate danger, we settled down to the long homeward flight across the hostile skies of Western Europe. During the hour immediately following on the flak burst, we had investigated the extent of the damage as best we could and this seemed to be mostly superficial. There were a good number of small holes to be seen, but the riggers would soon have them patched up – hopefully before we had to hand her back to her regular crew.

We had been airborne for the best part of five hours and the monotony was beginning to have an effect, so I decided to take a couple of wakey-wakey pills in order to ward off the drowsiness that was creeping over me. As I eased my oxygen mask aside in order to take the tablets, I got a strong whiff of a substance which I instantly recognised as 100-octane petrol. I at once reported this to the skipper and there followed an in-depth investigation into the accuracy of the damaged instrument panel. Three of the six fuel gauges were functioning normally, but it was probable that at least one tank had been holed. Bill and Reg then got involved in a discussion on the transfer of fuel from various tanks to others by means of the cross-feed system. Because we had undoubtedly lost an unknown quantity of fuel, there was also a need for fuel conservation and this was agreed and adopted.

We had not yet reached the enemy coast and the fighters were still amongst us, as was evinced by the number of aircraft to be seen spiralling down. One of these had bought it while flying close behind us on the starboard quarter. I was unable to see the fighter, but had seen the tracers strike home shortly before a wing began to burn brightly, casting a revealing red glow in our direction. We slid away from the danger and logged yet another casualty.

We were just crossing the French coast on the homeward leg when our starboard outer engine was showing signs of overheating. Bill ordered it to be shut down and feathered. Because we would be descending for the rest of the flight, the loss of the engine would not greatly affect the performance of an unladen Lancaster and we would also be saving on whatever quantity of fuel was still available to us. In truth, the need to shut down the engine was a blessing in disguise and would to some extent serve to conserve our precious fuel still further. All things considered, we should have sufficient in reserve to make it back to home base comfortably.

At last the English coast passed beneath us and we were soon within sight of the Bardney beacon. Because we were late, the circuit was almost devoid of aircraft as we joined it and Bill requested landing instructions over the r/t. As he was doing so, Reg reported that the starboard inner engine was showing signs of losing power. Bill at once requested landing priority, which was granted. As we hurriedly lined up for the runway, the faulty engine appeared to right itself, so we prepared for a normal three-engined landing.

We entered the funnel and lined up for the approach and as we did so I positioned my turret in the dead astern position in order to avoid the rudder effect. We were now flying at about 200ft and were about one mile from touchdown. Revs and pitches had been set on the three remaining engines and the order had been given for wheels and flaps to be lowered. Everything seemed to be OK from where I was sitting. She wallowed slightly as the flaps were applied and I could see that Bill was correcting the swing with the rudders as we continued our descent. The first indication I got that anything was wrong was when I discerned an abrupt change in the note of the three remaining engines. There was a lot of popping and coughing taking place and the aircraft at once swung sharply to starboard, seemingly out of control.

'What the hell's happening?' Bill shouted, as he struggled with the controls in an attempt to right the aircraft. 'I don't know, Skipper,' Reg replied in a voice that lacked the calm and confidence that it usually projected. 'We've got red lights showing all over the place.'

'Brace for a crash!'

In the eerie silence that followed the order, I watched the barely discernable horizon drop from my view as the nose dipped sharply downwards and our aircraft began its death dive and then righted itself for a brief moment as Bill fought desperately with the almost useless controls. In the few brief seconds that remained before impact, my thoughts raced furiously as I took stock of our situation. We were without power and the drag of the lowered undercarriage and wing flaps was adding to the problem. These could not be retracted because the hydraulics were no longer functioning. We could not bale out because we were too low. At less than 200ft, a parachute would not have time to open fully before its wearer struck the ground. There just was not anything we could do except sit tight and hope for the best.

'What a bloody awful way to die!' I thought, as I tried to prepare myself for whatever the event might bring.

Having sensed that the moment of impact must be near, I forced my back hard against the turret doors, just as I felt the Lanc give a sudden judder and I heard the sound of

tearing metal coming from I knew not where. We were still in a nose-down gliding attitude, but we had hit something a glancing blow – probably a tree or a building. Then I felt the tailplane shake and vibrate violently and I was forced away for the doors, only to be pounded relentlessly on either side of the head by the oscillating gun mounts. I instinctively raised my arms and clasped my hands behind my head in an effort to ward off the blows. Then we hit once more and I lost all sense of direction as I felt myself and my turret spinning through space at an alarming rate. There was just one more sickening impact amid all the turmoil and confusion. I felt a severe blow to the back of my head and this was accompanied by a sharp pain, the seat of which I was unable to determine.

There was so much pain. Then, the pain eased and there was nothing but nothingness; peace; silence; oblivion.

Air and ground crew on *N-Nan Nulli Secundus* on 103 Squadron at Elsham Wolds, showing 52 sortie bomb symbols and racy nose art. *(Bob Edie)*

OLD MAN LUCK

There were many stories from aircrew returning from the raid relating to the severe icing conditions causing engines to seize up and only restarting when the aircraft had been brought down to lower and warmer levels. Old Man Luck was still pointing us in the correct direction.

W/O Eric Jones, Leipzig raid, 20/21 October 1943

Late on in September 1943, raids were dispatched against targets in France and against the Dortmund-Ems Canal and to cities in Germany such as Hanover (twice) and Mannheim. On the night of 23/24 September, 627 bombers – 312 of them Lancasters – were dispatched against Mannheim and a diversion raid by 21 Lancasters and eight Mosquitoes was made on Darmstadt. At Mannheim 32 aircraft, including 18 Lancasters, were lost. Three nights of minor operations followed and then, on the 27th/28th, Hanover was the target for 678 bombers, 312 of them Lancasters. The bombing was very concentrated but the Pathfinders were misled by strong winds, and just four visual markers identified the aiming point, although some crews made as many as four runs over the city trying to identify the aiming point. Only 612 aircraft dropped their bombs and most fell on an area 5 miles north of the city. RAF crews were not yet expert with the new H_2S navigational radar, which showed up an expanse of water very well, but the Steinhuder See, a large lake that was used as a way point, had been almost completely covered with boards and nets. Millions of strips of 'Window' were dropped and a diversion raid on Brunswick by 21 Lancasters and six Mosquitoes was successful in drawing off some of the night-fighters, but no fewer than 39 bombers were shot down; ten of them Lancasters.

After a break in operations, over 350 aircraft took off on the night of 29/30 September for Bochum, home of the huge Bochumer Verein steelworks, producing 160,000 tons a month. Oboe-assisted Pathfinder aircraft marked the target and the bombing was concentrated. Four Lancaster IIIs and five Halifaxes failed to return. A total of 178 victories were credited to German night-fighters in September, and during October 149 RAF bombers were destroyed by *Nachtjagd*.

On the night of 1/2 October, 243 Lancasters and eight Mosquitoes of 1, 5 and 8 Groups set out for Hagen, one of the many industrial towns in the Ruhr. At Binbrook P/O Scott RAAF on 460 Squadron RAAF hauled his Lancaster off the runway for what was to be his 24th operation over occupied Europe. Over Holland the Australian's aircraft was hit by anti-aircraft fire and the cockpit canopy was shattered. With fragments of the Perspex lodged in an eye, his face covered in blood and only partial-sighted, Scott decided to press on to the target and dropped his bombs on the aiming point. The raid was a complete success, achieved on a completely cloud-covered target of small size. Two of the town's four industrial areas were severely hit and a third suffered lesser damage. An important factory, which manufactured accumulator batteries for U-boats, was among the 46 industrial factories that were destroyed and 166 damaged. Battery output to the *Unterseebooten* was slowed down considerably. Thirty thousand people were bombed out as over 3,400 fires, of which 100 were 'large', ravaged the town. One Lancaster failed to return and a second was lost when it crashed on return in the Bristol Channel with the loss of all seven crew.

On 2/3 October, 294 Lancasters were detailed to bomb Munich. Visibility over the target was clear but the initial marking was scattered. Eight Lancasters failed to return. The following night the target was the Henschel and Fieseler aircraft factories at Kassel, which made V-1 flying bombs. Just over 200 Lancaster crews were included in the force of 547 aircraft that were detailed. The route would take them in north of the Ruhr and out to the south. There was nothing unusual about the target and as usual crews were warned to expect a strong concentration of searchlights and anti-aircraft guns. The H_2S 'blind marker' aircraft overshot the aiming point at Kassel by some distance and the 'visual markers' were unable to correct the error because of thick haze which restricted visibility. *Nachtjagd* claimed 17 of the 24 heavies

LANCASTER OLD MAN LUCK

shot down for nine Tame Boar losses. On 4/5 October, just over 400 aircraft attacked Frankfurt. Clear weather and good Pathfinder marking resulted in extensive devastation to the eastern half of the city and in the inland docks on the River Main, both areas being described by the Germans as a 'sea of flames'. Tame and Wild Boars claimed 12 victories, though 11 bombers – five Halifaxes, three Lancasters, two Stirlings and one of three American-crewed B-17s – were lost.

Airborne Cigar (ABC) Lancasters on 101 Squadron equipped to carry out jamming of the enemy R/T frequencies made their operational debut on the moonless night of 7/8 October when over 340 Lancasters raided Stuttgart. Each Lancaster also carried a specially trained German-speaking operator who interfered with broadcasts on the German night-fighter frequency. The ABC-equipped Lancasters, which could be distinguished externally by three 7ft aerials – one under the nose and two on the upper fuselage – played their part on their debut, but any signal transmitted by a bomber could be used by a fighter to locate it and in time 101 Squadron, who flew more operations than most other squadrons as they were the only one equipped with ABC, would lose heavily, as the enemy sought to diminish their effectiveness. What really confused the *Jägerleitoffizier* (JLO) were the diversions by Mosquitoes to Munich, Emden and Aachen; only a few night-fighters reached the Stuttgart area and only then at the end of the raid. Just three Lancasters were lost.

On 8 October, just over 500 aircraft of the Main Force visited Hanover. At a certain position the Lancasters, Halifaxes, Wimpys and eight Mosquitoes turned towards Hanover, leaving 119 bombers as decoys still heading for Bremen. Unlike the two previous raids on Hanover, there was no 20-mile timed run from the Steinhuder Lake to confuse crews, and in clear conditions the initial H$_2$S-guided blind markers placed their yellow TIs very accurately around the aiming point in the centre of Hanover, which was brilliantly illuminated. The Main Force, following their usual procedure, bombed the first red TIs that they saw, which in this case were nearest to the aiming point. About 70 per cent bombed before these TIs were extinguished, with the result that an exceptionally concentrated attack developed with a 'creep-back' of only 2 miles, all within the built-up area. It was estimated that 54 per cent of the built-up area was destroyed by fire. *Nachtjagd* destroyed all 27 bombers lost.

Three more large-scale Tame Boar operations were mounted during October. On the 18th/19th, when Bomber

Lancaster III crew on 576 Squadron at Elsham Wolds before departing on an operation late in 1943. L–R: F/Sgt Len Sumak (mid-upper gunner); Sgt Jock Boston (rear gunner); Sgt Bob Hammond (WOp); Sgt Ted Roff (flight engineer); Sgt Cyril Rollins (pilot); P/O Louis Reese (bomb aimer); Sgt Jack Rutter (navigator). *(Len Sumak)*

Command attacked Hanover with 360 Lancasters, the target area was covered by cloud and the Pathfinders were not successful in marking the city. The raid was scattered, most bombs falling in open country north and north-west of Hanover. The JLO directed 190 twin-engined fighters into the stream 51 minutes before the Pathfinders arrived over the city and 14 victories were claimed for two own losses. The actual losses were 18 Lancasters. A 115 Squadron Lancaster crashed on return to Little Snoring in north Norfolk, injuring two of the crew. Another Lancaster ditched east of Aldeburgh. Eleven more landed with damage inflicted in the fighter attacks.

Bomber Command was stood down the following night and then on the 20th/21st, 358 Lancasters of 1, 5, 6 and 8 Groups were dispatched on the long trip to Leipzig 90 miles south-west of Berlin for the first serious raid on the city. At Syerston, Dick Starkey, who was from Barnsley and had volunteered for air crew in April 1941, received his commission two hours before take-off. This would be his first operation since joining 106 Squadron, though he and his crew had flown a 'Nickelling' op to France in a Wellington at 29 OTU:

59

Lydon's crew on
103 Squadron at Elsham
Wolds in late 1943.

against fighters. At this point of the operation 'Monica' pips (advanced warning of aircraft approaching – either friendly or enemy) were recorded and 'Jock' Jamieson the mid-upper gunner who was from Aberdeen instructed me to dive port. He had sighted a twin-engined aircraft which appeared out of the clouds as we also emerged. He quickly gave the evasion order and then fired a long burst at the enemy aircraft before it disappeared into the clouds. It must have been a shock to the night-fighter pilot when tracer bullets flashed past his aircraft, but under the conditions he could not do much about it. At this stage Joe Ellick the rear gunner, a Jew from Wallasey, also sighted the enemy aircraft but before he could bring his guns to bear it disappeared into the clouds on the starboard quarter up. On route the searchlights were ineffective because of cloud thickness and still without the ASI we bombed on ETA at a large glow on the clouds which we took to be the target. The return journey was uneventful but the appalling weather conditions did not improve until we descended over the North Sea and air temperature increased; the airspeed came back on the 'clock' much to my relief.[1]

About 220 night-fighters were sent up, returning with claims of 11 Lancasters destroyed for nine own losses. Sixteen Lancasters failed to return. Sgt Eric Jones on 49 Squadron recalls:

We entered dense cloud early in the flight and continued to climb in cloud up to our operational height. When flying in cloud for a long period and relying completely on instruments most pilots start to imagine all kinds of peculiar things. 'My left wing is low; the plane is flying left wing down.' 'My compass is giving an incorrect reading and we are getting more and more off track.' In fact, it took considerable concentration to avoid believing that all the instruments were incorrect. It needed the reassurance of the navigator to convince that everything was all right, 'Yes Skipper. Hold your course, we are on track, you are doing well.'

Our immersion pumps went unserviceable. Immersion pumps pumped petrol through to the engines when at height. We would be unable to get above 17,000ft and we were still in dense cloud with no sign of it breaking. This was the only time I 'boomeranged' (RAF jargon for an abortive sortie). We turned round and went back to Fiskerton. On our way home and by using H_2S we dropped our bomb load on the Dutch island of Texel. Some time previously we had

What a trip it turned out to be. We took off in Lancaster DV297 *O-Orange* and climbed to 20,000ft. The Lancaster was a beautiful aircraft to fly; definitely a pilot's aeroplane; it handled very lightly, could reach heights of 22,000ft fully loaded and could even maintain height on two engines. It had no vices, except for a slight swing to port on take-off and was nearly impossible to stall. If you put her in a stall position with nose up the airspeed would just about disappear before the nose gently dropped. We had no trouble on the first part of the trip but the weather over Germany was appalling, with cumulus clouds reaching over 20,000ft and we were frequently flying in and out of the tops in icing conditions. About halfway to the target the airspeed indicator suddenly dropped off the 'clock'. This left us in an awkward situation and I had to fly on other instruments to maintain the aircraft in a normal flying position. It reminded me of the Air Force adage – 'There I was at 20,000ft with sweet **** on the clock'. This is where I needed my instrument flying training, because the airspeed indicator was very important during corkscrewing, which was our main evasive action been told that the Germans had evacuated all the civilian population from the area and filled it with coastal defence weaponry. Out of a force of 358 Lancasters, 271 got to the target area but due to the atrocious weather it was unlikely that any damage was caused to Leipzig. There were many stories from aircrew returning from the raid relating to the severe icing conditions causing engines to seize up and only restarting when the aircraft had been brought down to lower and warmer levels. Old Man Luck was still pointing us in the correct direction.

On 22/23 October, 28 Lancasters and eight Mosquitoes of 8 Group set out to bomb Frankfurt as a diversion for over 560 Lancasters and Halifaxes heading for Kassel for the second raid on this city that month. Their targets were the Henschel locomotive-engine plant, the largest of its kind in Europe, and the Fieseler aircraft plant. It was the first trip for Bill Siddle's crew since their horrific crash on 7 September. Two of the crew had been so badly injured that they had been scrubbed from flying and were replaced by two new crew members. Clayton Moore, who had spent ten days in Rauceby Hall Hospital and then had a period of recuperation, recalls:

Lancaster CV340 (WS-Q) was our mount for the trip, so we did an NFT [night-flying test] on her before lunch. At the afternoon briefing we were told that our takeoff time would be just after 6 p.m. and that we would be attacking a city called Kassel, an industrial target deep inside central Germany. As such, we could expect to find it well defended, so it was with the now-familiar feeling of uneasiness mixed with excitement that we lifted off on a flight of about six hours' duration. According to intelligence, Kassel had a number of factories producing undisclosed types of 'secret' weapons and our job was to put them out of business. Lady Luck flew with us on the night. The marking was good and although we encountered cloud over much of the trip, the target was fairly clear on our arrival and fires were already to be seen burning beneath us as we went in. The defences were alert and heavy but our bombing run proved uneventful. We had some trouble with icing on the return flight, but we still made it back to our own base on time. In all our re-introduction to operational flying had turned out to be something of an anticlimax. We touched down at Bardney just before midnight and were pleased to learn that all of the squadron's aircraft had made it back safely, with the

exception of just one. F/O Albert Manning and crew, flying in EE188 (WS-B) had been ganged up on by no fewer than four twin-engined fighters and the mid-upper gunner had been killed and the tail gunner wounded during the battle.[2]

One hundred and twenty-five bombers were forced to return early on the Kassel operation because of severe icing en route, but Kassel was subjected to an exceptionally accurate and concentrated raid. They created 3,600 fires that were still burning seven days after the raid, the firestorm destroying 63 per cent of all of Kassel's living accommodation. The Fieseler factory was so badly damaged that the V-1 assault on Britain was delayed. At Frankfurt all the bombers returned safely but the RAF lost 43 aircraft on the operation to Kassel.

'November 2nd proved to be something of a red letter day for us,' recalls Clayton Moore:

because it was then that we were 'issued' with our own aircraft and its ground crew. Although intended for exclusive use by us, the exigencies of the service were such that 'our' Lancaster would sometimes be flown on operations by another crew if we were on leave or stood down. Similarly, we would be expected to operate in another aircraft if ours was out of action due to unserviceability. In squadron terms, the assignment of a particular aircraft to a crew marked the acceptance of that crew as one which could be relied on to take good care of it and also indicated the belief held by 'the powers that be' that the chances of the crew surviving were considered to be good. However, the allocation of ex-works Lancasters was usually reserved for crews of higher rank and experience than ours, so it was to be expected that our charge would be of the second-hand variety. After all, we had already subjected one Lancaster to a severe bending. Because the cost of a new Lancaster was estimated to be in the region of £75,000 at the time, we could hardly expect to be issued with a new one. We had managed to survive long enough to qualify for recognition, but our past reputation for recklessness precluded the granting to us of the ultimate accolade. So it was that the Lancaster we got was not new. In fact, EE136 WS-R was of near-veteran status, having been handed over to our keeping by P/O Jim Lyons and crew after the completion of a tour. Jim, an Australian, was then posted to a training unit for a period of rest, but was to be tragically killed in a flying accident a few weeks later.

Most of the afternoon was spent in inspecting our new charge and getting acquainted with the ground crew. It had emblazoned beneath the 'driver's window' in Olde English

Script the most unlikely name of *Spirit Of Russia*. Beneath the name were painted several rows of bombs, a record of the impressive number of trips she had so far completed. Although we had no way of knowing it then (otherwise we might have refused to fly in any other aircraft), WS-R was to complete 109 operational sorties before being withdrawn from active service and would ignominiously end her days as a practice hulk for fire fighting crews before being sold for scrap. The reason for choosing the name *Spirit Of Russia* was a mystery to us, it having been bestowed on the aircraft by P/O Lyons. One of the rumours circulating on the squadron at the time was to the effect that the cost of her production had been met through a collection made by the people of Leningrad, but recent inquiries made of the Russian Embassy in London have failed to confirm or deny this. The *Spirit Of Russia*, in common with most other aeroplanes, had her own characteristics, as we were to discover next day, when we took her up for an air test. She had a disconcerting habit of crabbing a few degrees to port in flight and this was a tendency that Bill had not encountered previously with the type. During takeoff, all Lancasters tended to pull to port because of the fact that all four propellers turned in the same direction and care had to be taken to compensate for this, particularly if taking off with a full load of highly combustible materials. In the case of WS-R, full rudder trim was necessary at all times and Bill foresaw difficulties during takeoff. Nonetheless, we found the *Spirit Of Russia* a delightful aircraft to fly. She was manoeuvrable and had plenty of power in her well-maintained engines. In the months that were to follow, we were to become jealously fond of our battle-scarred old warrior. In an attempt to find an answer to the question of WS-R 's waywardness, we began making some inquiries around the squadron, particularly of the ground crews. Among the rumours circulating, there was one to the effect that our Lanc had been shot up a few times and that two tail gunners had been killed by enemy action. It seemed that on one such occasion, the damage had been sufficiently severe as to warrant the fitting of a new tail section and the problem had been evident ever since.

On 3 November, Düsseldorf, after a lapse of almost five months, was earmarked for a raid by just over 570 bombers. The operation included a special force of 38 Lancaster IIs in 3 and 6 Groups equipped with the 'G-H' navigational device,

F/L R.A. Fletcher on 97 Squadron in the cockpit of *Sri Gajah/Jill*. Fletcher was shot down on 23/24 September 1943 on the raid on Mannheim and was taken prisoner.

which after a successful trial in the winter of 1943, had been withdrawn until enough sets could be produced to equip a large force of bombers. 'G-H' was a set that transmitted and received pulses from two ground stations. By plotting the point at which the two intersected, the aircraft's ground position could be plotted quite accurately. Aircraft without 'G-H' were to formate on a 'G-H' leader and release their bombs in unison. The Mannesmann tubular steelworks on Düsseldorf's northern outskirts was selected to test this precision device for the first time on a considerable scale. Five of the 'G-H' Lancasters had to return early and two were lost, while equipment failures in 16 aircraft reduced the numbers bombing on 'G-H' to 15, although these left a number of assembly halls burnt out. Photographs taken after the raid showed that half the bombs aimed by means of 'G-H' had fallen within half a mile of the aiming point. By October 1944, most of the Lancasters of 3 Group had been equipped with this important new aid. All 52 Lancasters, including 20 blind markers plus ten Mosquitoes that were detailed to carry out a feint attack on Cologne ten minutes before the start of the main raid, returned safely. Thirteen Oboe-equipped Mosquitoes that were detailed to hit Rheinhausen and two more, equipped with 'G-H', that went to Dortmund also returned without loss, as did 23 Stirlings and Lancasters that sowed mines off the Friesians.

A Victoria Cross award was made to Glasgow-born pilot F/L William 'Bill' Reid RAFVR on 61 Squadron, who put Lancaster *O-Oboe* down at Shipdham airfield after he was twice attacked by night-fighters before Düsseldorf was reached, but, despite severe head wounds, pressed on, bombed the target and brought the badly damaged aircraft back with the help of the crew.

On the night of 4/5 November, the RAF carried out mining of the western Baltic, with a Mosquito 'spoof' towards the Ruhr. At 1819hrs German radar picked up 50 to 60 RAF aircraft between Cap Griz Nez and the Westerschelde River at 23,000ft to 30,000ft. Their further course was south-east into the southern Ruhr area. As their speed at first was only about 250mph they were taken to be four-engined bombers, but later, taking headwinds into consideration, the defences identified them as Mosquitoes. Several night-fighters in the area of the western Ruhr were ordered to take off but the operation was abandoned after the approaching aircraft were identified as Mosquitoes. Meanwhile, at 1802–1840hrs 30 to 50 aircraft at heights between 3,300ft and 5,000ft flying at 200mph were picked up by German radar approaching the northern part of

west Jutland. Occupying two night-fighter boxes in Jutland, II *Jagdkorps* (2 JD) engaged 16 bombers and shot down four without loss.

During the next few nights, only minor operations were flown and then on the night of Wednesday 10 November, 313 Lancasters of 5 and 8 Groups were detailed to attack the rail yards at Modane set high up in the mountainous border region between France and Italy. At Fiskerton Sgt Eric Jones and the other crews that were 'on' that night were told at briefing that the idea was to close the tunnel and so prevent the movement of German troops into Italy.

'We were all very pleased with this arrangement,' recalls Jones:

A nice quiet 'stooge' across France, an undefended target and back home and another operation completed – no trouble at all. We were fourth in the line to take off and the squadron's operational effort was getting under way when there was a terrific explosion heard above the sound of the four Merlins each idling at 1,000 revs and a brilliant flash lit the sky. Some poor devil had crashed on takeoff and surely he and his crew must have 'bought it'. The runway must have been damaged; we could not see over that little rise in the centre of the runway. The trip was 'scrubbed' and back we went to dispersal cursing the character who had robbed us of that 'cushy' French trip. We were never again to get the chance of another French trip and it was to be Germany all the way.

Many years later, Eric Jones discovered that it was Lancaster *P-Peter* flown by W/O Ernest Webb:

Slowly advancing the four throttles until they were through 'the gate' the Lanc gradually gathered speed. Suddenly a tyre burst and Ernest became aware of his starboard wing looming large in his cabin window and almost immediately, without warning, his port undercarriage collapsed. *P Peter* slewed off the runway at speed narrowly missing the Watch Tower and finally finished up pointing in the direction from which it had just come. Small fires were already licking around the aircraft as Ernest dived out of his own sliding window. The fuselage had split into two sections and the rear turret had been thrown clear with the rear gunner still trapped inside. His flying boot was caught in the wreckage but, with a quick twist and a tug, Ernest was able to free him. By now the Blood Wagon had arrived complete with the Station Medical Officer. A head count showed that there were still two members of the crew missing. Ignoring the .303 ammo exploding all around the aircraft, Ernest and the MO dived back into the aircraft only to discover that the two missing men were nowhere to be seen. Eventually they turned up. Apparently, as soon as the aircraft came to rest, they scrambled out and ran at high speed into the darkness very much aware of the 4,000lb 'cookie', a number of 1,000lb bombs and many cans of 4lb incendiaries in the bomb bay. It was some hours later when there was a violent explosion and pieces of *P-Peter* were deposited over a wide area. All that was left of the aircraft was a single key. The five aircraft that managed to get airborne were, on their return, diverted to Dunholme Lodge.

At Bardney, Lancasters on 9 Squadron had taken off without incident. F/L Siddle's crew took WS-R to war on the raid on Modane, as Clayton Moore recalls:

Four hours into the flight, we arrived over some of the most picturesque scenery I had ever seen from the air. We found the target area bathed in brilliant moonlight and this was made all the more effective by the snow-covered peaks rising several thousand feet above us from the floor of the valley in which our objective lay. The Pathfinders had put down excellent markers, superfluous though they were in the circumstances, because the ground detail was plainly visible to us, with the steel bands of track glinting in the moonlight between the goods yards and sidings, at the far end of which was a rail tunnel leading into the foot of the mountain. The defences were so light as to be almost non-existent as we began our bombing run. Because of the lighting conditions, I could see for miles and of the many aircraft in sight, all were friendly. The bomb doors had been opened and Mike was giving directions, when I spotted the steam from a train making its way towards the station from about a mile out, so I at once broke into the conversation. 'Tail to pilot. Train approaching behind us. If we go 'round again, it should just be arriving when we bomb.'

'Good show, tail. Cancel the run-up, bomb aimer, we're going 'round again.'

As we wheeled between the towering peaks and returned down the valley, I watched the train steam into the crowded marshalling yards and come to a halt. Just why the driver

didn't realise that a raid was in progress remains a mystery, but there was no questioning the fate of his train and its cargo. We turned again and were soon on our second bombing run. As the bombs were released, I watched in fascination as one after another of them slammed into the long, ill-fated train. By way of a bonus, the last couple of two-thousand pounders burst in the entrance to the tunnel leading away from the yards.

Just as we were coming out of the target area, Bill reported an unidentified aircraft approaching on a reciprocal course on the port bow. I at once swung my turret to the beam and was just in time to see a Messerschmitt 210 go screaming past.

'Doesn't seem to be interested in us,' I reported.

'Keep an eye on him just in case,' Bill instructed.

'Roger. Cover the rest of the sky, mid upper.'

Leaving Gerry to keep watch, I followed the progress of the 210, which was plainly visible against the blanket of snow that stretched out beneath and above us at either side. He appeared to be in some difficulty, because he was losing height steadily. Then, the Messerschmitt did a steep turn to starboard and rammed headlong into the side of the mountain, throwing up large volumes of snow as it hit. There was no explosion or fire that I could see.

The four-hour return flight across France and the water passed without any problems, but we had used up a lot of fuel in negotiating the mountains around the target, so were obliged to lob down in Cambridgeshire, at Gransden Lodge, the home of 405 'City of Vancouver' Squadron RCAF. I had hopes of renewing one or two acquaintances from my training days in Canada, but was disappointed to find after debriefing that we had been amongst the last to land and most of the others had retired to their beds for the remainder of the night. In truth, a good number of 'cuckoos' had nested and there were no spare beds for us. However, we found that the crew room sported a large, four-sided fireplace in the centre of the room, so we all sat around it until dawn in a collection of easy chairs, warm, but snoozing fitfully as best we could.

After breakfast we flew back to Bardney only to find that we had been officially reported missing! Seemingly, Gransden Lodge control had not reported our presence to Bardney, while Bill had presumed that they would have. We managed to stop some (but not all) of the telegrams from being sent out to the next of kin. Unfortunately, the cable

to Canada informing my parents of their 'loss' was one that did get through and they were thus subjected to a lot of quite unnecessary grief and trauma, particularly since the cable cancelling the first one arrived ahead of the original. It wasn't until they received my letter explaining the mix-up that they were sure of just what had happened and that I was indeed safe and well.

On the night of Wednesday 17 November, the attack on the I.G. Farben factory at Ludwigshafen by 66 Lancasters and 17 Halifaxes of 5 Group, a purely H_2S blind-bombing raid without any TIs being dropped, was believed to be accurate. Most of the enemy fighters landed early because of misleading instructions broadcast from England to German night-fighter pilots and only one Lancaster was lost.

The following night, 18/19 November, the first of 16 major raids which have gone into history as the Battle of Berlin was mounted by 835 bombers, 440 of them Lancasters; another 395 aircraft, mostly Halifaxes and Stirlings, flew a diversionary raid on Mannheim and Ludwigshafen. Both forces were planned to cross the enemy coast simultaneously, the stream heading for the 'Big City' reaching the capital along the Havel River in eastern Germany and then splitting into groups to bomb Siemensstadt, Neukölln, Mariendorf, Steglitz, Marienfelde and other districts of Berlin. Berlin, however, was completely cloud-covered and both marking and bombing was carried out

Lancaster II DS689 which was lost with F/Sgt Malcolm Barnes Summers RCAF and crew on 7/8 October 1943 on the operation on Stuttgart. The aircraft crashed at Rachecourt-sur-Blaise with the loss of all the crew.

FIXES 1ST CLASS   2ND CLASS
Nº 49 SQUADRON
BERLIN 18/19 Nov 1943.

A/C Nº & LETTER	CAPTAIN	BOMB LOAD	CALL SIGN	FIXES	ETC	ATO ETR ATR	CLOUD	RISE	A/C Nº-LETTER	CAPTAIN
1st WAVE O	ADAMS		NMY	LANDED WARBOYS		1657 0131	LOCAL WEATHER TIME:-			
" A	DAY			TANGMERE		1658 0140				
" G	COTTINGHAM			DUNSFOLD		1700 0115	STATION VIS QFE WIND			
" N	EDY					1702 0112	SCAMPTON			
" C	HALES					1704 0135	FISKERTON			
" T	BLACKMORE					1705 0121				
" M	JUPP					1706 0116				
" H	HIDDERLEY					1710 0147	AERODROME STATE			
2ND WAVE F	REYNOLDS					1708 0129				
" P	TANCRED					1707 0137	FIXES QDM'S ETC			
" V	PALMER					1706 0139	A/C L	A/C L	A/C L	
" S R	FOSTER			LANDED CRANWELL		1711 0135	FIX TIME FIX TIME FIX TIME			
" J	BACON			WOODHALL		1714 0159				
" U	WEBB			GRANSDEN		1715 0229				
" K	JONES					1716 0142				
" Q	BARNES					1719 0148				
3RD WAVE E	SIMPSON			LANDED MANSTON		1718 0100				
" B	GEORGE			W. MALLING		1720 0139				
" D	ROANTREE					1731 0143				

OFFICER I.C. G/C GRINDELL DFC AFC

49 Squadron Operations board on the 18/19 November 1943 Berlin raid.

blindly. Eric Jones had not been to the German capital since his 'second dickey' in August and his crew had not been there at all.

'The very mention of the name Berlin,' he says:

created a certain tension and apprehension amongst all those present at briefing. There were to be a further three raids on Berlin in that week and we participated in all of them. *K-King* was to take a very active role in the 'Battle of Berlin' taking part in eleven of the raids.

Steve, when approaching the city on his bombing run, said, 'Sorry Skipper, dummy run, go round again,' and I put the Lanc into a 360° turn to come round onto the same heading. Well, we nearly finished our tour there and then and, I might add, some other poor devil's as well. When I had gone through 180° and facing the oncoming stream I was left in no doubt where the rest of the stream was. Lancs were visible all round and on a reciprocal heading. How we escaped a collision I will never know. Flying to the target one rarely saw another bomber unless they were being attacked by a fighter or coned by searchlights. To find the stream, fly

on a reciprocal course. I said to Steve, 'Don't you ever ask me to go round again. You make sure you line up correctly and we drop 'em first time.' We did and that was the first and last time we went round again.

Tens of thousands of incendiaries and 2- and 4-ton HE bombs were dropped on Berlin within a period of barely 30 minutes. More than 2,500 tons of bombs were dropped on Berlin and Ludwigshafen. With bomber formations simultaneously over two cities about 300 miles apart, the German night-fighters found it impossible to protect both targets adequately and just nine Lancasters were lost on the Berlin operation, but an effective Tame Boar operation mounted against the second force destroyed most of the 23 aircraft that failed to return.

After Berlin, Leverkusen was bombed on the night of 19/20 November, when bad weather on the continent prevented many night-fighters from operating and only five bombers were lost. Conditions were little better when, on 22 November, 764 bombers – 469 of them Lancasters – set off for Berlin again. Crews were detailed to fly a straight course in and out. The first Pathfinders arrived over the 'Big City' just before 2000hrs to find it covered by 10/10ths cloud. The forecast had been for clear conditions over the home airfields, broken to medium-level cloud over Berlin and low cloud or fog over much of the rest of Germany. Three of the five Lancasters equipped with the new 3cm H_2S Mk.III sets had to turn back after their equipment failed, but the two other aircraft's sets showed a clear outline and the blind markers accurately dropped four red TIs at the aiming point slightly to the east of the centre of the capital. Despite the risk of collisions, the rate of aircraft over the target was increased to 34 per minute. More than 2,000 tons of blast bombs and approximately 150,000 incendiaries were dropped on Berlin in barely 35 minutes. This was the most effective raid on the 'Big City' of the war. A vast area of destruction stretched across the capital, caused mainly by firestorms as a result of the dry weather conditions. Tuesday night leading into Wednesday, the west part of Berlin suffered more than in all the previous raids. On Wednesday afternoon, Mosquitoes flew over the capital and reported having observed over 200 giant conflagrations. *Nachtjagd* largely remained grounded due to adverse weather conditions and there were no diversions, but 26 aircraft failed to return. Eleven of these were Lancasters.

Twelve hours later the height of the smoke cloud over Berlin from 11 major fires still burning from the previous night was almost 19,000ft high, when 383 bombers – 365 Lancasters, ten Halifaxes and eight Mosquitoes – again made the long haul to the 'Big City'. This time the attack concentrated mainly on the western part of the capital with its three large rail installations. Major John Mullock of the Royal Artillery, attached to the Pathfinder Force as flak liaison officer, on his first trip to Berlin flew in a marker Mosquito captained by G/C 'Slosher' Slee DSO DFC. Mullock later reported:

> After the attack had been going for four to five minutes the flak was entirely barrage, spread over a vast area and wildly dispersed in height. No bursts were seen above an approx height of 23,000ft. There seemed to be no attempt to fire a barrage over any one part of the city as might have been expected. The rate of fire of most of the guns would appear to have been something in the region of five to six rounds per minute ... when the attack on Berlin had developed concentration and window cover obviously precluded the use of radar. The impression gained was that the defences were in a state of utter confusion and were firing blindly and wildly.[3]

Twelve Tame Boar crews shot down 13 of the 20 heavy bombers that were lost. From a distance of 30 miles the air crews could see the fires that had continued to smoulder since Monday. Finally in the fire glow over Berlin they were able to observe many details of the destroyed districts. The whole complex around the Wilhelmstrasse and Brandenburg Gate area, the Tauentzienstrasse, Potsdam Square, the Anhalter Strasse and many other streets were completely destroyed.

On the night of 25/26 November, 236 Halifaxes and 26 Lancasters attacked Frankfurt. Again, there were no major diversions. The JLO did not at first know whether Mannheim or Frankfurt was the real objective but eventually he chose Frankfurt. Flak was restricted to 15,000ft, cloud covered the target area and the bombing appeared to be scattered. Eleven Halifaxes and a Lancaster failed to return. With no respite, within hours crews were on the battle order again when the target for over 440 Lancasters and seven Mosquitoes was Berlin, and 157 Halifaxes and 21 Lancasters would fly a diversion on Stuttgart. At Oakington P/O A.S. 'Ted' Ansfield lay in a hospital bed recovering from burns he had received on an operation earlier in the week. Ted was an observer (a navigator trained as a bomb aimer and as operator of the air-to-ground radar used for blind marking when the target was under cloud) on 7 Squadron PFF. Casualties among Pathfinders at this time

were heavy and on 7 Squadron the 'chop rate' was less than six trips per crew. Ansfield listened to the roar from the dispersals where the station's Lancasters were being run up and tested, and it was enough to make him decide that he wanted to fly on that night's operation. He persuaded the MO to discharge him from hospital and let him take it easy in the mess. As he left, a quiet voice behind him said, 'Ansfield, briefing's at 15.00 hours – good luck.' Ansfield replied, 'Thank you, sir, thank you very much', and ran all the way to the briefing room. The crews were already assembling but he was just in time to get his name down on the list on F/O G.A. Beaumont's crew for the night's operation to Berlin in *F-Freddie*. As they prepared to go aboard their aircraft at dispersals in the winter darkness the 7 Squadron crews sang *Silent Night* – not because Christmas was only four weeks away but because the carol had become the squadron's departure theme song. Then, as each crew climbed aboard, the singing died away and soon the last voices were drowned in the vibrant crescendo of the engines as one by one the Lancasters started up and taxied out for take-off. It was six o'clock when *F-Freddie* climbed away from Oakington. Two hours later the crew were turning over Frankfurt and heading north-east for Berlin.

At Kirmington W/O J.E. Thomas DFC on 166 Squadron lifted his heavily loaded Lancaster III DV247 AS-F off the runway for Berlin. It was the crew's twentieth trip. The procession of bombers took 45 minutes to cross the coast. Both forces flew a common route over northern France and split nearing Frankfurt. At first the JLOs thought that Frankfurt was the intended target. Difficult weather conditions resulted in only the most experienced German crews being ordered to take off, and 84 fighters engaged the RAF formations. The outward trip, as far as Thomas was concerned, was uneventful, but over the 'Big City' they were caught in searchlights. Thomas was able to climb to about 28,000ft and got away from the beams but on the return and approaching the German-Dutch border, the Lancaster was attacked by a Bf 110, which riddled DV247 with cannon fire. Sgt Jimmy Edwards, the bomb aimer, had part of his left hand shot away and also suffered a gaping wound in his left thigh. Sgt W. O'Malley's voice announced over the intercom that he was trapped in his rear turret, whereupon Sgt A.V. Collins DFM immediately replied that he was going to help him. Thomas shouted that he was unable to hold the aircraft much longer and gave the order to bale out. Five men did so successfully but the two gunners were killed in the aircraft when it crashed. Sgt Bill Bell, the navigator, made three attempts to escape from captivity during 1944, but he was caught each time after being on the loose for a few days.

In all, 29 bombers were lost on the raid on the 'Big City'. Just two German night-fighter aircraft were shot down. On 7 Squadron's *F-Freddie*, Ted Ansfield heard F/O Beaumont warning the two gunners to keep a sharp lookout for fighters, Sgt P.J. Palmer, the mid-upper gunner, replying that all he could see was another Lancaster about 600yd away to port. The wireless operator at the airborne interception radar confirmed that the Lancaster and another – which the gunners could not see – were showing as a blip running in line with them on his screen. A moment later something else came on to his screen. He just had time to shout over the intercom, 'There's a smaller signal moving fast towards us', when there was a great crashing, hammering noise which swamped the roar of the engines and the Lancaster lurched as a stream of cannon and machine gun fire poured out of the darkness into them. The rear gunner saw the fighter and called to the pilot to dive to port, but it was too late. The damage was done. The Lancaster was mortally wounded. The four engines, the petrol tanks and the rear half of the fuselage was afire.

'Get the bombs off,' Beaumont shouted, but Ansfield had already thrown the jettison switch; nothing had happened. The electrical release gear had been damaged. He looked into his air-to-ground radar and from it fixed their position. They were 20 miles north-east of Frankfurt-am-Main, at 23,500ft. He tried to pass on the information but nobody heard him – the intercom wires to his helmet had been shot away. So he leaned out and shouted the position report to the navigator who drew the centre-section curtains aside to relay it to the wireless operator for transmission to base. However, the wireless operator was lying dead over his key. Ansfield went forward. The Lancaster was starting to go down and the pilot gave him a thumbs-down sign to bale out. 'Good luck, Ted,' he said, 'see you in hell!'

Down in the nose Ansfield found Sgt D. Ashworth, the flight engineer, crouching over the escape hatch. He looked despairingly at Ansfield. 'The ruddy thing's jammed,' he shouted.

'Put your foot through it,' Ansfield yelled at him and then, impatient to get the job done, he pushed Ashworth aside and kicked it open himself. 'Out you go and good luck – see you below.'

Ashworth, who was very young and very frightened, looked down at the black roaring void and backed away. 'I can't, sir,' he said.

'Don't argue, it's your only chance,' Ansfield snapped back. Reluctantly Ashworth sat on the edge of the hatch and hung his feet out. Ansfield gave him a push and he disappeared into the night. Ansfield checked his parachute and was about to follow him when above the noise of tortured engines and the roar of the flames further aft, he heard someone screaming, beseechingly, 'Oxygen! Oxygen!'

He crawled out of the nose back into the cockpit. Sgt Bill Meek, the navigator, told him it was the mid-upper gunner Palmer calling from the turret in the burning fuselage. 'The poor devil's in the middle of it. We can't get to him.'

Ansfield knew that oxygen wouldn't help Palmer. He could only assume that the gunner believed it might give him added strength to struggle out; so he leaned across the pilot and turned the oxygen lever to 'emergency'. They were diving now. Standing by the pilot, Ansfield saw the airspeed indicator going up towards the danger line and the altimeter unwinding fast from 20,000 to 19,000ft. He said, 'Can we pull her out together Skip?' Beaumont shook his head. 'Not a chance in hell – the controls are dead. Out you go. Meet you downstairs.'

Ansfield moved forward into the nose. There was a brilliant flash. The Lancaster's 5 tons of bombs and flares had exploded and the aircraft disintegrated. The next thing he knew he was lying on the ground in a forest in the darkness, numb with cold and his face badly frostbitten. The only sound was the distant barking of a dog.

For several minutes, as consciousness ebbed back, Ansfield lay there while his mind struggled to bring his situation into perspective. At first his last memory was a hospital bed. Then he remembered the trip he needn't have flown on, the fighter, the fire and the flash. He knew he hadn't jumped out and because of that he couldn't at first comprehend why he was alive.[4]

As the bombers returned home, fog covered much of eastern England, with many squadrons being diverted. As midnight approached, conditions at Fiskerton were quite severe. Radiation fog was 1,200ft deep and visibility down to 450yd. At 0015hrs it was decided to light FIDO (Fog Intensive Dispersal Operation) for 49 Squadron's returning crews. This was the first instance during the war that FIDO was used operationally to bring down one complete squadron of bombers (Graveley had used its system the previous week to land four Halifaxes). *H-Harry* and two more Lancasters touched down safely and were followed by *K-King* skippered by Eric Jones.

FIDO was a highly expensive way of dispersing fog. It was very simply a perforated pipe, one on each side of the main runway, for its full length. Raw fuel passing through the pipes was ignited and the resultant intense heat just burnt away the fog. Not many airfields were graced with this device so we were fortunate. When in use FIDO could be seen for miles. Not only by us, I always thought, but also by those damned intruders. Landing a Lancaster into the FIDO system seemed like a descent might be into Hades. It was only when the plane got reasonably close to the runway that a pilot realised there was indeed a space between the two strips of flame affording sufficient space to land. The process certainly kept pilots on their toes. No swinging off the runway in a crosswind unless they wanted to straddle those flames. Nevertheless it certainly got rid of the landing in fog problem. One operational night when the fog was very dense, we were the bolt hole for dozens of Lancs. Unable to get in at their bases they were diverted to Fiskerton. Lancasters were lined up on the runways not in use and on every available perimeter track. If 'Jerry' had got wind of this he could have enjoyed a 'Hey-day' or 'Hey-night'.

Four more Lancasters landed and at 0102hrs Sgt Roy Richardson RAAF flying *C-Charlie* entered the 'funnel'. Next in the stack behind *C-Charlie* was *A-Able*, flown by fellow Australian F/Sgt Clive Roantree who recalled:

We positioned ourselves to land immediately after *C-Charlie*. He would turn into the funnel, whilst we were on the down-wind leg and should be clear of the runway as we touched down. The two parallel bars of fire, one on either side of the runway, could be clearly seen with bars of flame at each end to stop the fog rolling into the cleared area. On our practice on 3 November we had found that after we turned into the funnel at 600ft and lined up with the runway, as we approached, the fire on the cross bar reflected on the Perspex windshield so that it was impossible for the pilot to see out. To offset the problem my flight engineer called height and airspeed as soon as we lined up on the runway at 600ft. For the inexperienced pilot it could be a frightening experience as it is not until the aircraft crossed the bar of flame at less than 100ft that it was possible to see clearly and then make a visual landing. Subsequently a shield was placed in front of the bar of flame to prevent windscreen reflection. On this night, with wheels down, pitch in fully fine

with 20 degrees of flap, we were at the end of the down-wind leg ready to make our turn across wind before entering the funnel, when there was a dull flash on the ground right at the beginning of the funnel. I knew that an aircraft had crashed and to my horror realised that it must be Richardson in *C-Charlie*. I continued the landing procedure turning across wind and there, right below us was an aircraft on fire! Giving the crew the order that we were going to overshoot, I called flying control, 'Hello Passout, Bandlaw Able overshooting – an aircraft has crashed and is on fire in the funnel – I say again an aircraft has crashed and is on fire in the funnel.' At the time I was completely unaware of Sgt Richardson's Lancaster crashing behind me in the approach funnel.[5]

Sgt M.O. 'Spud' Mahoney, the mid-upper gunner, recalled:

The cockpit was a sheet of flame, but I could just make out the skipper still in his seat … I went forward and could see his clothing was on fire … grabbing him by his parachute harness I pulled like hell, but the harness had been burnt part through and it gave way sending me tumbling back into the radio compartment. Regaining my feet, I went back into the heat again to try to get Roy out, but a stronger force seemed to be pulling me back; I then became aware of two figures holding me and one was shouting 'it's too late mate,

Lancaster III ED664 AR-A2 *Aussie* on 460 Squadron RAAF at Binbrook. Note the 'Advance Australia' badge and the 35 ops depicted on the bomb log by kangaroos with RAAF wings and a DFC ribbon above them. ED664 with F/Sgt Maurice Joseph Freeman RAAF and two of the crew were lost on 23/24 November 1943 on the operation on Berlin. The four others survived and were taken into captivity.

Pre-raid briefing at Wickenby for the attack on Berlin on 16/17 December 1943.

After Steve's call for a 'dummy run' over the target and turning to set course for home we hoped that no further problems lay ahead. Our wish was not granted. As the homeward trip progressed, Ken started to calculate very high winds. At first he thought he had got it all wrong but eventually established that we were running into a 100mph headwind and it was cutting our ground speed back to a comparative snail pace. I doubt if we were moving over the ground any quicker than 120mph. This situation had Ron doing rapid petrol calculations and as the homeward journey progressed it was obvious we were not going to make base and we might not even make England. Our luck lasted out and we crossed the coastline but it meant getting down onto the 'deck' as quickly as possible and this turned out to be an American base. That night the main Bomber Force was scattered all over Eastern England, obviously seriously disrupting any operational plans for the following night. Our American hosts, delighted at having a Lancaster in their midst, feted us with a cracking meal and excellent overnight accommodation (for what was left of the night). When we flew out the following morning we noticed that our night flying rations had been stolen from our Lanc, we suspected by American ground crew as some kind of souvenir. So who cared? They had looked after us so well.

it's too late … nothing can be done now.' The next thing I remember was being taken to the sick bay.

Richardson's was one of 16 Lancasters that crashed or were later written off after accidents at the end of the Berlin operation, one Lancaster having crashed after turning back with one engine out at the start of the operation.

The fifth heavy attack on Berlin within a fortnight was flown on the night of 2/3 December. Most of the Halifaxes were withdrawn from the raid because of fog which was forming at their airfields in Yorkshire, so 458 Lancasters, 18 Mosquitoes and just 15 Halifaxes were dispatched to the 'Big City' that night. There were no major diversions and the bombers took an absolutely direct route across the North Sea and Holland, and then on to the capital. Unexpected winds en route blew many aircraft off track and nullified the Pathfinders' efforts to make dead reckoning (DR) runs from Rathenow. Consequently there were gaps in the cloud covering the city; most of the bombing was scattered over a wide area of open country to the south. Many illuminated targets were provided for the night-fighters over the capital and Tame Boar crews claimed 40 kills. At least 32 bombers went down in the main air battle that was concentrated in the target area.

This was Eric Jones' crew's first raid on Berlin and their 12th sortie over Germany:

W/O J.A.R. Coulombe, the skipper of a Lancaster B.II on 426 'Thunderbird' Squadron RCAF, Linton-on-Ouse, recalled:

While over the target we were coned by 50–70 searchlights from 20.24 to 20.28 hours, during which time we were attacked five times by enemy aircraft and damaged by flak. The mid-upper gunner first sighted a Ju 88 on the port quarter down at 400-yard range and gave combat manoeuvre corkscrew port. The fighter immediately broke off his attack. No exchange of fire by either aircraft. The second attack developed from starboard quarter down and mid-upper gunner saw e/a at 400 yards, so gave combat manoeuvre corkscrew starboard. Again fighter immediately discontinued his attack and broke off at port beam down. No exchange of fire. The third attack came from the port quarter down at 400-yard range. Again mid-upper gunner gave combat manoeuvre corkscrew port and e/a broke off his attack to starboard beam down. Fourth attack developed from starboard quarter down at 400-yard range and mid-upper gunner once again gave combat manoeuvre corkscrew starboard and again the fighter discontinued his attack and broke away port beam down.

The fifth and last attack developed from port quarter down at 200 yards and mid-upper gunner gave combat manoeuvre again corkscrew port and at the same time opened fire. E/a came in to 60-yard range and broke away to port beam above, giving mid-upper gunner a sitting target. Tracer appeared to enter belly of e/a; sparks and tracer were seen to ricochet off fighter which dived steeply and was lost to view. During this attack our aircraft sustained damage to port inner engine and the R/T was rendered unserviceable. The rear gunner was completely blinded by the blue master and other searchlights throughout these five attacks. During all these attacks a Me 109 was sitting off at 1,000 yards dropping white fighter flares. Just as the Ju 88 opened fire on his last attack a Fw 190 was seen by the pilot and flight engineer off on the port bow up at 400 yards coming in for an attack. The cannon fire from the Ju 88 caused the FW to break off his attack to the port beam and down at 100-yard range. He was not seen again. The mid-upper gunner claimed this Ju 88 as a 'probable'. The port tyre, port outer tank and hydraulic system were damaged.

At the beginning of the attack heavy flak was fired in a loose barrage up to 22,000ft around the marker flares and was predicted at seen targets through gaps in the cloud. Searchlights were active in great numbers and took every opportunity the weather offered for illuminating the bombers. After the raid had been in progress half an hour, and soon after the appearance of fighter flares, the ceiling of the barrage was lowered and the flak decreased, although individual aircraft were heavily engaged when coned. Forty bombers, 37 of them Lancasters, were shot down by night-fighters. Fifty-three aircraft were damaged by flak. Five Lancasters on 460 Squadron RAAF, including two that carried press correspondents, were lost. Captain Nordahl Grieg of the Free Norwegian Army, representing the *Daily Mail*, and 40-year-old Australian Norman Stockton, of the *Sydney Sun*, were killed. A 50 Squadron Lancaster that carried Lowell L. Bennett, a 24-year-old war correspondent employed by the *Daily Express*, also failed to return to Skellingthorpe. Ed Murrow, who also flew on the raid, returned safely and his report of the first Berlin raid that month appeared in the morning edition of the *Daily Express* under the banner headline: 'BERLIN – ORCHESTRATED HELL OF LIGHT AND FLAME.'

Sgt Stanley Miller of Scarborough, who was killed on the night of 16 December 1943 when LM385 crashed in fog at Caistor, returning to Kirmington from Berlin. All of his crew died.

On the night of the 3rd/4th, over 520 bombers flew another direct route to Berlin before turning off to bomb Leipzig. Several of the 24 bombers that failed to return were shot down by night-fighters in the bomber stream before the turn was made. The night-fighters were directed to Berlin when the diversionary force of nine Mosquitoes appeared over the German capital and it was believed that only three bombers were shot down in the Leipzig area. More than half of the bombers that went missing on this raid were shot down in the defended area at Frankfurt on the long southern withdrawal route. The Pathfinders found and marked Leipzig accurately and the raid was adjudged to be the most successful on

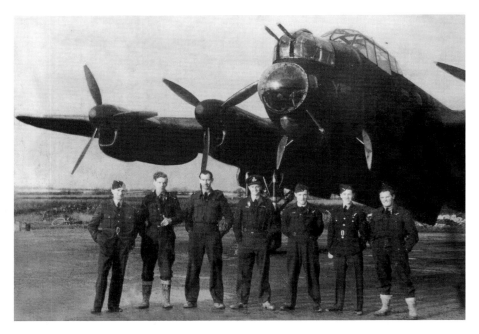

Twenty-year-old P/O Richard Anthony Bayldon's crew on Lancaster III DV293 *Y-Yoke* on 9 Squadron at Bardney, who were lost over Berlin on 16/17 December 1943. All seven crew were killed.

The planned route was straight in and out again over Denmark but the enemy fighters, which were supposed to be sitting on fog-shrouded airfields in Holland, Belgium, northern France and Germany, were airborne and the first intercepted the stream of Lancasters over the Dutch coast. The German controllers had plotted the course of the bombers with great accuracy; many German fighters were met at the coast of Holland and further fighters were guided in to the bomber stream throughout the approach to the target. More fighters were waiting at Berlin and there were many combats. Widespread mist and fog at 150–300ft over the North German plains reduced the overall effectiveness of the fighter defence, and 23 aircraft, mostly Bf 110s, had to abandon their sorties prematurely; yet 21 Lancasters were shot down and four lost in collisions over Berlin. Oberleutnant Heinz-Wolfgang Schnaufer, *Staffelkapitän*, 12./NJG1, shot down four Lancasters over Holland to take his total to 40 victories. Berlin was cloud covered but the Pathfinder sky-marking was reasonably accurate and much of the bombing fell in the city. By this stage of the war sustained bombing by the Allies had made more than a quarter of the capital's total living accommodation unusable.

this distant inland city during the war. The raid caused considerable damage, particularly to the Junkers factories in the old World Fair exhibition site.

The Main Force was prevented from making any more major bomber operations until the middle of the month, when Berlin was again the target for the Lancasters on the night of Thursday 16 December. Over 480 Lancasters and ten Mosquitoes were on the battle order for the 'Big City'. During briefings that afternoon, crews were told that Bomber Command had been waiting to mount a raid on Berlin when the weather was so bad that the fighters would be grounded and they would have an easy trip. This was to be it. W/C Jimmy Bennett, with two tours behind him already, had arrived at Waltham three weeks earlier to form 550 Squadron and he chose to fly that night with 'Bluey' Graham and his crew. Jimmy Bennett recalled:

On their return to England crews encountered very low cloud, as Eric Jones on 49 Squadron at Fiskerton recalls:

Our take-off was early, about 4.30 in the afternoon and even then visibility wasn't very good and it was plain we were not going to be in for a very pleasant journey. There was no high cloud and at times we could see dozens of aircraft around us. The clouds below cleared slightly over the city, we dropped our bombs and got away again. There was some fighter activity but we were not bothered.

We had experienced brushes with electrical thunderstorms on a few occasions but well into the homeward journey we ran into dense cloud and I tried to get above it but with no success. Initially, it was fairly smooth going and I was, by now, quite accustomed to such flying conditions. But the going got rougher with the aircraft becoming difficult to hold on a steady course and we were flying into the heart of an electrical storm. I knew that the cloud tops would be well above the capacity of the Lanc and that the cloud base could be as low as 600ft. We didn't know whether it thinned out to the right or to the left so the only thing to do was to keep going until we flew out of it. We then experienced St Elmo's Fire and this occurs when the whole aircraft becomes charged with static electricity. The propellers become giant Catherine wheels and every bit of Perspex in the plane (all the windows) are framed in a blue flashing light. A scaring but fantastic experience. I had never seen anything like it before. In these conditions it was nothing for the plane suddenly to lose a few hundred feet and regain them just as quickly. We must have passed through a series of these storms without realising it; we were in cloud for what seemed like 'forever'. Ken eventually said that we were over the North Sea and should start dropping off height. This was a relief because I don't recall any icing up and to get to lower levels would help eliminate this further hazard. Once again there was no time for ration eating and smoking over the North Sea – it was going to take all my concentration to get us all safely back to Fiskerton. I was well aware that the safety height over Lincolnshire was 1,500ft (the height to which one could safely descend in cloud) and to break cloud below this height without knowing one's exact position was asking for trouble. So I was back with Ken and his 'Gee' fixes and when he said we were over the coast I continued to lose height below 1,500ft. I was very conscious that Lincoln Cathedral must be at least 500ft above sea level so I was putting implicit faith, once again, in Ken's navigational skills. Suddenly, we were out of the cloud and very low with perfect visibility, and the 'Drem' systems of Lincolnshire's airfields, for miles around, formed a very welcoming scene. One operational night and sometime after debriefing, we heard of reports coming in of engines freezing up and only restarting when the aircraft reached the warmer air levels at lower altitudes.

Jimmy Bennett recalls:

Coming back the cloud started to increase again and it was clear that by the time we reached England it would be almost right down to the deck. Bluey decided to come down through the cloud over the North Sea. In conditions like that it was always wise practice. Lincolnshire may have been fairly flat, but other places weren't and there were always a few of what we called 'stuffed clouds' around, clouds which contained something hard, like a hill. We dropped down into the mist but Bluey picked up the outer circle of sodium lights at Waltham, stuck his port wing on them and followed them round until he found the funnel and put her down. We rolled along the runway to the far hedge and we were already aware that planes were coming down all around us, landing at the first opportunity, so we decided it would be a lot safer to leave the aircraft where it was and walk the rest of the way. As we headed across the airfield I remember looking up and seeing the red starboard wing-tip light of a Lancaster and thinking that it shouldn't be there. The rule was to go round in the other direction. Then there was an almighty crash and a Lancaster hit the ground.

Clearing the winter snow.

When we got to the debriefing room, Nick Carter (the station commander at Waltham) told me what was happening and asked me to go with him to a village somewhere just outside Louth to help him sort out some of the mess from that night. When we arrived we found one of those wartime wooden buildings in the charge of an Irish nursing sister and she had had to turn it into a temporary mortuary. The crash crews were going all around North Lincolnshire and there must have been the bodies of 40 or 50 lads laid out in there, down both sides of the room, all covered in service blankets. It was a terrible sight, a sight I will never forget. It was impossible to recognise some of them. All we had to go by was their identity discs.

Later Jimmy Bennett discovered that one of the four 100 Squadron crews lost that night was that of his friend and the squadron commander, 22-year-old W/C David Holford DSO DFC*. On operations he had one idiosyncrasy. After the normal 'All set boys? Here we go!' he would sing, 'I've got spurs that jingle, jangle, jingle as we ride merrily along' until 'Wheels up!' was ordered. Holford's Lancaster had come down at Kelstern. He had nursed his badly damaged Lancaster back but chose to remain airborne until less-experienced crews had

Eric Jones, pilot of
K-King on 49 Squadron.

landed. On making his approach, the Lancaster clipped a small hill; David Holford was thrown from the cockpit, suffering two broken ankles. He was not found until dawn, after lying in the snow, by which time he was dead. At the time Bennett and his wife were staying at the Ship Hotel in Grimsby. Also staying at the hotel at the same time was David Holford and his wife, a WAAF intelligence officer at Elsham Wolds. They had been married for only a matter of weeks. Bennett had the terrible job of breaking the news to her that her husband was dead.

D-Donald on 405 Squadron, piloted by F/O Burus Alexander McLennan RCAF, crashed in a potato field near Ely after the Lancaster ran out of petrol while attempting a second landing at Graveley. WO2 S.H. Nutting DFM RCAF, the rear gunner, was one of only two men who initially survived the crash. His recollections of events after the crash are dim:

We were stacked up over Graveley trying to get down and after one unsuccessful attempt to land on the beam, I think we went to come in again. We were very low, perhaps no more than 50ft off the deck. When the aircraft crashed and the tail broke off, I was strapped in securely, braced against the guns. When my head struck the gun-sight, I was knocked out. I recall climbing out over the guns. Further on the aircraft was burning. Various fail-safe devices were blowing up.

Nutting then went to look for the other crew members. He found two, one of them dead, the other, McLennan, barely alive:

When I went to move him I saw the bare bone where his leg was all but severed at the knee, so I left him. The next thing I remember was going into hospital with McLennan in the back of a van. He died, I was told, shortly after we got there.

In all, 29 Lancasters either crashed or were abandoned when their crews baled out. South of the Humber, in north Lincolnshire, 1 Group lost 13 aircraft. 166 Squadron at Kirmington lost two Lancasters and 14 men when the returning bombers found north Lincolnshire shrouded in low cloud and fog. At a minute before midnight LM385, which had taken off from Kirmington eight hours earlier, emerged from low cloud and struck high ground at Caistor. There were no survivors. The pilot was Sgt Stan Miller from Scarborough. He had been married barely four months and was later to be buried in his home town. At around the same time, a second Kirmington Lancaster, *S-Sugar*, piloted by Sgt Arthur Brown, crashed in open land between Elsham and Goxhill. Again, there were no survivors on the seven-man crew.

After a lull in operations, on Monday night, 20/21 December, 650 bomber crews were detailed to attack Frankfurt. One of them was captained by P/O Dick Starkey on 106 Squadron, who was flying his third op:

'Jock' Jamieson had been granted compassionate leave and his replacement was a sergeant whose crew had already completed their first tour and he had to complete his last trip by flying with other crews. We took off in JB534. At 120mph the wheels came off the ground and as the speed built up I started a flattish turn out of the Drem system with the inner wing barely 50ft off the ground. We had no trouble on the outward journey and flew at 21,000ft. The target was covered by a lot of cloud so the ground-markers were hidden and the Germans had lit a decoy fire south of the city. About ten miles north of the target on our return journey we were fired upon by cannons and machine gun fire from what we presumed was a night-fighter. Joe Ellick immediately instructed me to corkscrew as enemy tracer came from the port quarter. I did so and after one complete corkscrew resumed normal course. I could tell we had been hit around the port main plane and prayed we would not catch fire. However, after breaking off the first attack the fighter attacked again almost immediately and the aircraft was hit again. I corkscrewed but no fighter was sighted by either gunner so we resumed course. The fighter was never seen and although Joe attempted to open fire on three occasions, his guns failed to function.

When we resumed course it was evident that the aircraft had been severely damaged because it started to shudder violently and I had great difficulty controlling it. The vibration transferred to my body as I fought to maintain control. Joe Ellick reported that the port fin and tailplane rudder was extensively damaged and a large part had disappeared. As for the main plane we could not see any damage but knew there was some.

Soon after the attack Colin Roberts, the navigator, who was from Sheffield, instructed me to change course, but on applying rudder and aileron the aircraft began to bank steeply and I had to put her back on an even keel by using automatic pilot – manual controls were ineffective. The shuddering continued over the North Sea and I gave instructions to prepare for any eventuality. However, we managed to remain airborne and after approximately two and three quarter hours we approached base. George Walker, the wireless operator, who was from Northampton, informed the control tower of the condition of the aircraft and that it was essential to make a right-hand circuit and also that we must land immediately. Permission was given to circuit at a height of 800ft and other aircraft were ordered to maintain their height until we landed. On our approach down the funnel of the Drem system we began to drift to starboard and I dare not counteract this because to make a turning to port would have been a disaster. After an anxious couple of minutes we touched down on the grass in darkness fifty yards to the right of the flare path, and as the mid-upper gunner left the aircraft he knelt down and kissed the ground, having completed a memorable 30th operation.[6]

Eric Jones and crew on 49 Squadron.

Dick Starkey had been doubly lucky. A diversion operation to Mannheim by 44 Lancasters and ten Mosquitoes did not draw fighters away from the route to the target until after the raid was over. The German control rooms were able to plot the bomber force as soon as it left the English coast and to continue plotting it all the way to Frankfurt so that there were many combats on the route to the target. The Pathfinders prepared a ground-marking plan on the basis of a forecast giving clear weather, but at Frankfurt they found up to 8/10ths cloud. The Germans lit a decoy fire site 5 miles south-east of the city and also used dummy target indicators. In a period of barely 40 minutes, 2,200 tons of HE and incendiary bombs were dropped on Frankfurt. Some of the bombing fell around the decoy but part of the 'creep-back' fell on Frankfurt, causing more damage than Bomber Command realised at the time. Altogether, 41 aircraft failed to return.

As far as Clayton Moore was concerned, the six-hour trip to Frankfurt was mainly 'run-of-the-mill', with the opposition being what he had come to expect:

I observed a few aircraft being shot down and it was apparent that the Luftwaffe was out in force, but we didn't encounter any of their fighters. Flak and searchlight activity was heavy, but we completed our bombing run without incident and had an untroubled return flight. The Frankfurt raid was my eleventh sortie. At the de-briefing, I learned that just one squadron aircraft had so far failed to return and I noticed that none of the P/O E.J. Argent's crew was in the room. I became concerned because Dick Lodge, our regular navigator, now fully fit after the crash and eager to get back into harness had flown with Argent. A message came through from Group stating that WS-O had come down in the sea off Great Yarmouth, that the crew were in the dinghy and were being picked up by ASR, but that an as yet unnamed member of the crew was dead. It was to be mid-morning before the picture became clear. Argent's Lancaster had been attacked simultaneously by a Me 210 and a Ju 88. The aircraft had suffered severe damage during the combat, but had managed to escape. Unfortunately, the 19-year-old Canadian tail gunner, Sgt Vincent Knox, had been killed during the exchange. The remaining six members of the crew, including our navigator, were unhurt and had survived the enforced ditching. However, all were suffering from the effect produced by their four-hour exposure to the rigours of the North Sea in mid-winter.[7]

On the night of Thursday, December 23rd we carried out a further attack on the 'Big City' in the *Spirit Of Russia*. Berlin, together with Essen and the industrial towns of the Ruhr, was looked upon by most crews as the ultimate test of one's nerve and skill because of the strength of the defences. The 'Big City' differed from the Ruhr targets in just one respect: A trip to Essen or Cologne, although it could prove hectic in terms of the opposition to be met, usually lasted little more than four hours, of which about ninety minutes would be spent in enemy air space. Berlin, being inside Germany, demanded long hours of concentration, alertness and the suffering of severe cold on the part of the crews involved. In addition to the heated reception one could expect from the city's considerable defences, there was always the guarantee that the fighters would be in close attendance for most of the six or more hours spent over Germany and the occupied areas of Western Europe.

Despite the trepidation I felt at being ordered to visit Berlin for the first time, the outward flight passed without incident. The presence of fighters within the stream was savagely demonstrated by the number of bombers that I saw spiralling down in flames along the route to the target. The fearsome spectacle of a fully-laden bomber meeting such a fate was one never to be forgotten. The slow, agonising, downward spinning, flame-engulfed hulk; the pieces breaking away; the holocaust when it hit the ground. I observed and reported six such disasters during our flight to the target.

On reaching Berlin, we found the city to be almost totally blanketed in cloud which was effectively illuminated by the searchlights beneath it. This presented the area as a sheet of opaque Perspex across which the bombers could be clearly seen as they crawled like army ants across the brightly lit cloud tops. The gunners were sending up a box barrage and I saw a couple of single-engined fighters lurking beneath us as we began our bombing run. At last the bombs were gone and we had taken our worthless picture of the cloud cover before turning north into the night. Suddenly the gathering darkness glowed red as another Lancaster stopped one a few hundred feet above us and began its death dive. The skipper took immediate evasive action, throwing us into a steep dive to port. But the stricken Lanc followed us and Bill asked me for instructions on its location and what further action to take in order to avoid a collision. I watched and waited unable to decide on which way we should go. The doomed aircraft was obviously out of control and it was difficult to predict which direction it would take next. The port wing was blazing furiously; also a large section of the fuselage aft of the wing. It was directly behind us and getting uncomfortably close despite our dive and I could make out details of it as it began to break up about a hundred yards astern.

Fearing the danger of the aircraft exploding, I realised the need for immediate action on my part if we were to avoid being brought down with it. 'Level out to starboard, NOW!' I shouted into my microphone, having detected a slight course alteration to port by the stricken bomber as it continued its death throes. Even as I spoke, the fire in the after part of the fuselage flared and I could clearly see inside the cockpit two dark figures silhouetted against the flames as they struggled to make their escape. The *Spirit Of Russia* winged over to starboard at once and I fought against the forces of gravity in order to stand up and watch the wretched Lancaster as it disappeared from sight beneath my turret. As we began to level out, the falling aircraft again came into view, permitting me to see that the port wing had broken free, leaving the still burning fuselage to spin drunkenly downwards. Once clear of the danger, we resumed course, while I watched the ill-fated Lanc continue its downward plunge. The fire had spread from the fuselage to the starboard wing by this time and more bits were beginning to break off, but I didn't see any parachutes come out, nor had I been able to read the aircraft's markings. 'Poor devils,' I thought as I watched it strike the ground far below us, 'there but for the grace of God go us!' Four hours later, we touched down at Bardney.

F/L J.H. Clark and crew on 460 Squadron RAAF at Binbrook. Third from left is 20-year-old P/O Roberts Dunstan, the Australian rear gunner. When he was 17 he had joined the Australian Imperial Forces as an engineer. He went into battle with the 6th Division and on 15 January 1941 was hit in the right knee by a shell fragment. His right leg was amputated after five operations, and after seven months in hospital he was discharged with an artificial leg. He persuaded the authorities to let him join the RAAF despite his disability and was accepted in June 1942; at the end of May 1943 he joined 460 Squadron. His first operation was an attack on Düsseldorf on 11 June, and he finished his tour with an attack on the same target on 3 November. He would take his crutches with him in the air and when he had to go aft would crawl there on one leg. Dunstan was commissioned in August 1943. During a raid on Kassel in the following October, a fighter attacked his aircraft and a shell smashed through Dunstan's turret and tore his sleeve. The turret itself was so badly damaged that the crew had to cut away the wreckage before they could get him out. Dunstan was awarded the DSO at the end of his tour, the only Australian air gunner to gain this award.

Although the Battle of Berlin was proving quite costly to the hard-pressed aircrews of Bomber Command and the Pathfinders, Nine Squadron suffered no losses on the trip, although a total of fifteen heavies were knocked down.

In the early hours of Christmas Eve, 379 aircraft took off for Berlin. Once again the Main Force Halifaxes were rested and only seven took part. This raid was originally planned for a late afternoon take-off but a forecast of worsening weather over the bomber stations caused the raid to be put back by seven hours to allow the bombers a return in daylight. Two Lancasters collided while taxiing; another Lancaster crashed out of control and two more were involved in a collision over Lincolnshire, both crashing and killing everyone on board. At the target there were few searchlights and no fighters, partly because they encountered difficulty with the weather and partly because the JLO was temporarily deceived by the Mosquito diversion at Leipzig. Marking was poor mainly because of H$_2$S failures, and the bombing was predominantly in the suburbs. Fighters appeared in the target area only at the end of the raid and could not catch the main bomber stream. Fifteen Lancasters in all were shot down. Two were claimed by Oberfeldwebel Karl-Heinz Scherfling of 12./NJG1. Three were destroyed by Oberleutnant Paul Zorner of 8./NJG3, who had been credited with 16 victories including five on his first two nights of operations in a Bf 110 equipped with SN-2 at the end of November.

LAC Ray Meredith, a flight mechanic, airframes in 58 MU at Skellingthorpe, recalls:

From our domestic site early one winter morning in December 1943 after a heavy overnight frost I remember a peaceful Christmas scene. The pine trees were white with hoar and Lincoln Cathedral stood majestically on top of the hill with its spires glinting in the early morning sunshine. This tranquillity didn't last long, as in the distance the sound of Merlin engines could be heard as the station's Lancasters returned from a deep penetration raid over Germany. One by one the aircraft circled the airfield and made their landing approach. Just before touching down each pilot throttled back the engines before gliding over the threshold of the main runway. The fading sound of spluttering engines could be heard in the distance as each Lancaster came to the end of its landing run, only to be quickly replaced with the sound of the next aircraft home.

Some of the damaged aircraft had badly injured or dead aircrew aboard. Generally, returning aircraft automatically made their way to their own Squadron dispersal, irrespective of damage and casualties sustained. However in some cases, the station's ambulances met an aircraft as it landed, to take off the badly wounded to the station's sick quarters or the nearest hospital. I remember one Lancaster that was towed into the hangar after it had been in a mid-air collision the previous night. The starboard outer engine and its sub-frame were completely missing. The three bladed propeller sheared off and cartwheeled along the wing embedding itself in the fuselage, one blade missing the navigator by inches. The aircraft returned with the other two blades making their own 'V-for-Victory' sign behind the cockpit. Later, another aircraft returned without its main starboard fuel tank and the wing panelling underneath. It was another casualty of the crowded sky in the target area. In the upper skin of the wing, above the tank was found the perfect stamped out profile of a 500lb GP bomb. A metal stamping press could not have done it better and it was a miracle that no one was killed. When we took delivery of aircraft with major damage it either taxied around to our hangar from its dispersal or was towed tail-first by a Davy-Brown tractor. Sometimes the sight which met us on entering damaged aircraft was very upsetting. Remnants of torn blood-stained flying clothing were found and smears of dried blood around crew positions could be seen. These were constant reminders of the dangers faced by aircrew in the night skies high above Germany.

We mainly worked in a large T2 hangar which could accommodate two Lancasters at a time. Outside the hangar was our flight hut and technical stores. Hangars were vast acoustic chambers mostly clad in corrugated metal sheeting with massive sliding doors at each end. A number of Windy drills at work and the sound of riveting, which was all done by hand I hasten to say, built up into a crescendo of sound which was deafening. Inside, the aircraft stood in various stages of repair like patient horses waiting to be groomed, some with trestles under every jacking point for tests on the undercarriage hydraulics and to allow checks on the angles of incidence of the mainplane and tailplane with a clinometer, to make certain the longitudinal spat was perfectly level. The hangar floor area was kept spotlessly clean, while around each aircraft was a crisscross of hoses, supplying the Windy drills, interlaced with numerous power cables for inspection lights and electrical services. There were large rectangular oil trays placed on the floor beneath the engines and any other point where oil might have dropped or be spilt by accident. Protective covers were also draped over the main wheel tyres, to protect the rubber from being soiled, especially if work was being carried out on either of the inboard engines where oil spillage was often unavoidable.

Nos 50 and 61 Squadron ground crews' main task was to make the aircraft ready for the next operation by carrying out daily inspections, servicing and repairing minor operational damage. Each ground crew had a pride in and an affinity with their aircraft which ensured all the work done was of the highest standards. However, it was the 'Flying 58th' who were the surgeons at Skellingthorpe. We attempted to make whole again the badly damaged aircraft that returned from bombing raids. How some of them got back, God only knows, they were in such a dreadful condition. A tail-fin and rudder missing, a whole engine gone, a fuselage or mainplane raked with cannon shells or shrapnel from anti-aircraft guns and all too often the blackened and charred skin of the aircraft indicated there had been a fire on board. Sometimes we worked all day and night in order to get severely damaged aircraft into the air once more.

The Lancaster was made up of several sections and working from nose to tail along the fuselage, the D1 which formed the front gun turret and the cockpit was followed by the D2 centre section which housed the navigators table, the W/T and the flight engineer's panel. Behind that came the D3 section housing the mid-upper gun turret and radome beneath, and lastly came the D4 containing the flare chute, rear gun turret and leaflet chute. Each fuselage section was bolted together at two adjoining bulkhead frames with numerous nuts and bolts. The tailplane came in four sections including the elevators, the tail-fins and the rudders. The port and starboard mainplanes were made up of a leading edge, trailing edge and wing tip all bolted to the main section. Ailerons and flaps were then added along with four engine sub-frames and the undercarriage. One of the most onerous tasks was the removal and replacement of the trailing edge to the mainplane. This involved undoing dozens of bolts holding it to the rear mainplane spar. Each mainplane was held in position to the centre-section with four huge high tensile steel bolts. For this operation the mainplane was suspended on two slings with a spreader bar between them. The replacement of the trailing edge was a nail-biting exercise for the operator of the Coles crane when lowering it into the correct position. Of all tools employed in the major surgery we performed on those aircraft, the Coles crane was the maid of all work and without it, many of the tasks would have been impossible. The assembled aircraft was magnificent and a tribute to the makers A.V. Roe. It would be wrong of me not to mention the support we received from their repair depot at Bracebridge Heath near Lincoln. We and the other airfields close by were wholly dependent on their workforce for Lancaster spares. We could only work as fast as they could produce parts for our damaged aircraft.

I will never forget the sound and smell of Merlin engines being started up and the scenes of activity in the various dispersals as the ground crews fussed over their aircraft. After the last aircraft had clawed its way into the air and departed with the sound of its laboured engines fading

Lancaster III EE136 *Spirit of Russia* on 9 Squadron which had flown 93 sorties when it was re-assigned to 189 Squadron in October 1944. By the time it was taken off ops, the last on 2/3 February 1945, EE136 had 109 bombs painted below the cockpit.

away, the airfield felt strangely silent. Rural tranquillity descended once again on the surrounding woods and fields and as the group of WAAFs and airmen strolled back to their billets, rabbits came out of hiding and scurried across the main runway before disappearing once again into the long grass that edged this great expanse of concrete, while in the distance ground crews could be seen tidying the dispersals in readiness for their aircraft's return or cycling around the perimeter track to the cookhouse for their evening meal. Some of the aircraft would not return from the night's operation, but of those that did, it was inevitable that one or two would suffer severe airframe damage and would become 58 MU's patients in the next morning.[8]

On the night of Wednesday 29 December the next raid on Berlin took place when 712 aircraft – 457 of them Lancasters – took part. Clayton Moore recalls:

We again took off in *Spirit Of Russia* for what was to be the squadron's last sortie of 1943. Although it was to be my thirteenth trip, it proved to be reasonably trouble free, despite the feeling of fear born of superstition that stayed with me throughout the flight. Again, all of the 9 Squadron crews returned safely, thus indicating the degree of efficiency attributable to the squadron's air and ground crews. However, command's tally amounted to twenty

FIDO in use at Graveley.

Pathfinders had arrived. In another moment they had dropped the target indicators, great shimmering Christmas trees of red and green lights; you couldn't miss it. It would be impossible to miss such a brilliantly marked objective. Bright flares started going off under the clouds that would be the Cookies of the plane ahead. *V for Victor* started the bombing run. 'Left, left, steady, now right a bit, steady, steady, Cookie gone.'

V-for-Victor shot upwards slightly. 'Standby!' came the voice of the rear gunner, Bob [Sgt C.E. Shilling]. 'Corkscrew, starboard!' he called. The pilot instantly went to starboard and dived headlong down.

A stream of red tracer whipped out of the dark, past the rear turret and on past the wing tip missing both, by what seemed inches. Then the fighter shot passed.

'A Me 109,' shouted Bob.

They went on corkscrewing over the sky and so the fighter was finally shaken off and a normal course resumed. The dark shapes of Lancasters could be seen all over the sky against the brilliant clouds below. The attack only took fifteen minutes.

Then came, 'Standby – Ju 88, starboard corkscrew,' from the rear gunner again.

The aircraft went into a dive and the tracer from the Ju 88 missed. From there on the journey was uneventful – the searchlights of the English coast sent out a greeting of welcome.

As the pilot of *Victor* circled over their base, the WOp called ground control. '*V-Victor*,' he said. A girl's voice came over the intercom. '*V-Victor*, prepare to pancake.' From the WOp, '*V-Victor* in the funnels.' – the girl's reply, '*V-Victor*, pancake.'

The aircraft landed and gently ran down the runway and turned off onto the perimeter track. '*V-Victor*, clear of flare path.' The ground crew met them with questions: 'How was it?' The reply: 'A piece of cake.'

The crew got out and the pilot had a look around the aircraft. One small hole through the aileron. Into the debriefing by bus where a cup of hot tea laced with rum is waiting. Each pilot signed on the board as he came in.

aircraft [including 11 Lancasters] lost, which was about the going rate for such a raid, but the overall trend was moving steadily upwards. 1944 was just one day and twenty minutes old when *S-Sugar* lifted from the Bardney runway bound for yet another raid on Berlin – my third. On this occasion I was flying as stand-in tail gunner with another crew. The most outstanding feature of the trip was a large hole in the cloud covering much of the capital. Through this we glimpsed a lake which we were able to identify, thus making sure that we were in the right place at the right time.

Sgt Ben Frazier, a *Yank* magazine staff correspondent, flew on the raid on Lancaster III ED888 *Victor Squared* on 576 Squadron at Elsham Wolds, piloted by F/O Gomer S. 'Taff' Morgan. ED888 was a veteran of 57 ops and had a DFC ribbon painted on its fuselage to celebrate its 50th trip. On his return, Frazier wrote an article entitled 'Night Plane to Berlin':

Suddenly the whole city opened up. The flak poured through the cloud, it poured up in streams of red as if shaken from a hose – it went off in bright white puffs. The

4

CHOP CITY

The 'Big City' now referred to as 'Chop City' by some. God, the losses over Berlin! Seem to remember that over four raids at the beginning of the year, 158 kites bought it. A slugging match with masses of night-fighters all along the route with even more over Berlin itself: and the heaviest concentration of flak in all Germany.

Campbell Muirhead

It snowed on Sunday 2 January 1944. At Syerston 26-year-old F/O Harold 'Johnny' Johnson, air bomber on F/O Vic Cole's crew on 106 Squadron, learned that, having just returned from an eight-hour round trip to the 'Big City' when Berlin was raided by 421 Lancasters and 29 Lancasters were lost, they were 'on' again that night. They were one of the 'gen crews' at Syerston, having flown 25 trips since their first on 9 July 1943. Vic Cole, who came from Farnham in Surrey, had been employed by Shell before the war. 'Johnny' Johnson had been born in Stepney where his father was a cycle maker and plumber for the London County Council and had joined the Metropolitan Police in 1937. PC Johnson and another constable tried to enlist as pilot trainees and fight in the Spanish Civil War, which they thought would be over by the time they had completed training, but news of their proposed venture reached Sir Philip Game, the Police Commissioner, and they were summoned to Scotland Yard where they were suitably admonished. Game was ready to discharge them but they remained in the force. When war was declared in September 1939, 'Johnny' was stationed at Rochester Row police station. He could not volunteer for the RAF because the police was a 'reserved occupation', and on Saturday 24 August 1940 he was on duty in Downing Street when the first raid on London took place. Finally, in 1942, when policemen could be considered for air crew, Johnny was selected for pilot training. The medical was at the recruiting centre at Lord's cricket ground in the famous Long Bar. All the recruits stood in line and were ordered 'shirts up – trousers down' as the MO stood ready with a stick. Suddenly a Cockney voice bawled out, 'Arry, if he gives you that bloody cup, you've won!'

'Johnny' went to Canada for pilot training but eventually washed out because he could not judge height during landings and he retrained as a bomb aimer. He crewed up at 19 OTU

Kinloss and soon formed the opinion that there was not a better pilot anywhere than Vic Cole. 'Vic always went in as high as he could and would then glide the Lancaster home. Nine times out of ten we would be first back.' The rest of the crew included Alfie Bristow, the navigator; Bill Haig, the WOp/AG; Eddie McColn, the 'excellent' flight engineer; and Malcolm 'Parky' Parkinson, the always-alert rear gunner.

Fatigue mixed with anger caused severe rumblings and ructions at briefings on many stations in Bomber Command. Once more the target was Berlin. And there were to be no

Lancaster PB410 on 97 Squadron in flight. This aircraft was scrapped in January 1947 after seeing further service on 12 Squadron. *(RAF Museum)*

'Vic' Cole's crew on 106 Squadron at Syerston celebrate after completing their tour with a trip to Magdeburg on 21/22 January 1944. L–R standing: F/Sgt W.P. 'Bill' Haig (WOp); F/O V.L. Cole (pilot); Sgt E.H. Woods (flight engineer); F/O H. 'Johnny' Johnson (bomb aimer); F/Sgt M.D. 'Parky' Parkinson (rear gunner); F/Sgt T. Ross (mid-upper gunner). Kneeling is F/O A.E. Bristow (navigator). Cole, Johnson and Bristow were awarded the DFC, Haig and Parkinson the DFM. Eddie McColn, the regular flight engineer, was sick and missed the trip. (Johnny Johnson)

diversions. A long, evasive route was originally planned but this was changed to an almost straight in, straight out route with just a small 'dog leg' at the end of it to allow the bombers to fly into Berlin from the north-west, to take advantage of a strong following wind from that direction. It would be another midnight take-off and runways would have to be cleared of snow to allow 383 Lancasters, Mosquitoes and Halifaxes to take off. The weather was foul throughout, and cloud contained icing and static electricity up to 28,000ft, but clearer conditions were expected at the target.

'On approaching Hanover,' recalls 'Johnny' Johnson:

with the cloud we were in thinning, 'Parky' Parkinson ordered Vic to 'corkscrew starboard'. Without hesitation he carried out this operation but not before all hell broke out. We were under attack by a Me 110, which had come out of the cloud with us and was underneath us. We were an easy target and he opened fire with his *Schräge Musik* guns. He missed the bombs but we were holed everywhere and one or more shells must have hit the No.2 tank in the port wing. The jettison toggle dropped out and so did all the petrol but there was no fire. Next Vic called out 'Prepare to abandon aircraft – I can't get this bastard out of the dive.' I jettisoned the escape hatch but the 'G' force took over and I could not move. I knew that I was going to die and in that terrifying moment I had a vision of my Mum and Dad together with my fiancée Enid. We were circling to such a degree that we were more or less transfixed but as a last resort Vic put the auto pilot in and it pulled the aircraft out and his corkscrew tactics lost the Me 110. We slowly levelled out at 6,000ft before beginning a slow climb and Vic ordered the bombs to be jettisoned. I said 'No – we've got a 4,000lb "cookie" on board' and we were too low. I would take a chance on 8,000ft but I was not about to do so at 6,000ft. We finally dropped the 'cookie' and the rest of the bombs at 10,000ft and I had remembered to fuse them.

Next 'Parky' Parkinson called out, 'Skipper we're on fire at the back.' The WOp/AG was belting out 'Mayday, Mayday'. I got a fire extinguisher and climbed over the main spar but there was no smoke and the interior light was on. I could see that the rear door was open. John Harding, the mid-upper gunner, had baled out! We found out much later that he broke his ankle on landing and was captured. Eddie McColn estimated that we had less than an even chance of making it back to base but Vic decided to try and get us home. Alfie Bristow quickly assessed a heading for base and I went up into the mid-upper turret as Vic knew we would probably be attacked on the way home. We came back fairly low over the North Sea. We couldn't afford to ditch because the back door was open, all the hatches were gone and we were full of holes. Everything was wide open for water to come in. We were very fortunate that we were not attacked again. Vic flew very carefully, hit the Lincolnshire coast and flew straight to Syerston. We landed and as we taxied off the runway all four engines stopped! The first to get to us were our overworked, underpaid, seldom mentioned ground crew – God bless them – with the blood wagon not far behind. But we didn't need him.

Next day the station commander sent for Vic and wanted to know why he hadn't continued to the target! At that moment the engineering officer came in and told the Group Captain that our Lancaster had been SOC [signed off charge] – it had a broken back.

There were two nights of rest for the bomber squadrons and then on the night of 5/6 January, 348 Lancasters and ten Halifaxes raided Stettin. Clayton Moore recalls:

Alan Falconer and his crew on 166 Squadron at Kirmington with 'Ronnie', their WAAF MT driver.

We took the *Spirit Of Russia* to the Polish port on what was to be the longest operational flight we would ever undertake – nine hours and 35 minutes of danger, darkness and extreme cold. The route was a devious one and included a feint at Berlin, which together with a spoof raid on Berlin by [13] Mosquitoes, succeeded in fooling the German defences into thinking that the capital was again to be attacked. We found Stettin to be mostly clear of cloud cover and were able to press home the attack without meeting serious opposition. Because of the lateness of our take-off, dawn had broken by the time we got out over the North Sea on the homeward leg.

Stettin was W/O Eric Jones' 23rd op and his crew's 21st:

I wonder what the crew would have thought if we had known at the time that we were one of only six Lancasters engaged on mine-laying that night and that the Main Force of 348 Lancasters were bombing the city of Stettin. This was our longest penetration into enemy territory and we were to land 9 hours and 10 minutes after take-off (with a full fuel load of 2,155 Imperial gallons the Lancaster could stay airborne for just over 10 hours so this trip was pushing things to the limit). Each Merlin engine burned about 50 gallons an hour and a considerable amount could be added to this for take-off and climb. This was a 'Gardening' operation. In plain language a mine-laying exercise in enemy waters and the water in question this time was the harbour approach into Stettin. Mines were dropped on parachutes to minimise the impact when they hit the water

Tommy Heyes' highly decorated crew on 100 Squadron, who completed their tour with a raid on Berlin on 28/29 January 1944. They had crewed up at 28 OTU before they were joined by their engineer at 1656 HCU at Lindholme, being posted to 100 Squadron at Waltham at the end of August 1944. Eleven of their 30 ops were to the 'Big City', including the raid on 16/17 December when the squadron lost four in crashes. This photograph was taken on the day after their final trip. L–R: Tommy Heyes DFC AFC (pilot); Peter Ashden DFC (engineer); Sid Emmett DFM* (navigator); William Kondra DFM RCAF (bomb aimer); Glynn Jenkinson (WOp); Ken Kemp DFM (mid-upper gunner); Jock Ross (rear gunner). *(William Kondra)*

and it also meant going in quite low to drop them. We were routed in over the most northerly cape of Denmark right up to the Swedish coast and then we turned due south for about a 200-mile run up to the target. At some point we must have crossed over into Swedish territory because, suddenly, all hell was let loose with the Swedes shooting off with everything they had, but only up to about 10,000ft. It was quite a sight, quite like flying over a bright red carpet with nothing coming anywhere near us. We learned later and I don't know how true it was, that they did exactly the same for the Germans. They were a neutral country of course and they were telling us to stay up there. Come any lower at your peril. We witnessed the Aurora Borealis (the Northern Lights), way to the north, shafts of light like searchlights piercing the night sky.[1]

After Stettin we went to Brunswick nine nights later and then there were four more Berlin raids which took us up to the end January 1944. I had now completed 28 operations and the crew 26 (except for those members who had been sick). Almost half our tour had been taken up with raids on Berlin and it was incredible that we had survived. Going into breakfast one saw but tried not to notice that a certain table was empty, maybe sometimes two but these would soon be filled with crews straight from training. The crews of Hodgkinson, Brunt, Cottingham, Petty and countless others had all gone missing. I knew these four pilots; they were all from 'A' Flight and I trained with Hodgkinson and Brunt. But

it was never going to be us, not the crew of *K King*, not after we had come so far.

On the raid on Brunswick on 14/15 January, 498 bombers set out but most of the attack fell either in the countryside or in Wolfenbüttel and other small towns and villages well to the south of the city. 'Once again,' recalls Clayton Moore, 'we had an uneventful flight, although we saw a lot happening all around us to indicate that some of our friends were having a rough time of it.' The German defences picked up the attacking force only 40 miles from the English coast and many night-fighters entered the bomber stream soon after the bombers crossed the German frontier near Bremen. Eric Jones recalls:

'Stooging' out of Germany I heard the chatter of our machine guns. I had received no call from Jock in the mid-upper or Peto in the rear so it had to be Steve up front. There had been no call for a 'corkscrew'. There wouldn't be for a frontal attack. I couldn't see anything so I assumed that the attack was coming from below. It was all over in seconds. Looking out on my starboard quarter I saw a German fighter flying from left to right and trailing smoke. From his position in the nose of the aircraft Steve was able to follow the fighter all the way down and he claimed it as his very own. At debriefing, on our return, I substantiated his claim. Steve was later awarded the DFM and although I never saw the citation I suspect that the German fighter had something to do with it.

By the time the heavies crossed the Dutch coast on the return flight no fewer than 42 Lancasters had been shot down. *Nachtjagd,* operating Tame Boar, freelance or Pursuit Night Fighting tactics to excellent advantage, seemed to have rendered 'Window' counter-productive.[2]

'Losses in the Battle of Berlin,' recalls Clayton Moore:

had been averaging about 6 percent of late and this figure was disturbingly high for anyone hoping to complete a tour. There was no question of the Germans not having adapted to the effect that 'Window' had given earlier. For one thing, they had changed the wavelength of their radar systems, so that we were now obliged to drop strips measuring almost twice the size of the originals. The increasing efficiency of the German air defences was borne out by the loss rate suffered on the Brunswick raid. This had been more than 7 per cent. Although the 'chop rate' was steadily mounting, so too was the degree of success we were having in hitting the enemy where it hurt. The techniques being developed by the innovative Pathfinder Force were beginning to show promise and a lot of crews were showing an interest in joining AVM Bennett's 'Blue eyed Boys'. Bill called a crew meeting to propose that we might also make the move. The vote went in favour of the proposal and the application was lodged the next day. Little time was wasted in processing it and acceptance was confirmed almost immediately. PFF would be an entirely new ball game. Emphasis would centre on a crew's ability to navigate within precise limits, with timing to within a tolerance of plus or minus ten seconds at the target being the maximum error allowed. We also learned that each member of the crew would receive an automatic bump up in rank on completion of the training.

The straight in, straight out routes, which so often characterised previous raids on Berlin, were abandoned on the night of Friday 20 January, when 769 aircraft – 495 of them Lancasters – returned to the 'Big City'. It was a fine day and the bombers took off in the late afternoon. Over Germany the bombers ran into the cloud of a cold front and Berlin was completely cloud covered. The timing of the blind markers was reported to be 'excellent' and a good concentration of sky-markers was maintained throughout the attack, but the crews of H$_2$S aircraft thought that the attack fell on the eastern districts of the city. Conditions were particularly favourable to night-fighters since a layer of cloud at 12,000ft illuminated from below by searchlights provided a background against which aircraft could be silhouetted. About 100 twin-engined fighters were sighted over Berlin and night-fighters shot down all 13 Lancasters and 22 Halifaxes that were lost on the raid.

On Saturday night, 21 January, an 83 Squadron PFF Lancaster crew, captained by F/O Ken Hutton, who was from Tasmania, were on the battle order, having just flown their sixth trip to Berlin. Hutton's 26-year-old navigator, P/O Geoffrey Breaden, and 21-year-old F/O Ron Walker, the mid-upper gunner, also came from Tasmania. Breaden recalled:

During the pre-flight briefing for the raid, tension started to rise while we all awaited the unveiling of the target for tonight. We all felt that our luck was being stretched by so many consecutive sorties to this heavily defended spot. Imagine the almost audible sigh of relief when the target of Magdeburg was revealed to us. Our task this night was

Lancaster I NG264 IQ-B *We Dood It Too* on 150 Squadron at Hemswell in the winter of 1944. This crew had flown 20 operations on 550 Squadron at North Killingholme before going to Fiskerton, where they helped re-form 150 Squadron and went on to complete a full tour of 36 operations. L–R: Len Buckell (mid-upper gunner); Ken Brotherhood (engineer); Bill Mann RAAF (navigator); Gordon Markes (pilot); Danny Driscoll (rear gunner); Vernon Wilkes (bomb aimer); Frank Petch RAAF (wireless operator). The aircraft survived the war and was operated by 10 MU before it was scrapped in March 1947.

'Primary Blind Marking' and our time on target was shown as zero minus five! I could hardly believe my eyes. Five minutes alone over a target in Germany before anyone else arrived. Suicide, I thought. As the navigator, I felt that I would be signing my own death warrant, as it were! In the event we contrived to time our arrival for zero hour, 23.00Z.

Favourable weather permitted sending 648 aircraft in four waves to Magdeburg, 60 miles west of Berlin. Winds were stronger than forecast and the outward route was not dissimilar to that of the night before. A feint by 22 Lancasters and a dozen Mosquitoes who bombed Berlin was largely ignored and the JLO ordered Tame Boar to assemble at a beacon between Hamburg and Cuxhaven. Later he ordered them to Hamburg and then to Leipzig, just south of Magdeburg. The Bf 110s and Ju 88s struck between Cuxhaven and Lüneburg, south of Hamburg, with devastating effect. The night-fighters remained with the bomber stream all the way to Magdeburg, where most of the bombing is believed to have fallen outside the city. The stronger-than-forecast winds brought some of the bombers into the target area before the Pathfinders' zero hour and 27 Main Force aircraft bombed without delay. The Pathfinders later blamed the fires started by these aircraft and some effective enemy decoy markers for the scattered bombing that followed. Five bombers were shot down over the target. About 60 aircraft were damaged by flak and fighters, and some of these did not

BQ-Q *Queenie* and F/O Godfrey Arnold Morrison's crew on 550 Squadron at Woodbridge after landing there following combat with a night-fighter on the operation on Berlin on 30/31 January 1944. The bomb aimer baled out and was taken prisoner. Both the mid-upper and rear gunner were killed. F/L Morrison DSO and his crew were killed on the operation on Dortmund on 22/23 May.

7,000 feet. So coming back we flew right across the heart of Europe at tree-top level, shooting at any light along the way.

There were four nights of rest for the Main Force following the Magdeburg débâcle and then on Thursday 27 January, 515 Lancasters and 15 Mosquitoes were detailed for Berlin again. Half the German night-fighter force was duped into flying north by a diversionary force of Halifaxes laying mines near Heligoland, so only a few enemy fighters attacked the bombers before Berlin was reached. The target was cloud covered and sky-marking had to be used. This appeared to be accurate but the strong winds blew them rapidly along the line of the bombers' route and bombing was 'spread well up and down wind'. This raid was the first time that the flare force marked with 'supporters' from non-Pathfinder squadrons. Until now the 'supporters' had all been from 8 Group. Supporting the flare force meant arriving at the target at the same time, but flying at 2,000ft below them to attract the flak and enable the PFF to carry out a straight and level run. After drawing the flak the 'supporters' then re-crossed the target to drop their bombs. Twenty-eight of the most experienced crews in 1 Group acted as 'supporters', two of which were shot down.

Gunners on *S-Sugar*, a Linton-based 408 'Goose' Squadron Lancaster B.II piloted by W/O2 J.D. Harvey, claimed an enemy fighter. The e/a was identified by the rear gunner, F/Sgt S.E. Campbell, as a Me 110 and was first sighted below and slightly to port, range 600yd, commencing to attack. Campbell instructed Harvey to turn port towards the dark side of sky. The fighter appeared to be trying to position himself to fire rockets. The rear gunner opened fire at 300yd with long bursts of approximately 300 rounds. He observed his tracer entering the Me 110's starboard wing, hitting the engine and knocking off one of the rockets. The fighter broke off down to port and Campbell instructed Harvey to climb starboard, do a banking search and then resume course. Campbell again sighted the Me 110 right below at a range of 800yd and opened fire with another long burst of 300 rounds, his tracer entering rear of the fighter's cockpit. A large blue flash appeared in the cockpit and every light came on. The Me 110 started to weave, going over to port quarter down and back again underneath the bomber and commencing to attack with all its lights on. Campbell again opened fire with a long burst of 300 rounds, his tracer this time entering the fighter's cockpit. The Me 110 caught fire, rolled over and went down out of control, disappearing beneath the clouds with flames completely enveloping the fuselage. This

make it home. A total of 58 bombers were shot down – Bomber Command's heaviest loss of the war so far.

F/O Len Sumak DFM RCAF, a mid-upper gunner on 576 Squadron who flew on the raid on Magdeburg, could see a train pulling into Magdeburg and then all of a sudden the train started going in reverse. 'He knew what was coming,' recalled Sumak:

In my tour we hit cities like Stuttgart, Leipzig, Augsburg, Frankfurt, Schweinfurt, Brunswick and Berlin a total of eleven times. On one raid on Berlin I got credited with shooting down a Focke-Wulf 190. We had just dropped our bombs. Gunners were always told to keep their eyes open and shoot at anything that didn't look right. Right away I saw a Focke-Wulf fighter coming at us. I notified the skipper to take evasive action as the tail-gunner and I started firing away at the Jerry. The fighter got to within 200 yards on the port side before he gave a burst. Meanwhile I kept shooting until I actually froze on the triggers. I got him finally, but as I was still firing, the skipper shouted, 'For God's sake, Len, stop! You already got him!' Then another fighter came at us and we started firing again and once again we took evasive action. At that point we were at 21,000 feet, but after it was all over – we were coned in searchlights too – we were down to about

was seen by the rear gunner, mid-upper gunner and WOp/AG. Then a glow appeared beneath the clouds, which was assumed to be the fighter hitting the ground.

All told, 33 Lancasters failed to return.

A maximum effort was ordered immediately against Berlin by 677 bombers on the night of 28/29 January and a full range of diversionary operations was put into action. The cloud was well broken and the routes to the target went north over northern Denmark on both the outward and return flights, but the JLO still managed to concentrate large numbers of fighters over the target and 26 Halifaxes and 18 Lancasters were shot down. Two other Lancasters were lost with their crews in a mid-air collision over Alsace.

Crews had just one night's rest before the bombing of Berlin resumed on the night of 30/31 January, this time by a force of 534 aircraft for the third raid on the capital in four nights.

It was the start of a new moon period and a quarter-to-half moon was expected during the outward flight. Also, the only diversionary raids were by 22 Mosquitoes on Elberfeld and five more on Brunswick. While the outward route for the Main Force was again a northerly one, it was not as far north as the previous raid on Berlin. The night-fighters were unable to intercept the Main Force over the sea and the bomber stream was well on the way to the target before they met with any opposition. At the target the bombing was made through complete cloud cover. Twin-engined Tame Boars wreaked havoc, continuing their attacks until well into the return flight, accounting for all 33 bombers that were shot down.

F/L Thomas H. Blackham from Dunoon, Scotland, a pilot on 50 Squadron, was flying Lancaster DV368 *S-Sugar* and had already been to Berlin four times. He was attacked by a fighter whose approach had not been seen or detected. The hydraulics and oxygen supply were damaged by shell splinters. F/Sgt J. Shuttleworth, the Australian rear gunner from Brisbane, was wounded and slumped unconscious in his turret. The port fin and rudder and the tailplane were shot up in the attack – the hole later found was large enough for a man

Right: Lancaster B.I
W4964 WS-J *Johnnie
Walker/Still Going
Strong!* on 9 Squadron
flew 106 ops between
April 1943 and October
1944. It is pictured
here with Doug Melrose
and crew.

Far right: *O-Oboe* on
460 Squadron RAAF
at Binbrook with the
port inner christened
Christopher J. and the
port outer, *Betty.*

to crawl through. The mid-upper turret had also been hit and the gunner wounded in the head. They were later to discover the left tyre had burst and a cannon shell had gone through the port tailplane. It also holed the outer petrol tank, but the self-sealing there had held. There were cannon shell holes all along the fuselage, but despite this carnage, they carried on and bombed the target.

Sgt Charles Richard Ernest Walton, the 28-year-old flight engineer from Birmingham, went back with an oxygen bottle for the rear gunner who was trapped in his turret. He feebly waved to him, his face covered in blood. He tried to work the dead-man's handle to release him but because of the lack of oxygen, Walton kept passing out. The bomb aimer, Sgt Stewart James Godfrey, from Paisley in Scotland, went back to find out what was happening and when he too failed to come back, the WOp, Sgt Sidney Charles Wilkins, went aft – and he too passed out. It was left to the 24-year-old Welsh navigator, P/O David Gwynfor Jones, to help. He found Godfrey, brought him round and then went back to sort out the engineer, but then Jones too passed out. The WOp, who then came too, gave a running commentary to the pilot: 'The navigator is down; no it's the flight engineer, the navigator is up, no he's down, the engineer is kicking him, yes the nav's on his feet …'

The engineer came around and the navigator got back to his seat. They were now about 30 minutes from the French coast and nearly out of oxygen, so Blackham got the aircraft down to 4,000ft to cross the coast. They flew over the North Sea with their wheels down; it took the engineer 20 minutes to pump them down by hand. The rear gunner remained trapped until the aircraft was about to land, when Sgt Herbert George Ridd, the 29-year-old mid-upper gunner, also from Wales, hacked the doors free with an axe and pulled him out. Despite a burst tyre they landed safely, although petrol spilled out of ruptured fuel tanks. Shuttleworth had an operation on his damaged eye, as well as on a fractured forearm.

There was a remarkable escape for one Lancaster and its crew on 550 Squadron at North Killingholme. *Q-Queenie*, flown by F/O Godfrey Arnold Morrison of Scarborough, was attacked over the outskirts of Berlin by a night-fighter. Morrison managed to evade and dropped his bombs at the aiming point only to be attacked by a second fighter. *Q-Queenie* was hit repeatedly by cannon fire; both rear and mid-upper turrets were put out of action and their gunners killed. The escape hatch was blown off and with it went the navigator's charts. To make matters worse, the controls were damaged and Morrison needed help in holding the rudder. The crew were desperate to find an airfield and finally they did spot landing lights and were in the funnel when they realised that they were on the wrong side of the North Sea. They spotted their mistake just in time and Morrison managed to nurse the damaged *Q-Queenie* back to East Anglia, where he needed most of the 1,500ft runway at the emergency airfield at Woodbridge to bring the Lancaster to a halt. The squadron records later spoke of Morrison's 'exceptionally' good landing in view of the fragile condition of the aircraft. F/L Morrison DSO and his crew were killed in Dortmund on 22/23 May.

Harsh weather curtailed flying into February. On 10 February, when there was no flying, a parade was held at Kirmington in blizzard conditions. 'It is noteworthy,' said the records, 'that the flight commanders had no idea how to look after their men.' By now work had been completed on the sodium lighting ring around Kirmington and, on a good night, homecoming crews were said to be able to see the airfield from 50 miles away. It was also decided to use red lights to illuminate Pelham's Pillar, the monument standing on top of the Wolds a few miles south of Kirmington, as an additional beacon.

Replacement crews and aircraft were on their way and existing crews were told that they would not be required for operations for nearly two weeks. On 12/13 February, ten Lancasters on 617 Squadron carried out the third operation on the Anthéor Viaduct, 15 miles west of Cannes on the coastal railway line leading to Italy. As on the two previous raids, the 90ft stone arches curving back across the beach at the foot of a ravine resisted all attempts to destroy them. Two low-level Lancasters flown by W/C Leonard Cheshire and S/L Mick Martin were damaged by anti-aircraft guns, which mortally wounded F/L 'Bob' Hay, the bomb aimer on Martin's aircraft.

Above: Sgt Cliff Fudge DFM, gunner on Lancaster R5868 PO-S on 467 Squadron RAAF, is toasted in cocoa on return from Berlin on 16 February 1944. Fudge had turned 21 years of age over the target at midnight. His skipper, P/O J. McManus RAAF, is second from right.

Left: A 460 Squadron RAAF Lancaster at Binbrook with flying kangaroo art.

Martin had headed for Ajaccio in northern Corsica after one of his crew confirmed that the island was in Allied hands – he had read it in the *News of the World* the Sunday before – but he finally put *P-Popsie* down at Elmas Field in south Sardinia, where a doctor was available. They buried Bob Hay in Sardinia. The other Lancasters landed back at Ford where the weather threatened to prevent them from returning to Woodhall Spa.

Due to bad weather, it was not until the night of Tuesday 15 February that the offensive on Berlin could recommence, and just over 890 aircraft were dispatched. This was the largest force sent to Berlin so far and the largest non-1,000-bomber force dispatched to any target. It was also the first time that more than 500 Lancasters (561, of which 226 were from

Late in 1944 some Lancasters began receiving the 'Village Inn' tail warning radar, which was also used for gun-laying. The equipment was fitted to the rear of the aircraft below the turret and gave coverage of 85° either side of the turret and 45° in elevation. The antenna was not linked to turret movement and when a target was located the antenna 'locked on' to the target return. The rear gunner was provided with a semi-transparent screen on to which a target spot was projected through a collimator and prism. Detection of an enemy fighter was typically 1,000–1,500yd and the range was given to the rear gunner by the navigator as the fighter approached. The gunner simply lined up his sight graticule with the target spot until the range was within 400yd, when he would open fire – whether he could see anything or not.

5 Group) and more than 300 Halifaxes were sent on a raid. In addition, a series of diversion and support operations was mounted, including one by 24 Lancasters of 8 Group on Frankfurt on Oder.

Bill Siddle's crew, now on 83 Squadron PFF at Wyton in Cambridgeshire, flew their first PFF op in *N-Nan*, as Clayton Moore recalls:

For the first time, I was seated inside the modified and greatly improved FN 120 tail gun turret. Although the space available to me was still somewhat limited, the layout of the Perspex canopy was much improved, thus making possible a less restricted view of the sky around me and

this had been further enhanced by the removal of most of the Perspex from the front of the turret. I had been issued with the standard pilot's seat pack parachute, so the small seat cushion (normally a feature of the FN 20) had been removed, together with the minute piece of armour plate intended as a means of affording some degree of protection to the 'family jewels'. Because there were within the air gunners' fraternity some who complained that the plate was not in any case of sufficient dimension to give complete protection, its removal was not entirely regretted.

The flight was to be the first of six trips as 'supporters' and it was with some trepidation we learned at the briefing that our target was again to be the 'Big City'. Because we were to go in with the leaders of the pack, our take-off was set for the early time of 21.30 hours and our load would be made up entirely of high explosive and incendiary bombs. All things considered, there was no denying the fact that we were being thrown in at the deep end for this one. The outward flight proved no more traumatic than any of the previous Berlin raids in which I had been involved. Again, weather conditions were a problem, mainly because of the cold, and we were forced to come down below our designated operational height when condensation trails were seen to be forming behind us. These were a dead give-away to any fighter that happened upon them and both Gerry and I had strict orders to report these to the skipper whenever they appeared, whereupon Bill would bring us down a hundred feet or so. I was also having problems with my microphone due to the condensation of my breath. This quickly turned to ice, thereby making conversation with the rest of the crew almost impossible. Because of the importance of me being able to give clear orders during a possible attack, I had to dislodge the ice at frequent intervals during the flight by giving my oxygen mask a sharp rap with my gloved hand. At a later date, the vexing problem was to be overcome by the extension of the heated clothing to include a small heater within the mask itself, an improvement which was to earn the gratitude of many.

This was the first time we had gone in with the primary marker aircraft and we found the German capital unusually quiet as we began our run-up. As on previous occasions, there was almost total cloud cover in the area, but this time there was nothing to indicate that the city had a defence to offer. Darkness reigned, there being no searchlight beams to be seen anywhere in the area and this, together with the marked absence of a flak barrage, led one to believe that we were nowhere near our intended objective. I was later to learn that the Jerries often 'played possum' in this way, hoping that the Pathfinder crews would go on to drop their marker flares over one of the many decoy targets that had been laboriously and skilfully constructed in the vicinity of the more important German cities. Berlin was not without one of these and there was no doubt that more than a few crews had in the past been fooled into dropping their markers on these 'cardboard' cities. But our marker crews were not to be duped this time. The H_2S set was behaving well and Dick was confident that we were on track and on time. We already had the bomb doors open when we saw the

Above: An all-Australian crew on 103 Squadron at Elsham Wolds.

Left: F/L Gomer S. 'Taff' Morgan's crew on 576 Squadron at Elsham Wolds. On 29 December 1943, Sgt Ben Frazier, a *Yank* staff correspondent, flew on the raid on Berlin in a Lancaster on 576 Squadron flown by Morgan.

first target indicators go down just ahead of us. That was the signal for all hell to break loose around us.

Suddenly the city beneath took on the visual impact that we had all come to expect of it. On came the scores of searchlights as if by the throwing of a single switch and I again sensed the feeling of naked exposure as the sky around and beneath us became flooded with the all-revealing illumination that I hated so intensely. In almost the same instant, the flak barrage opened up with deadly accuracy and I again heard the familiar 'whump' of the shells bursting near us and the metallic rattle of the shrapnel against the aircraft as we flew steadily onwards on the bombing run. This was the part of a raid that we all feared the most. There was nothing we could do except sit it out and hope to God that our luck stayed with us. Dead astern, somebody's luck let them down and their aircraft suddenly exploded in a ball of white heat and coloured marker flares, indicating that it had been a Pathfinder. Off to starboard, two others were going down, one in flames, the other with no sign of fire, but apparently out of control as it spiralled down in the direction of the opaque cloud tops. In all probability the controls had been damaged, or maybe the pilot had been killed or injured. Whatever the cause, I was too busy to reflect on it. I had seen a twin-engined fighter silhouetted against the clouds as it passed from port to starboard a couple of hundred feet beneath us, so the situation required that I exercise the utmost vigilance, otherwise we too might join our unfortunate friends in their death dive.

At last the bombs were released and the snapshot had been taken for posterity, thus freeing us for evasive action as we winged over to head for the comforting darkness beyond the city. There now lay ahead of us the three-hundred mile flight over enemy territory that would take us out across Holland to the comparative safety of the North Sea. Indications were that it wasn't going to be an easy trip. Even though we were in the forefront of the attacking force, the German fighters were already in the bomber stream, as was evinced by the number of aircraft we could see going down on either side of us and astern. I was reporting such sightings every few minutes and it was clear that the Jerries were having a right old turkey-shoot. Bill had issued an order for all non-essential crew members to assist in scanning the sky around us for signs of a possible attack and both Gerry's turret and mine was constantly on the move. Our vigilance paid off and we spotted two fighters during the return flight. Gerry let go a burst at the second one (a Me 109), but we managed to lose him in the darkness before he could launch an attack on us. I was later to have a word with Gerry on the matter of his having opened fire first. This was a policy favoured by a lot of gunners, but it was one with which I disagreed strongly, not only because the sight of his tracer bullets had served to betray our position to the enemy. The seven-hour, forty-minute flight ended with touchdown just after 5 a.m. on the 16th. On arrival at dispersal, we inspected the aircraft and were not surprised to find that *Nan* had picked up a couple of flak holes, but nothing serious had resulted.

The JLOs plotted the bomber stream soon after it left the English coast but the swing over Denmark for the approach flight proved too far distant for many of the German fighters. A record 2,643 tons of bombs was dropped through thick cloud over the 'Big City' by 806 aircraft out of the 891 dispatched. The aiming point was marked by red and green stars and the 'blind backers-up' were ordered to keep it marked throughout the raid with green TIs. The attack lasted for 39 minutes. Nine Lancasters and six Halifaxes acted as primary 'blind markers', dropping their flares two minutes before the arrival of 11 special Lancasters acting as 'backers-up' and equipped with H$_2$S. They dropped their markers at the rate of one every two minutes and were followed by three Lancasters and 11 Halifaxes flying in pairs. After these came the visual 'backers-up', 20 Lancasters, dropping flares at double that rate; their supporters, 58 Lancasters and the Halifaxes; and finally, the Main Force, divided into five waves of an average

F/L Knute Brydon, the Canadian bomber leader on 101 Squadron (seated with greatcoat), debriefs F/Sgt Ross (far left, seated) at Ludford Magna early in 1944 after a raid on Berlin. At far left in the background is the Met Officer, F/O Stan Horrocks, who was interested in all the route weather information. At the extreme right, holding a cigarette, is F/O Arthur, the deputy squadron engineering officer. The remaining air-crew members are from W/O Jack Laurens DFM's crew (missing in action 19/20 February 1944). *(Imperial War Museum)*

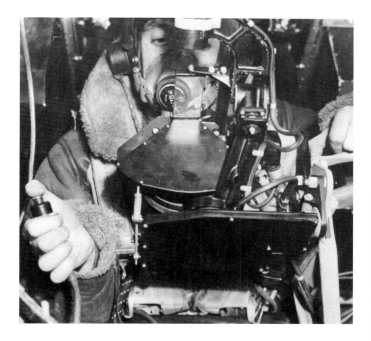

number of 140 aircraft. 'Window' was dropped throughout the attack until supplies were exhausted. The attack was remarkable for its precision, though no glimpse of the city was seen. The last arrivals were able to report the glow of large fires and a column of smoke rising 30,000ft into the murky air.

After leaving Denmark there were 22 attacks by enemy fighters and upwards of 50 sightings. In the target area 30 fighters were seen and there were 13 attacks. Eight attacks and 15 fighters were encountered on the return. *Q-Queenie* on 460 Squadron was attacked by a Ju 88 and a Me 210. P/O Robert William Burke DFC RAAF, who was from New South Wales, had just celebrated his 22nd birthday and he and his crew were on their ninth op. In the action the Ju 88 was shot down and the Me 210 damaged, but *Queenie* had been damaged in the starboard wing and rudder, and the starboard inner propeller blade had been hit and the engine holed by cannon fire. The starboard outer engine was also holed and there was a jagged hole in the fuselage. Burke got *Queenie* back to Binbrook but he would not live to see his 23rd birthday. He and five of his crew were killed on the night of 9/10 April 1944, when his Lancaster crashed in Denmark on a 'Gardening' trip.

Forty-three Main Force bombers failed to return. Six other aircraft were lost on the return over England. This raid marked the close of the true 'Battle of Berlin'. A raid on the 'Big City'

was ordered on each of the following three days – 16, 17 and 18 February – but each time the operation had to be cancelled because of unfavourable weather conditions. Only one more raid took place on the German capital – and that would not be for another month. Thirty-five major attacks were made on Berlin and other German towns between mid-1943 and 24/25 March 1944; 20,224 sorties, 9,111 of which were to Berlin. From these sorties, 1,047 aircraft failed to return and 1,682 received varying degrees of damage. At the start of battle, Sir Arthur Harris had predicted that Berlin would 'cost between 400–500 aircraft' but that it would 'cost Germany the war'. He was proved wrong on both counts.

Above: F/O J.B. Burnside, a flight engineer on 619 Squadron on 14 February 1944.

Left: F/L Walmsley at his Mk.XIV Stabilized Automatic Bombsight (SABS) on 14 February 1944. By mid-1944 this became standard equipment on most operational heavy bomber aircraft in Bomber Command.

5

ROUND-THE-CLOCK

We used to come home with all sorts of holes in the Lancaster, particularly when we were doing Main Force trips. We did Munich twice, Stuttgart twice, we did Darmstadt, all long-distance penetrations, lots of aeroplanes, 500/600 aeroplanes on the target at the time. You were as scared about the bombs on the aircraft above you hitting you, as you were about the Germans. On one operation, a Pas de Calais run on the buzz-bomb sites, when we got back our ground crew counted 400 holes in our Lancaster and then gave up counting. This was flak damage. On another trip to the same area we actually watched a bomb go through the wing of a Lancaster in front of us. It just folded up and went down. Also in daylight, a raid on buzz-bomb sites, the mid-upper gunner went, 'Christ!' A bomb had dropped between our main-plane and tail-plane, just came down beside us from the chap on top.

Sgt John Langston, navigator on 630 Squadron at East Kirkby

At 2333hrs on the night of 19 February 1944, S/L Barry Douetil taxied *T-Tommy* out at Kelstern on the bleak Lincolnshire Wolds, took off and climbed through 10/10ths snow clouds with clear skies at about 12,000ft. At Cleethorpes he joined the mass of over 800 bomber aircraft making for Leipzig and its aircraft assembly factories, part of a true round-the-clock offensive in concert with the American air forces. (In just one week of sustained operations Bomber Command and the USAAF dropped 19,000 tons of bombs on the Reich but 224 American and 157 British bombers failed to return.) Douetil headed for two searchlights that made a 'gate' through which he flew, and his navigator then set course over the North Sea towards Denmark, then down to Zwolle in Holland and across Germany between Hamburg and the Ruhr. It was hoped that this route and a 'spoof' feint over the North Sea by an OTU force and a Mosquito 'spoof' force carrying on due east to Berlin, as the Main Force turned sharply south to Leipzig at Brandenburg, would keep German fighters back. Unfortunately the JLO did not take the bait and another Mosquito 'spoof' on Dresden failed. A forecast steady headwind turned into a tailwind in excess of 100mph and it caused chaos. Many crews arrived at the target early and then orbited the area waiting for Pathfinder markers to go down. Some crews did not wait and, using their H₂S, bombed before zero hour. Pandemonium ensued when the markers did go down, as several hundred bomber crews raced in from all directions and tried to bomb at once before leaving the area as quickly as possible.

Dick Starkey on 106 Squadron recalled:

This raid was made during the Battle of Berlin and as the two cities are fairly close to each other the German night-fighter commander expected another raid on Berlin, to which city we appeared to be heading, and for the first time Dornier bombers flew above the bomber stream on either side dropping flares to suspend in the air and form a 'lane' towards Berlin. They were also helpful to the night-fighters. The turning point to fly south to Leipzig was ignored by several crews who for some unknown reason decided to take up the time by flying towards Berlin where they were caught in the 'lane' of flares which never seemed to go out and hung there for what seemed ages. We saw all this happening as we flew towards Leipzig. I had decided to risk flying over the city and go round again on the east side by which time Pathfinders would be dropping markers on zero hour. We could bomb on them and hopefully be away. On the return journey we saw many combats but were not troubled ourselves.[1]

At 0243hrs near Langenhagen airfield in the region of Hanover, still on course for Leipzig, Barry Douetil and the crew on *T-Tommy* was attacked twice by a Focke-Wulf 190. On the second attack cannon shells hit the Lancaster's starboard wing, setting it alight. Barry Douetil recalled:

I could feel the impact of the hits through my hands on the control column and at the same time could hear the rear gunner's four guns firing back, but by this time I could also hear the noise of the flames over the noise of the engines. Following the usual drill, whilst I was still controlling the aircraft, Sgt John Gill the flight engineer undid my seat belt and clipped the chest type parachute pack on to the harness; we did not fly with parachutes attached. A short while afterwards the aircraft went over onto its back and I fell on to the roof which seemed to give way and I found myself at about 20,000ft falling rapidly through the darkness.[2]

The Main Force lost 78 bombers – 44 of them Lancasters – Bomber Command's worst casualties so far. About 20 of these were victims of flak, while the rest were shot down by a very efficiently deployed Tame Boar operation by 294 aircraft (17 only of which were lost). Two Lancasters on 103 Squadron collided on the return to Elsham Wolds. One of the aircraft managed to crash-land at the base without injury to the crew but five men on board the second aircraft were killed. *Q-Queenie*, a 166 Squadron Lancaster, crash-landed at Manston on its return to Kirmington, all the crew suffering injuries.

The crew of *J-Jig* on 550 Squadron at North Killingholme had a fortunate escape. On the outward leg the Lancaster, which was being flown by F/L Jack Simon Gustave Crawford of Gloucester, was attacked by a Me 109 which damaged the starboard inner engine but was then shot down by the Lancaster's gunners. The flight engineer, Sgt John Reginald Powell of Morecambe, feathered the damaged engine and a small portion of the bomb load had to be jettisoned in order that the Lancaster could maintain height. Crawford flew *J-Jig* on to Leipzig, dropped what remained of his bombs and turned for home, only for the aircraft to be hit again, this time by flak, and it took the combined strength of the pilot and flight engineer to handle the controls. Crawford ordered the other members of the crew to prepare to bale out as he and his flight engineer battled to try to maintain the precious height they would need if they were to get home. It was a long haul back but *J-Jig* made it, finally being guided in by searchlights to Manston in Kent.[3]

On 20/21 February, the Stuttgart operation got off to a bad start when two Lancasters swung out of control on take-off and crashed in flames. Over 590 bombers reached Stuttgart where the target was cloud covered and the bombing scattered. Night-fighter activity was reduced because the bomber stream flew a North Sea sweep to keep fighter attacks to a minimum, and a diversionary feint towards Munich successfully drew

A 4,000lb 'cookie' arrives at dispersal ready for loading.

the German fighters up two hours before the Main Force flew inland. Just nine bombers were brought down. Three Lancasters and a Halifax crashed in England on their return. Two further effective Tame Boar operations were directed against Bomber Command raids before February was out. The first was on the night of the 24th/25th, when 209 Tame Boar crews destroyed all except two of the 33 Lancasters and Halifaxes that were lost raiding a ball-bearing factory at Schweinfurt. For Eric Jones, who was commissioned in January and on 15 February had been awarded the DFC, the raid was his 29th operation and for four members of his crew, their 27th:

I had high hopes of at least Ron, Steve and Jock finishing their tour along with me but I wasn't too sure about Ken, Pat and Peto. They had missed out on quite a few trips. I suspect that we were all apprehensive about the approaching end of our tour and at even this late stage our luck could still run out. At briefing we were told 'Knock out the ball-bearing factories and the whole German war machinery will be out of action.' Everything ran on ball-bearings. The Americans were bombing it by day and we were to bomb it by night.

Lancaster crew on
460 Squadron RAAF in
front of *R-Robert*.

Anyway the wings didn't drop off and we were out of range of those guns and we eventually landed safely at Fiskerton.

Reporting to the Flight Office the next morning I was told that our tour was completed. I just couldn't believe our luck. I always understood that the commitment was 30 operations. For the four of us it was the end of nightly journeys over Germany. Later I was to learn that a pilot completing 200 hours of operational flying constituted a tour. Actually I had completed 190 hours 35 minutes including that 'boomerang' to Leipzig so, somewhere, someone, had got their sums wrong and I could have done the full 30 ops within the 200 hours. Perhaps they thought that with 12 trips to the 'Big City' included in the tour it was sufficient to call it a day. I was not asked to volunteer for a second tour which would have meant a further 20 ops, possibly with the Pathfinder Force. I was informed, as were Ron, Steve and Jock, that we would be 'screened' which meant being posted to training unit to instruct future bomber crews. Ken, Pat and Peto were to stay behind and complete their tours with other crews. I could have volunteered to continue with a second tour with all the crew sticking together but I was firmly convinced that luck had been the biggest factor in our survival. If we could have foreseen the fate of Peto and Pat perhaps other decisions would have been made. It was to be some weeks before I heard that Pat and Peto had gone missing and that Ken had completed his tour.[4]

On the night of 25/26 February, nearly 600 bombers carried out the first large raid on Augsburg, which was made in clear weather conditions. Dick Starkey on 106 Squadron recalled:

The main target was the Daimler Benz factory where engines were assembled for Messerschmitt aircraft. It was the first time an operation was planned with two waves of aircraft bombing two hours apart. We were in the first wave and the weather over Germany was clear with a starlit sky, no cloud and snow on the ground. It was our second visit to Bavaria, having bombed Munich on a previous operation, and as you gained experience it was possible to identify cities by the position of searchlights and anti-aircraft guns. Augsburg and Munich are both in Bavaria and as we flew east towards them both defences were in action, especially the searchlights. We identified one cluster of searchlights ahead of us as 'Augsburg' and confirmed with the navigator's H₂S that this was the target. However, as we approached the

I must say how much we all admired the Yanks in their Flying Fortresses. They flew in daylight and their losses were horrendous. [Two hundred and sixty-six B-17s had raided the factories the previous day.]

734 bombers attacked Schweinfurt, the attack being split into two raids with a two-hour gap between each raid; the first time a major attack had a two-hour gap between waves. The raid was considered a success. When a piece of spent shrapnel managed to find its way through the underside of our Lanc and into the cockpit between my legs it was a good indication the anti-aircraft guns had got our range. We had completed our bombing run and we were on our way home so with one thought in mind – 'get out of the guns' range' – I stuck the nose of *K-King* right down and sped post-haste towards the ground. The speed built up very quickly and we were well over 300mph when I started to pull out of the dive. The controls were as heavy as lead and only lightened as the speed fell away. The maximum permissible speed for the Lanc was around 350mph and I must have been pretty close to it.

two minutes before zero hour when markers were dropped by the Pathfinder Force nothing happened, which was strange; a point I conveyed to the crew. 'Jock' Jamieson, the mid-upper gunner, immediately came on the R/T and said the markers were being dropped on the beam of the port side. I looked over my left shoulder and there they were dropping on the other city which of course was Augsburg; we had mistakenly identified Munich for Augsburg. I turned towards the target which was several minutes' flying away and knew we should be over the city when the first wave was already on the return journey. I thought about all the defences concentrating on one aircraft – ours – at the same time we would be on our own on enemy radar. Should we drop the bombs straight away and skirt the city to get on the return track or risk it and bomb the fires in the city which were now well alight? I decided to carry out the operation and approached the target. Fortunately the flak was not very accurate, although some of it was too close for comfort; I managed to evade the searchlights and the bomb aimer dropped the bombs from our solitary aircraft twenty minutes after the Main Force had gone. It was an uncomfortable few minutes as we flew over the city and we were relieved to get to the other side.[5]

John Sanders, a pilot on 617 Squadron, recalling the raid on Augsburg, said:

We climbed up from the airfield like we always did and as we were heading for the coast of France the sky was still light behind us. Where the sun was setting, there was a golden tinge to the sky. Obviously a German fighter, coming up from the coast, must have seen us silhouetted against the sky. From down to my left, he fired a short burst of cannon fire. He must have been a pretty good pilot to be able to do that, coming from the opposite direction and hitting us with about six or seven cannon shells. One of them exploded right on the side of the cockpit at my eye level. There was a fairly thick bar across the windscreen on that side, but the shell blew a hole in the Perspex. So if I turned my head sideways, I could see the instrument panel. I managed to straighten up. The next thing I did was to feel my face. There was nothing wrong. I told the flight engineer to check everything in the aircraft. He found that various cannon shells had hit the aircraft and done some superficial damage. But there was no leak of petrol so we decided to press on. Just as we

were approaching the target and getting ready to line up, there was an almighty blast behind me. A fighter's cannon had hit us in the tail and flung the aircraft nose down. We went screaming into a dive and when I looked round all I could see was a wall of fire. The back of the aircraft was just yellow with flames. I managed to pull the aircraft out of the dive with great difficulty, feet braced against the rudder pedals and pulling like mad.

While Sanders wrestled with the bomber, some crew members tackled the fire, putting it out with their extinguishers. Another went to the aid of the rear gunner, who had collapsed in his turret because his oxygen supply had been severed. The navigator, a Glaswegian solicitor before the war, who had to plot by dead reckoning and the stars since his equipment had been hit in the attack, worked out the best course for home. The intercom had been destroyed when cannon shells ripped through the wiring so the navigator wrote out the course on a note which he passed to Sanders. Sanders continues:

I was finding the aircraft most difficult to handle. I could not keep it level. If I loosened the pressure on the wheel, it just slammed against the instrument panel. I had to hold it back but I could feel something grating all the time. I sensed that something was going to break.

By wrapping his arms around the wheel and bracing his feet against the rudder pedals he could just keep the nose up.

F/L Ernie Wharton (back row, third from right) on Brian Frow's crew on 7 Squadron PFF at Oakington. Ernie was a rear gunner who had completed his first tour on 97 Squadron and shot down two enemy fighters. After his second tour he flew further ops on SOE. The cocker spaniel did 22 ops with Frow's crew!

From June 1944 onwards, 1 Group experimented with the privately produced Rose-Rice turret developed by Rose Brothers of Gainsborough, which was much roomier than the FN types and was equipped with twin .50in guns, as a way of increasing the Lancaster's defensive firepower.

He now had a journey home of more than four hours in this position:

Every so often I would get the flight engineer to reach across my shoulder and take over the wheel while I had a rest because it was getting very tiring. All of a sudden I could hear somebody speak to me. So the navigator and I were able to talk. It was much simpler to fly and pass instructions after that.

The Lancaster crossed the English coast and Sanders reached Lincolnshire:

As the aircraft got lower, it got heavier and heavier on the wheel because the air was denser. It was really hard work.

There was a solid sheet of cloud at 2,000ft. The navigator put me in position, saying, 'Right, you're lined up to start the descent.' As I came out of the cloud, there was the runway in front of me, absolutely perfect. I told the wireless operator, who had the Very pistol, to keep loading reds and firing them to let them know that we were coming in without radio. There were other aircraft landing so control had to know that we were a plane without lights or any way of talking. I got the aircraft lined up and the runway was coming up nicely. Then I realised, to my absolute horror, that I was not going to make it. As hard as I pulled back on the stick, I could not get the nose up any further. By my own judgement, I could see that we were going to be short of the runway by about half a mile. I thought to myself, 'Good grief, after all this, it isn't going to work.' Then I suddenly remembered that in a Lancaster, the first ten degrees of flap increase the lift. So I tried it. I popped down ten degrees of flap and the nose came up just enough. I put the wheels on the end of the concrete, but once the tyres bit on the runway, there was no way I could get the stick back and bring the tail down. The poor old Lanc kicked up at an alarming angle, but eventually it crashed onto the tail wheel. And we were down safely. I taxied into dispersal and all the crew bundled out of the aeroplane. I got into the crew bus and it was the first and last time in my life that I was kissed by another man. The mid-upper gunner threw his arms around me and said 'You made it.'

Guided by accurate ground-marking by the Pathfinders, the Main Force aircraft were able to release more than 2,000 tons of high explosive on to the beautiful and historic centre of Augsburg, devastating the whole area, wiping out 3,000 dwellings and damaging a further 5,000 houses in a conflagration of fire and explosion. Ninety thousand people were bombed out and left the city in droves. The flak defences were weak but 165 twin-engined Tame Boars claimed 19 bombers. Altogether, 23 aircraft were lost and on the return two Lancasters crashed, killing 11 crew and injuring three others.

A measure of the determination of 1 Group at this time to make sure every possible bomber could fly was demonstrated at Kirmington early on the morning of 27 February. Heavy snow was falling and with ops likely that night the whole of the station staff, including the air crew, was turned out at 4.30 a.m. to begin snow-clearing operations. By midday the runways had been cleared, but then a planned raid on Munich was scrubbed.

The station records commented, 'With adequate equipment and manpower it is possible to keep a bomber station open in these conditions.' However, it was stressed that those involved in the operation should be fortified with a hot drink every hour.

Main Force raids followed to Stuttgart and to aircraft factory and railway targets in France and Belgium. On the night of 2/3 March, 15 Lancasters of 617 Squadron using the Stabilized Automatic Bombsight (SABS) set out to bomb the GSP machine-tools factory and the adjacent BMW aero-engine repair complex on the outskirts of Albert, 15 miles north-east of Amiens. All of the crews had been trained to group eight bombs within 50yd of the target on Wainfleet Sands, computed to a height of 15,000ft, and they were expected to continue practising until they could consistently group them closer to a maximum of 20yd. The target was covered with camouflage netting on which roads and buildings had been superimposed. The weather in the target area was unfavourable, with a great deal of cloud, severe icing and static electricity. Leonard Cheshire located the factory by identifying the surrounding landmarks in the light of the flares accurately dropped by the squadron's flare force. He carried out the usual 'shallow' dive approach but his stores hung up. S/L Les Munro then laid his incendiaries and spot fires nearly on the site and 12,000lb high-capacity bombs and 1,000-pounders obliterated both factories. Two nights later, the needle-bearing factory works at La Ricamarie at St-Étienne, near Lyons, was the target for 15 Lancasters of 617 Squadron. It was the smallest and most difficult target yet. It was in a narrow valley with 4,000ft hills on each side and the factory, in the middle of a built-up area, was only 40yd by 70yd. Met forecast good weather but unbroken cloud in the target area caused Cheshire to abort the operation.

Another attempt was made on 10/11 March with 16 Lancasters, four of them acting as markers. The cloud in the valley was broken and Cheshire found, after five dicey runs, that he could see the factory only at the last moment. On his sixth run, he judged the distance, dropped the nose of his Lancaster and the incendiaries landed on the roof of the main factory building but then bounced off and ignited about 100yd beyond. He summarised this result to the second marker, who used the burning incendiaries as a datum but released his load too early so that it undershot. The third marker's load repeated the behaviour of Cheshire's initial effort. The fourth marker came in on a much steeper dive and his markers effectively stayed in the middle of the factory buildings. Cheshire then instructed the bombing Lancasters to aim at this 'central

460 Squadron RAAF crews in front of AR-Q at Binbrook.

marker'. Photo reconnaissance revealed that only the wall round the factory remained; the rest had been completely destroyed and there was no damage outside.

Another 86 Lancasters of 5 Group carried out moonlight raids on three other factories in France. At Clermont-Ferrand, 80 miles west of Lyons, 33 aircraft bombed the Michelin rubber works once again. Another 53 aircraft attacked targets at Châteauroux and Ossun. Dick Starkey on 106 Squadron was one of those who flew on the Châteauroux operation:

The raid, on a factory making powered gliders for the Germans, was to be nothing like the high-altitude bombing of German cities; it was to be made by 22 aircraft bombing individually between 7,000 and approximately 11,000ft flying in full moonlight. The raid would be controlled by a Master of Ceremonies (W/C Baxter, CO of 106 Squadron) flying a Lancaster who would open the attack by placing a red spot flare on top of a large hangar and then call in the attacking force with the first wave carrying a 4,000lb bomb and nine 500lb bombs; the first and last bomb fitted with delayed-action fuses to explode several hours later. The second wave of bombers would carry incendiaries to set the factory alight. We were to fly from our base, another squadron from Coningsby would also be on the raid, and both squadrons had to rendezvous at a cross-road two miles north of the target, which would be marked by the Master of Ceremonies five minutes before zero hour. The first wave of attackers was called 'Apples' and the second wave 'Pears' with the Master of Ceremonies called 'Big Stiff'.

The Met forecast was for cloud cover at fairly low level en route over France, clearing well north of the target to give us visibility to identify landmarks such as rivers. However, the cloud cover extended a lot further south than Paris and we did not clear it until a few minutes before 'Big Stiff' was due to mark the cross-road. As ETA approached we waited for him to call up, which he did on time saying he had found the rendezvous and would be dropping the marker. We waited for him to do this because our ETA had not come up with a corresponding identification of the cross-road. To our relief when the marker dropped we were only three miles from it and by that time we were down to our bombing height – 7,200ft. I commenced circling the cross-road and could see the factory a couple of miles down the road. The planes continued to circle as the MC ordered and this gave the night shift at the factory sufficient time to evacuate before we commenced our attack. The first wave bombed on

time in perfect conditions without any opposition, except for one gun which was firing near Châteauroux a few miles south. I was surprised when the first aircraft dropped its 4,000lb bomb on the airfield adjacent to the factory with the other bombs also bursting on the field. We were the second aircraft to attack and as the target was only 250yd wide and half a mile long running by the side of the road, we came in on a diagonal run across it. It was perfect for Wally Paris, the bomb aimer, who could see where he had placed his bombs; the 4,000lb bomb exploded on a workshop and destroyed it as the photograph proved and the 500lb bombs straddled all the buildings; it was a perfect bomb run.

After bombing we turned 180 degrees and ran up the side of the road observing the result of the rest of the attack; after the first wave had done their job the target looked gutted with smoke and flames but the second wave went in to fire it with their incendiaries. One crew on the return journey talked continually with their R/T switched on and although they were ordered to be quiet they kept on for some time. One aircraft was lost on the raid as we wondered if it was the crew who 'broadcast'.[6]

It began snowing for days and runways had to be cleared and ice removed from the wings of aircraft. On the night of Wednesday 15 March the main operation was on Stuttgart by 863 aircraft – 617 of them Lancasters. In an attempt to avoid contact with night-fighters the Main Force split into two parts, flying a roundabout route over France nearly as far as the Swiss frontier before turning north-east to approach Stuttgart. Clayton Moore recalled:

The seven-hour flight proved uneventful. In our role as supporters, we were this time required to assist the 'backers-up' – the crews whose job it was to replenish the original marker flares before they had time to burn out. This meant that we were not in the forefront of the attack and we found the German defences somewhat overloaded with attacking aircraft by the time we arrived over the target. On this occasion, the weather was good, as was the marking, and all 83 Squadron aircraft returned to base safely.

The deception plan had worked until just before the force reached the target when 93 1 JD crews were fed into the bomber stream. Bomber Command lost 37 aircraft and six more were written off in crashes and collisions.

Lancaster B.I DV397 QR-W, which operated on 61 Squadron from 30 November 1943 to the night of 24/25 March 1944, when it was one of 44 Lancasters that failed to return from Berlin. P/O Denis Carbutt and five of his crew were killed.

On 16/17 March, 15 Lancasters on 617 Squadron and six H$_2$S-equipped Lancasters on 106, there to drop parachute flares, carried out a precision attack on the Michelin tyre factory at Clermont-Ferrand. The briefing orders were to destroy three of the four large factory buildings but to leave the workers' canteen intact. The aiming point was first accurately marked with red spot fires. These were then overlaid with green target indicators to emphasise the aiming point. Six of the 617 aircraft carried 12,000lb 'blockbusters' and each and every one was a direct hit on the workshops, which ceased production. All the aircraft from both raids returned safely. Daylight reconnaissance by a Mosquito revealed that the workers' canteen just beside the workshops was undamaged.

On the night of 18/19 March, 846 aircraft – 620 of them Lancasters – were dispatched to Frankfurt and Lancasters of 5 Group, including 13 on 617 Squadron, carried out an accurate raid on a former French state gunpowder factory at Bergerac, 50 miles east of Bordeaux. The target was marked from 5,000ft and the aircraft then bombed from 18,000ft. For 15 seconds it looked as though the sun was coming up underneath; the ground was just one great orange flash which lit up the sky for miles. No bombs fell outside the works and all aircraft returned without loss.[7] In Frankfurt part of the *Nachtjagd* force was sent north when the JLOs were deceived by the appearance on radar of 98 aircraft which were going to lay mines in the Heligoland area, but another force of night-fighters in Germany met the Main Force stream just before Frankfurt was reached. The Pathfinders marked the target accurately and this led to heavy bombing of eastern, central and western districts of Frankfurt. The target was identified by TI reds and red/yellow flares covering an area 2 miles square. The later phases of the bombing were scattered but extensive destruction was caused in the city.

F/L Bill Siddle's crew took *U-Uncle* for the five-hour trip. Clayton Moore recalls:

Again our role was that of supporters and we would be going in with the leaders once more, as we had done on the earlier Berlin trip. Everything went more or less according to plan. We had no difficulty in locating the target and the markers looked good as they went down. After we had disposed or our bombs and turned from the target area, all the indications were that it was going to be a well concentrated raid. The Main Force pounced at once and the pot was really boiling as we set course for the return flight. During the homeward part of our flight over enemy territory, it became increasingly evident that the German fighters were once more amongst us. A lot of aircraft could be seen on fire and both Gerry and I were kept busy reporting sightings. Off to port I saw a four-engined bomber which had strayed over a built-up area as it spun earthwards, completely enveloped in flames. As it neared the end of its fall, the fierceness of the flames that surrounded it served to illuminate the ground beneath it and I watched as it plunged into the saw-toothed roof of a large factory, where it started an equally large fire. As we forged on through the darkness, searching for a possible attacker, the quiet voice of Clem, the wireless operator, came over the intercom.

'I've picked up an aircraft on Fishpond,' he informed us. Fishpond was an extension of H$_2$S and gave a reasonably clear indication of what was happening between us and the ground. Such detail showed up on a small screen in the radio compartment and a good operator could usually differentiate between friend and foe by turning detective. Because of its superior speed, the blip produced by a fighter was easily distinguishable from that of a more ponderous bomber.

'I think it's a fighter. He came up fast behind us.'

'Where is he now?' I asked.

'Underneath us, but slightly to port and he's slowed to our speed.'

I stood up in my turret and peered out over the gun sight, but couldn't see him.

'Instructions, gunners', came from the cockpit.

'Corkscrew port, GO!' I ordered.

As we winged over into the dive, I again stood up in the turret in an attempt to catch a glimpse of the fighter, which I figured would be popping up into view somewhere to starboard.

'There he goes!' shouted Gerry. 'High on the starboard beam and he's trying to turn in on us. Up starboard, GO!'

Gerry had given the order for evasion, so I left him to get on with it, at the same time thinking that it might be a good idea if we were to show some fight this time. There was little doubt that the sight of a shower of tracer bullets coming in his direction would have a demoralising effect on the Jerry pilot who, after all, had spotted us and was out to get us. I recalled that more than a few gunners were of this belief and that some of them insisted on having more than the prescribed percentage of tracers in their belts for that reason. Having a go once it was obvious that the fighter had seen you made sense. I wrenched my turret around just in time to see the Me 110 go streaking past into the darkness. It was standing on one wing and had side-slipped into the night before 1 got time to open fire. Thankfully, we didn't see the Messerschmitt again and it was to be much later that I realised it must have been a *Schräge Musik* aircraft with upward-firing cannon and that it had probably been on the point of opening fire on us when we began our dive. It was only when I learned of the existence of such an aircraft that I came to appreciate the vigilance of Clem in having saved our bacon that night.

Twenty aircraft – 12 Halifaxes and eight Lancasters – were lost and two Lancasters crashed on return.

When, on 22/23 March, 816 bombers revisited Frankfurt the bombers took an indirect route to the city, crossing the Dutch coast north of the Zuider Zee and then flying almost due south to Frankfurt. This and a mine-laying diversion operation confused the enemy defences for a time but there was a heavy concentration of searchlights at Frankfurt and 25 Lancasters and seven Halifaxes were lost. A Lancaster which crashed in Suffolk on return resulted in the deaths of five of the crew. The marking and bombing were accurate and damage to Frankfurt was even more severe than a few nights previously. On 23/24 March, 20 of 5 Group, including 617 Squadron, dropped delayed-action bombs on the Sigma aero-engine factory at Vénissieux in the Rhône Valley, south of Lyons, and 70 miles south-west of Geneva without loss, but flare marking in hazy conditions was imperfect and the Dam Busters were ordered to return to the factory two nights later. Marking was much improved but the raid by 22 Lancasters was largely unsuccessful, with bombs being scattered. After an intensive practice bombing programme was carried out, 19 Lancasters – including 15 of 617 Squadron – returned to finish the job on 29/30 March. This time the bombing was concentrated around very accurate markers but a noticeable increase in defences was reported by the returning crews.[8]

'Bomber' Harris had realised that blind marking was unlikely to prove successful against German cities, principally Berlin, so he waited patiently for clear conditions over the 'Big City'. Finally, on Friday the 24th, 811 aircraft – 577 of them Lancasters – were detailed for the raid and diversionary and 'spoof' raids were laid on. During the afternoon ND625 was being bombed up at Kirmington when the full bomb load was jettisoned on the ground while checks were being carried out. The duty rigger, LAC Bishop, moved quickly to kick away a number of incendiaries which had ignited. His actions certainly saved the Lancaster and probably the four others on nearby dispersals.

Dick Starkey's crew on 106 Squadron had flown on eight operations to the 'Big City'; three of them out of four nights at the end of January when they were in the air for 24 hours out of 96. 'It was also a period of changing situations. You could be watching *Alexander's Ragtime Band* featuring Tyrone Power, Don Ameche and Alice Faye in a cinema in Lincoln at 8 o'clock one night and the next night at 8 o'clock be over Berlin.'

By now the crew had their own Lancaster – ND535 *Q-Queenie*, as Dick recalls:

Our ground crew at Metheringham who looked after *Q-Queenie* were of the best. They did their job magnificently and nothing was too much trouble. I remember asking them if they could find a leather cushion for the pilot's position and the next time I flew it was in place. I think that they took it from the flight commander's aircraft. They also named the aircraft *Queen of Sheba* and painted a picture of a nude lady just under the pilot's window. When we returned from an operation the crew shot a line to them saying that the flak over the target was so hot she came back tanned.[9]

Take-off at several 5 Group stations was delayed for two hours because of fog. A 20mph wind was expected but soon the 70-mile long bomber stream ran into very strong winds, which scattered the aircraft and made navigation tricky. Over the Baltic the jet stream was so strong that aircraft were registering ground speeds approaching 360mph. When crews made landfall on the Danish coast many navigators realised that they were much further south than they should have been, as 100-knot winds were experienced that night, instead of the anticipated 60-knot tailwind. The bombers on the diversionary raid west of Paris had already turned for home, and, when the Main Force crossed the German coast east of Rostock and bypassed Stettin, the JLO became convinced that the target was Berlin. The attack was very scattered at first and some of the TIs were seen burning about 10 miles south-west of the target. The raid developed into a 'terrific overshoot' because of the strong winds but, equally, many crews who overshot turned back to bomb. The bomber stream was spread out over 50 miles of sky and stretched back for another 150 miles. Single-engined Wild Boar and twin-engined

Tame Boar fighters took advantage and they and the flak claimed 44 Lancasters and 28 Halifaxes. F/L Tom Blackham on 50 Squadron was attacked over the target by a Ju 88. Bert Ridd, the Welsh mid-upper gunner, and Bill Dixon, the rear gunner, concentrated their fire on the fighter as it shot at them. Strikes were seen and the Ju 88 was claimed as 'damaged'.[10]

So far Vic Cole's crew on 106 Squadron had beaten the odds and their ops had been mostly without incident, but this operation was 'the scariest time', recalls 'Johnny' Johnson:

We were on our way back from Berlin when the bomb aimer said he could see a light in the sky and there was a Me 110 crossing in front of us. I went into the front turret and the mid-upper turned his turret round that way. Obviously the fighter saw us and soon got onto our tail. 'Parky' took over control of the aircraft – once the rear gunner is shooting at an aircraft he directs the pilot. All we heard from 'Parky' was, 'Keep her straight and level, skipper.'

The gunners wanted an even platform. It was all right for him, but just sitting there was scary. It was like waiting to be shot down. I thought, 'This is it!' On went my parachute and it seemed ages before I heard, 'Keep her like that.'

As soon as it was in range, there was the noise of six machine guns in unison followed by, 'It's on fire, dive starboard.'

When we levelled out, we saw the two bale out and the skipper waved to them as they were going down. Because there were no other witnesses we were only given a 'possible' and not 'aircraft destroyed'. More or less we attacked it, rather than it attacked us – very lucky. In all, we would go to Berlin twelve times, dodging the massive banks of searchlights, the AA shells and night-fighters, the latter paying more attention to us than was healthy – usually on the way home.

The Berlin trip on 24/25 March would be Dick Starkey's crew's ninth operation to the 'Big City':

The outward route was over the North Sea to Denmark then south-east over the Baltic Sea, crossing the German coast and continuing south-east before turning south through the target. The trip was to be one of the worst we encountered because of the strong winds. On the way out over the North Sea Colin Roberts, the navigator, who was from Sheffield, was finding winds with speeds far in excess of those in his

Armourers hoist a 4,000lb 'cookie' into the bomb bay of a Lancaster on 166 Squadron at Kirmington.

flight plan and coming from a more northerly direction than predicted at briefing. We were 'wind finders' that night and I remember the navigator advising me that the wind speed was unbelievable – approaching 100mph – and should he broadcast his findings back to Bomber Command. I said if he was satisfied with his calculations he must transmit them back to England. A number of aircraft were detailed as windfinders on every raid and when the navigators had calculated the actual wind speed and velocity they were transmitted back where an average wind speed was calculated from those sent back by aircraft and then relayed to the Bomber Force to use on their journey.

I ordered my navigator to work from his own calculations and ignore the wind speeds being sent back to us because they were far too low. By the time the Danish Coast was crossed we were many miles south of track as a result of the high wind speed from the north. At that time nobody had heard of the Jet Stream, but many years after the raid and on reflection, Bomber Command met this phenomenon on that night. The Force was scattered over a very wide front as we approached Berlin well before zero hour. Some captains ordered their navigators to work to the winds broadcast from England and found themselves hopelessly off track; others navigated on their own findings and were reaching points well in advance of ETA but they were not as far off course as the others.

We arrived over the target early and I decided to risk going round the city on the eastern side, by which time the PFF markers would be going down and we could start our bombing run. The activity in the sky over the city was awesome and frightening, as were all raids on Berlin. The sky was full of sparkling flashes as anti-aircraft shells from twelve hundred guns burst in a box barrage which was sent up every two minutes, containing the equivalent of an ammunition dump. I estimated that anyone getting through that would be very lucky indeed especially as the aircraft had to be flown straight and level with bomb doors open during the bombing run and take photographs after dropping the bombs. There were also hundreds of searchlights, making two cones over the city which the bombers had to try and evade.

The fighters no longer waited outside the perimeter of the target where they were in little danger from their own flak, because we were now severely damaging their cities. They flew amongst us in this area of death ignoring their own safety, meeting the anti-aircraft fire in order to get amongst us, and many a bomber was shot down when most vulnerable with bomb doors open. When we were on our bombing run with two other Lancasters, whose bomb aimers had chosen the same markers as my bomb aimer, a twin-engined fighter flew past our nose with cannon and machine guns firing at one of the Lancasters; there were tracers flying all over the sky as my gunners and the others in the third aircraft joined the targeted Lancaster to return

The flight engineer on *R-Roger* on 576 Squadron relaxing in the afternoon sunshine at Elsham Wolds in 1944. In the background is *Mike Squared,* then on the strength of 576 Squadron.

the fire. However, we lost another aircraft that night as the stricken Lancaster turned over on its back and went down in flames; we did not see anyone escaping because we were concentrating on the bombing run.

The Luftwaffe were now using single-engined fighters in the battle, generally over the target, and as I took a quick glance down at the fires I saw twelve of them circling up line astern towards the bombers whose bellies were red from the reflection of the flames below. The searchlight cones held two bombers like moths round a candle; the pilots were tossing their aircraft all over the sky but they were held like stage artists in a spotlight. The next move was from the fighters who came in and inflicted the *coup de grâce*, the bombers plunging down in flames before exploding and cascading in balls of fire to splash among the inferno below. A pilot had to take whatever action he could to get across the target area, and one practice was to fly near a coned aircraft and hope the action against it would help him get across. This wasn't always possible because, although the brightness was less intense, they could be seen. When a raid was at its peak with 800 aircraft bombing in a twenty-minute period, the illuminations had to be seen to be believed. The Target Indicators' red and green chandeliers 200ft in length cascaded down with a shimmering brightness; flak was bursting, filling every part of the sky with twinkling bursts and as you flew towards them there was no escape; you thought you would never get through it.

After bombing the target I gained height to 25,000ft and with relief at surviving the anti-aircraft, searchlights and night-fighter defences, but we had another fight on our hands before we reached England. The strong headwinds and nightfighters had not finished with us. It soon became apparent that our ground speed was very slow and we did not appear to be making much progress. As we crawled our way west to the next change of course which was to take us north-west between Hanover and Osnabrück, the navigator was continuously amending his air plot to try and keep us on course, but we were being blown south of our intended track. It soon became apparent that the conditions were getting worse and because of the effect of the wind on navigation found ourselves further west than the point where we should have turned north-west to fly between Hanover and Osnabrück. Instead we amended our course to fly between Osnabrück and the Ruhr, making sure we kept well clear of the latter area.

We had seen many aircraft shot down since we left Berlin, proof that the force was well scattered and aircraft were being picked off. As we looked over towards the Ruhr we saw many more, who had wandered over that area, shot down, so they had flown into the two heaviest defended areas in Germany, Berlin and Ruhr, in one night.

I was concentrating our efforts to get to the coast without further trouble when a radar-controlled searchlight was suddenly switched on just below the aircraft; these searchlights had a blue-white beam and more often than not hit the aircraft at the first attempt. The searchlight crew knew they were near us because the beam started creeping up in front of the aircraft. I put more power on and raised the nose to maintain our position above the beam, but it still continued creeping towards us. I was just on the point of putting the nose down and diving through it when it was switched off – talk about a dry mouth. If the searchlight had found us it would have been joined by others and as was the customary practice a night-fighter in the vicinity would have attacked us as we were caught in the beam.

Our last turning point was near the Dutch border and although our ground speed was very slow, the intensity of the defences had slackened off and for the first time in the raid, fighter activity had ceased. Maybe they had landed to refuel because we were approaching their airfields in Holland. We did not have any further trouble and eventually reached the North Sea coast where I pushed down the nose of the aircraft and did a very fast descent to 2,000ft, to the relief of the crew who were thankful to have the raid almost behind us.

As we flew towards the English coast, the wireless operator received a signal ordering us to divert to Wing, an OTU near Luton. It was a dark night and normally as you approached the coast you saw the odd searchlight but we did not see one light and I was surprised when the navigator told me, that according to his calculations, we had already crossed the coast, and gave me a course to Wing. We were by then well inland with navigation lights on, flying at 2,000ft, but could not see a thing. Suddenly a searchlight switched on to us followed by two more; they could not have been practising because they could see the lights of our aircraft. I cursed as they held us, thinking back to the hundreds we had evaded over Germany only to be caught in the beams of a searchlight battery in England. I was told afterwards that the lights were operated by a crew of ATS girls. We eventually landed at Wing, after a flight of seven and a half hours.[11]

Lancaster B.III ND356 *G-George* on 100 Squadron at Waltham piloted by Squadron Leader Hugh Grant-Dalton was attacked from below by a *Schräge Musik* fighter on the operation on Stuttgart on 15/16 March 1944.

Seventy-two missing bombers was the price paid by Bomber Command for the delivery of 2,493 tons of bombs on Berlin this night. Four of the missing Main Force Lancasters were on 115 Squadron at Witchford, which had dispatched 18 aircraft. It was the final raid of the 16 in which 166 Squadron participated and four aircraft were lost, bringing the squadron's losses for the raids to 17 of the 270 dispatched, a loss rate in excess of 6 per cent. It was the highest missing rate in 1 Group and cost the lives of 109 men, while another 24 became prisoners.

The raid was the last of Clayton Moore's six trips to Berlin:

Our mount for the night was *Q-Queenie* [now on display in the RAF Museum at Hendon]. A visit to the Big City was always a daunting prospect, but this was to be the worst. Our problems began on the outward leg, when we found that the information on wind speeds supplied by the Met Office was far from accurate. Dick Lodge, our navigator, was quick to realise that something was wrong and he directed us to carry out a series of time-wasting dog-legs in order to compensate for the strong winds that were pushing us on to the target ahead of time. Despite this, we still arrived too early, only to find the raid in full progress, since most of the other crews had not questioned the error.

It was after we had dropped our bombs and turned for home that the real trouble started. Now we were heading almost directly into a wind (later to be recognised as the Jet stream) which Dick estimated to be more than 100mph – almost twice the predicted speed. As a result, our rate of progress was almost halved and we were being blown off course by the hurricane which was attacking us from a north-westerly direction. Fortunately, the skies were reasonably clear above us and Dick was able to take a few star shots with the sextant, thus enabling him to plot our position. From this he was then able to get a fairly accurate estimate of the true wind speed. The clear skies would also have provided the defending fighters with an advantage, except for the fact that they too were faced with the same problems that we had. Not so the searchlight and flak batteries, however. For them, the conditions were ideal and the degree of co-operation between the two forces was to be admired. Because of the mix-up, many bombers were straying over heavily defended areas where they were being shot to pieces by the ground defences.

Dick was only one of the many navigators who had calculated the correct wind speed and the information had been radioed back to Group. But, because the information was so unprecedented, each of the numerous reports was considered exorbitant, so was ignored. Winds of such speeds had never before been encountered over Western Europe. As a consequence, the entire force was strung out all over central Germany. Those crews which, like us, had elected to ignore the broadcast winds and work to their own findings had a chance. The considerable remainder were in deep trouble. Lost, well behind schedule and with fuel stocks running dangerously low, they blundered on in bunches, straight into the waiting ground defences, where they were picked off one by one. When the tally was finally arrived at, it was learned that a total of 72 bombers [44 of them Lancasters] had been lost during the action. 83 Squadron had not contributed to this number however and we all looked upon this as a reflection of the undoubted efficiency of the crews (the navigators in particular) rather than luck.

The carnage we witnessed that night was uppermost in our minds when, two nights later, we were briefed to do a raid on Essen, right in the heart of the Valley itself. We took with us JB402 (OL-R). She had been on the squadron for four months and was to continue in service until failing to return from a trip to Mailly-le-Camp in May. This was to be our last trip in the role of supporters, it being the sixth such duty that the skipper had completed. The Happy Valley had a reputation for being inhospitable to the extreme and there was always an audible intake of breath throughout the briefing room when the curtain was drawn aside to reveal the tape running to and from one of the several industrial towns or cities in the Ruhr. Essen was looked upon as being a particularly nasty place since it contained within its boundaries the Krupp Works, a most important part of the German war effort and one which they fully intended to defend to their utmost.

Although short in duration, a visit to the industrial heart of Germany was guaranteed to provide a good deal of entertainment of the kind that most of us (myself included) found undesirable. It followed that I wasn't looking forward to the experience. Take-off was set for 21.20 hours and the flight was to take just four and a half hours. Despite my dread and diffidence, the trip turned out to be something of an anticlimax. We found the target defences to be no more impressive than others we had visited and we didn't see a single fighter during the whole of our flight. This was due largely to some clever flight planning by 'Butch' Harris and his boffins. We had gone in over Holland, thus giving the impression that we were aiming at a target in central Germany. The German controller had just managed to assemble his forces in the area of our supposed track when we turned south and headed for our true objective, thus throwing the airborne defences into a state of turmoil. Our attendance, plus another by the Americans earlier in the day, served to create a lot of problems for Herr Krupp who, it is alleged, suffered a stroke on inspecting the damage next day. 83 Squadron again survived unscathed from its ninth consecutive loss-free attack on Germany.

Over 700 aircraft, 476 of them Lancasters, took part in the raid on Essen and only six Lancasters and three Halifaxes failed to return. Clayton Moore thought that their luck would run out one night soon:

The equation was simple: $20 \times 5\% = 100\%$. The Essen trip had been my 22nd and I still had another eight to do to complete the tour. I was already into borrowed time. $30 \times 5\% = 150\%$. We had a first-class and efficient crew, but the odds against us were just too great.

This was never more so than on the night of Thursday 30 March, when crews were detailed for a 'maximum effort' on Nuremberg. F/L Siddle's crew were stood down, having completed four trips in the previous two weeks. The plan was to send the large force on what was virtually a straight line, which was ideally suited to Tame Boar interception. The AOC 5 Group, AVM Sir Ralph A. Cochrane, did not favour a dog-leg route because he believed it would only lengthen the flying time to the target and in turn greatly increase the risk of night-fighter interception. Of the original force of 782 heavy bombers that took off, 57 aborted with unserviceability problems. 'The forecast winds were not at all accurate and our navigator instructed the pilot to dog leg on at least two occasions,' recalls Sgt Len Whitehead, who flew a tour of operations between 13 December 1943 and April 1944, flying as mid-upper gunner on both Sgt Martin's and F/O Pitch's crews on 61 Squadron at Coningsby:

However, it was not as bad as it had been the week before on 24 March when 72 were lost on Berlin. When we had clear sky and moonlight we knew we were in for trouble. It was normal for the gunners to report to the navigator when we saw an aircraft go down and give such details as to whether we saw any baleout, if it was flak, or if a fighter etc. As regards the Nuremberg raid there were so many aircraft going down that the skipper told us not to report, just concentrate on looking out for and avoiding attacks. It was difficult not to look at those going down but we knew we had to keep a careful watch. We would sometimes see some tracer fire and then a small flame which would quickly grow until it lit up the whole aircraft and then frequently would finish with a terrific explosion. Sometimes we were rather puzzled because we did not see any tracer fire and no flak, just the aircraft catching fire. I did not learn the answer to this for many years. It was of course *Schräge Musik*. When using this no tracer was present so that it did not give away the form of attack. When an aircraft went down, or jettisoned the bombs, there would be a long line of incendiaries burning with a bright silver and further illuminating the aircraft above. There is little that I can say about the Nuremberg raid itself; we were just sitting ducks.

At North Killingholme 550 Squadron lost two Lancasters that night. The squadron had contributed 17 aircraft towards the raid and both the crews lost were experienced men.

F/Sgt Arthur Harrington Jeffries CGM and crew were on their 19th trip and four died, including the pilot, when they were shot down early in the battle by batteries near Liège. The three survivors were thrown out as the Lancaster blew up, Sgt S.A. Keirle sustaining very serious stomach, ribs and leg injuries. Jeffries is remembered by men who served with him as one of North Killingholme's 'character' pilots. He always seemed to be wearing a large brown Irvine jacket and 'couldn't give a damn about anything', according to Ted Stones, who was a member of his ground crew:

One cold winter day he and I were walking past the guardroom when the Station W/O 'Lavender' Yardley stepped out. He said that he thought a flight sergeant ought to be setting an example to the men under him. 'I am,' replied Jeffries. 'It's bloody cold and they ought to keep their hands in their pockets!'

The second 550 aircraft lost was flown by Sgt Charles Grierson Foster RNZAF, who died along with his entire crew when they were shot down by a night-fighter near Schweinfurt, the city bombed by mistake by many of the aircraft sent to Nuremberg that night.

At Ludford Magna 101 Squadron had six aircraft shot down and another lost in a crash in England. Its aircraft were strung out along the stream and 47 of the men on board died. The loss of seven aircraft out of 26 sent speaks for itself. John Wickman, a fitter-armourer at Ludford, remembers the following day on the airfield: 'There were a lot of empty dispersals. Almost a

Lancaster III LM418 on 619 Squadron, which was flown on the 30/31 March raid on Nuremberg by Sgt John Parker. He took off from Coningsby at 2213hrs and, on return, crash-landed at Woodbridge and the Lancaster was consumed by fire. Parker and his crew were killed on the operation to Kiel on 23/24 July 1944.

Z-Zombie on 408 'Goose'
Squadron RCAF
being refuelled at
Linton-on-Ouse on
24 April 1944.

before or since, has seen a greater rate of aerial destruction. By the time the Main Force reached the target area they had lost 79 aircraft, a figure exceeding the Leipzig total of six weeks earlier. Altogether, 512 aircraft bombed in the Nuremberg area; 48 crews bombed Schweinfurt in error and other bombers released their loads over German cities when they realised they were lost.

S/L Arthur William Doubleday DFC RAAF was first back at Waddington and Sir Ralph Cochrane called him up to the control tower. 'He asked, "How did it go?" I said, "Jerry got a century before lunch today." He didn't quite – he got 95.'

The Nuremberg raid was another disaster for 166 Squadron at Kirmington, with four of the 20 Lancasters sent failing to return. The first of the 166 Squadron Lancasters to go down that night was flown by P/O Walter Henry Burnett, who was killed, but three of his crew escaped before the Lancaster crashed near Bonn. The only one of the four pilots to survive was F/L F. Taylor, whose aircraft was shot down by a Ju 88 near Weidenhahn, two crew members being killed. The third to crash was the Lancaster flown by F/Sgt Roy Fennell. It came down on an airfield near Giessen and only one of the crew survived. The fourth 166 Squadron Lancaster to be lost was piloted by F/L Gordon Procter and was shot down by a night-fighter near Fulda; all eight men on board died.

At Coningsby 61 Squadron had dispatched 14 Lancasters. The following morning only ten crews reported bombing the target, three aircraft returned with damage and two failed to return. Air-crew casualties amounted to 16 men killed and four wounded in returning aircraft. Lancaster ME595 QR-R was flown by P/O Donald Paul and crew for this operation. As the aircraft flew past night-fighter beacon *Ida*, near Cologne, on the outward leg, it was attacked by a Ju 88 night-fighter. Over the next 15 minutes vigorous corkscrewing by the pilot and good work by the gunners managed to stave off two further attacks. However, during the defensive manoeuvres the aircraft had lost altitude and suffered severe damage to two engines which had to be shut down. Still losing height and down to 10,000ft, the skipper had no alternative but to get rid of his bomb load and turn for home. Later, after throwing everything moveable out of the aircraft in order to maintain height and flying speed over the North Sea, Paul managed to land safely at RAF Manston. Amongst the failed-to-return casualties was Lancaster DV311 QR-P flown by S/L 'Ted' Moss DFC. Nuremberg was the crew's 20th operation and they were regarded as the 'Gen Men' or 'Old Sweats' of the

third of the squadron had gone overnight and I can still picture all those weeping WAAFs around Ludford that morning. But new aircraft and replacement crews arrived and within a day or two it was as though nothing had happened.' Such was life and death on a bomber station. Five weeks later, 101 Squadron would suffer badly again when five aircraft were lost in the attack on Mailly-le-Camp.

Nuremberg, as far as the *Nachtjagd* was concerned, was the 'night of the big kill'. The 220 miles from Liège to the last turning point was littered by the blazing remains of 41 Lancasters and 18 Halifaxes. It is unlikely that a single hour, squadron. However, all the crew were killed when their aircraft was shot down by Hauptmann Fritz Rudusch flying a Me 110 night-fighter near Rimbach, north-west of Fulda. Another victim was veteran Lancaster R5734 QR-V, piloted by Australian P/O James Arthur Haste. *V-Victor* was detected by Major Rudolf Schoenert flying a Ju 88 night-fighter fitted with Naxos radar which could pick up H_2S radar signals, and the German night-fighter followed QR-V for over half an hour. His persistence finally paid off when he saw his quarry heading for home at 20,000ft. Unseen, Schoenert positioned his Ju 88 underneath the Lancaster and fired cannon shells from his *Schräge Musik* weapons into *Victor*'s right-wing tank. The Lancaster caught fire and crashed near Namur in Belgium. There were no survivors.

Mickey the Moocher, another 61 Squadron aircraft, ran into difficulties while flying over the North Sea towards the north Norfolk coast. It encountered stormy conditions and was struck by lightning on the front turret, causing Aussie skipper P/O J.A. Forrest to lose control and the aircraft plummeted towards the sea. In a blinded and shocked state the pilot ordered the crew to bale out while he tried to pull the aircraft out of the dive. Forrest managed to gain control again at an altitude of only 1,000ft and immediately countermanded his bale-out order. Unfortunately it was too late to save the lives of the WOp/AG and mid-upper gunner who had already parachuted into the sea.

Meanwhile, *Q-Queenie* had just flown past the *Otto* night-fighter radio beacon near Frankfurt, when it was raked by cannon fire from a Bf 110 and two Ju 88 night-fighters. Despite suffering severe damage to his aircraft and with four badly injured crew members aboard, P/O Denny Freeman managed to return to England and crash-land the aircraft at RAF Foulsham in Norfolk. Although this was only his third operation, Freeman displayed great courage after the night-fighter attack and was awarded an immediate DFC upon his return to Coningsby. Freeman's wireless operator, F/Sgt Leslie Chapman, was awarded the CGM for repairing his damaged radio and getting vital navigational radio fixes. Both men were to die later while flying ops with other crews.[12]

The Air Ministry 39-word communiqué broadcast by the BBC said, 'Last night aircraft of Bomber Command were over Germany in very great strength. The main objective was Nuremberg. Other aircraft attacked targets in western Germany and mines were laid in enemy waters. Ninety-six of our aircraft are missing.' In a subsequent communiqué the Air

Z-Zombie on 408 'Goose' Squadron RCAF at Linton-on-Ouse on 24 April 1944.

Ministry amended the number of missing aircraft to 94, but this took no account of the bombers that came down in the sea on the homeward flight or the 14 bombers that crashed or crash-landed in England. It was the worst Bomber Command loss of the war. More air crew had been killed in one night than in the whole of the Battle of Britain. Nuremberg was – as Winston Churchill recorded in his history of the war – 'proof of the power which the enemy's night-fighter force, strengthened by the best crews from other vital fronts, had developed under our relentless offensive'.

Nuremberg brought, for a brief period, the virtual cessation of heavy attacks because without 'radical and remedial action' the prospect of sustained and massive long-range operations deep into Germany was impossible. After Nuremberg the night-fighters were largely given a three-week respite before the next big raid, though Aachen was bombed on 11/12 April by over 340 Lancasters, and Cologne was hit by almost 380 aircraft on the 20th/21st.

Clayton Moore recalls:

April the seventh signalled a brief and much welcomed respite from the battle fatigue that was undoubtedly having such an adverse effect on me and the other members of the crew. At this time we were granted another week's leave and for me at least, it couldn't have come at a more convenient moment. Although it was seldom expressed, I felt certain that fear was lurking in the minds of us all and that rest was the only effective cure. Tempers had been on a short fuse of late and I had been experiencing concern over the effect that the strain of battle was having on the members of our crew.

Lancaster II on 408 Squadron at Linton-on-Ouse in May 1944.

It was an inescapable fact that thousands of bombs fell into mostly urban areas and there were many French casualties. Just one Lancaster failed to return from the Lille operation. Night-fighters were active over Denmark, and nine Lancaster mine-layers, including three on 460 Squadron RAAF at Binbrook, were shot down.

It was not just enemy action which killed the young men and destroyed Lancasters. At Kirmington on 10 April, as Lancasters were taking off for an attack on the railway yards at Aulnoye, four aircraft were already airborne when *F-Fox* came to grief. It seems a tyre may have burst, for the bomber, carrying 14 1,000lb bombs, swerved off the runway and within a minute or so caught fire. The crew scrambled clear before at least nine of the bombs on board exploded, completely destroying *F-Fox*, damaging four aircraft nearby and leaving a crater 50ft wide and 15ft deep which later needed 500 tons of rubble and soil to repair. Seventeen aircraft were unable to take off and those that did had to land away after the raid as Kirmington was officially out of action. That night Aulnoye was attacked by 132 Lancasters of 1 Group and 15 Pathfinder Mosquitoes, and 287 bombs hit the railway yards and the engine shed, putting 30 locomotives out of action, but 340 houses were destroyed or damaged and 14 civilians were killed. All seven Lancasters that were lost were shot down in 46 minutes by Hauptmann Helmut Bergmann, *Staffelkapitän*, 8./NJG4, operating from Juvincourt, (his 17th–23rd kills). Another 148 Lancasters and 15 Mosquitoes hit the corner of the railway yards at Laon for the loss of one Lancaster, and 180 Lancasters of 5 Group attacked the railway yards at Tours, also losing one Lancaster.

By six o'clock the next morning 100 men were at work repairing the crater at Kirmington and within four and a half hours the runway was reported to be usable, and the station was fully open again by eight o'clock that evening. The blast from the bombs had also blown in the doors of the nearest hangar and it was later found that 500 panes of glass had been broken in Kirmington village. Seventeen aircraft were able to operate that night when 352 heavies and Mosquitoes raided Aachen for the loss of nine aircraft, but there was to be a sequel for in mid-May it was found there was major subsidence at the point on the runway where the explosion had occurred and Kirmington was closed for flying until more substantial repairs could be carried out.

The next Main Force operations took place on the night of 18/19 April, when rail yards at Rouen, Juvisy, Noisy-le-Sec and Tergnier were bombed by four separate forces and 168 aircraft

On Easter Sunday, 9 April, scores of Lancaster, Halifax and Stirling crews carried out air tests in readiness for the night's Main Force raids on the goods station at Lille-Délivrance close to the border with Belgium and the rail yards at Villeneuve-St-Georges near Paris. La Délivrance was surrounded mainly by housing areas and commercial premises but it was felt that by using a combination of the new low-level marking by 11 Oboe-equipped Mosquitoes and a Master Bomber the bombing would be accurate enough to avoid unnecessary casualties. A further 103 Lancasters of 1 and 5 Groups were detailed to lay mines off Danzig, Gdynia and Pillau in the Baltic. In total, 697 sorties would be flown on the night of 9/10 April.

dropped mines off Swinemünde and in Kiel Bay and off the Danish coast. The Dam Busters provided 19 Lancasters and all four Mosquito aircraft of the Marker Force for the attack by 209 Lancasters of 5 Group on the Juvisy marshalling yards in the southern suburbs of Paris near Orly Airport. Three Mosquitoes of 8 Group carried out high-level 'Windowing'. Flares were dropped to the north of the target but illuminated the Arc de Triomphe. Cheshire dived to back up this initial marking but his stores hung up. However, his observation at low level made him decide that the markers were perfectly placed and he instructed the other two marker Mosquitoes to refrain from back up. The Lancaster force was given bombing clearance as it approached the target. The 617 Squadron Lancaster force spearheaded the bombing, each carrying a load of 14 1,000lb bombs or a mixed high-explosive and incendiary load. Their bombs burst laterally along the narrow bottleneck of the marshalling yard, successive loads supplementing and adding to damage already wreaked on the target. It appeared to the watching Mosquito crews that this initial bombing had torn the heart out of the target. The remainder of the bombing spread this wholesale damage to the rest of the area, but Cheshire reported 'two wild sticks' in his debriefing report. Sadly, these loads, a mix of instantaneous fused and delayed-action bombs, caused distressing casualties in the civilian areas in which they fell.

Some 273 Lancasters of 1, 3 and 8 Groups claimed a concentrated attack on the railway yards at Rouen with much destruction being caused. Mindful of the poor results of the 10/11 April raid, marking instructions for the return raid on Tergnier included the use of 24 Lancasters of 8 Group for visual marking, illuminating, support and back up. The Master Bomber, 23-year-old W/C John Fraser Barron DSO DFC DFM, a New Zealander, one of three visual markers on 7 Squadron, controlled the raid using VHF radio. The attack began as planned but again the Mosquitoes suffered poor serviceability of their Oboe equipment and of the eight detailed to attack, three only released their TIs, creating an MPI (mean point of impact) further to the west of those of the previous raid. Fifty railway lines were blocked but here too bombing fell on housing areas south-west of the railway yards and caused a number of French casualties. At Noisy-le-Sec the marshalling yards were so well hit that they were not completely repaired until six years after the end of the war, but 750 houses were destroyed and more than 2,000 damaged, and 464 French civilians were killed. Aircraft losses were relatively light. A Lancaster was lost on the Juvisy marshalling yards operation

and a Lancaster came down on the return. No aircraft were lost over Rouen but three Lancasters were shot down on their return over eastern England by Me 410A-1 intruder aircraft of KG51 *Edelweiss* at Soesterberg airfield in the Netherlands.

By mid-April, 83 Squadron had transferred from 8 Group to 5 Group and had joined 97 Squadron as the Group's Special Marker Force at Coningsby. Clayton Moore recalls:

It was widely rumoured that the move was something to do with the impending invasion of Western Europe. Whatever the reason, thousands of irate aircrew, ground staff and WAAFs were loaded into a large convoy of trucks and charabancs for the long journey north, while the pilots, flight engineers and station bikes flew first-class in the squadron's Lancasters. I and the other members of the crew were looking forward to a foray into the village in order to explore the boozing facilities, but the war intervened on the first available evening [on 20/21 April, when 247 Lancasters of 5 Group and 22 Mosquitoes of 5 and 8 Groups visited the marshalling yards at La Chapelle] for such pursuits. Instead, at about the time that the local barmaids would be draping the towels over the pump handles we were racing down the runway, bound for Paris. We were riding *Queenie* again and she was laden this time, not with markers, but with high

P/O John Dennis Carter and crew on Lancaster B.III JB405 *Hellzapoppin* on 12 Squadron at Binbrook in April 1944. Carter and six of the crew were killed and one taken prisoner on the operation on Mailly-le-Camp on 3/4 May.

Lancaster B.I
R5868 *S-Sugar* on
467 Squadron RAAF
taxiing out at an
American base.

Lancaster B.I
R5868 *S-Sugar* on
467 Squadron RAAF
taxiing out at an
American base.

explosive bombs and parachute flares. On this trip, the marking was to be done by Mosquitoes and the flares we were carrying were to be used to illuminate the marshalling yards. At the earlier briefing it had been forcefully pointed out that the dropping of bombs on any of the residences surrounding the yards would result in the most severe disciplinary action being taken against the crew or crews responsible. It followed that our brief called for exceptionally accurate bomb aiming from a height well below our usual altitude of about 20,000ft. The purpose of the many flares to be dropped was to ensure that everybody got a good view of the area, thereby avoiding serious damage to the surrounding dwellings – broken windows excepted, of course. Apart from the need for precision, weather conditions had to be good over the target, since cloud cover would have caused a late cancellation. But the Met men had got it right and we found Paris almost totally clear of cloud when we arrived just after midnight. Enemy opposition was moderate, although there was a lot of light flak coming at us as we swept in low over the brightly lit city on our bombing run. The flares had turned the night into day, making ground features plainly visible. As a result, such landmarks as the Eiffel tower, the Arc de Triomphe and other tourist attractions were easily recognised. We got it right the first time across the city and I watched our bombs as they blasted a path of destruction throughout the length of the rail yards.

Three other rail targets in France and Belgium were bombed that same night, and 357 Lancasters and 22 Mosquitoes raided Cologne. All told, 1,555 sorties were flown – a new record. Four bombers failed to return from the raid on Cologne and eight bombers were lost in the attacks on the rail targets. On six nights

– 21–28 April – 1,407 night-fighter sorties were dispatched, which resulted in claims for 135 bombers destroyed.

'Next day,' continues Clayton Moore:

we got the news that all the squadron aircraft had made it back and that we had won yet another aiming point. There was some bad news, too. Somebody 'up top' had decided that the Paris trip, together with all future 'softening up' targets would only count as one-third of an op. There was even talk of basing a tour on the number of hours flown on operations, instead of the number of trips. The news raised a lot of eyebrows – and tempers. As one plaintiff put it, attacking a target was like going over the top with bayonets fixed and few swaddies were ever expected to do that too often. 'They're getting blood,' he stormed. 'What more do they want?'

Sgt George Woodhead, the mid-upper gunner on W/O A.J. Higgs' crew on 156 Squadron at Upwood, would agree:

It had been a very bad winter for Bomber Command. There were some very tough raids. We did nine on Berlin alone. One of the worst was the night we were hit by bad winds and were blown right over the Ruhr defences. And on another our aircraft, *V-Victor*, wouldn't climb above 17,000ft and we arrived half an hour after everyone else and came back across Denmark in daylight, but we got away with it. That aircraft had done 80 raids when we got it but I believe it later went to a training unit and survived the war.

On the night of Saturday 22nd April when 596 aircraft – 323 of them Lancasters – of all groups except 5 Group were dispatched to Düsseldorf our Lancaster, *C-Charlie*, was hit by flak while flying at 19,000ft over the target. Sgt William A. Webb, the rear gunner, was killed instantly but the other six all escaped by parachute. I was the last out of the plane. I landed ten miles from the target, the wireless operator and skipper six miles and eight miles away. I believe the bomb aimer, navigator and flight engineer who landed in or near Düsseldorf were lynched. I know I nearly was myself. After I was captured I was taken into Düsseldorf and we stopped at a crossroads and they were bringing dead and injured out of a building. When we reached a police control centre one of the wardens saw me and went berserk. He grabbed me by the throat and tried to throttle me but was dragged away by the chief warden. I spent four days at a nearby airfield and then with five other downed RAF men was taken to a PoW centre. A dozen guards were needed to ensure our safe passage, yet there were still some ugly moments. With hindsight, it is easy to see why this sort of thing happened.

At Düsseldorf 2,150 tons of bombs were dropped, mostly in the northern districts of the city, which caused widespread damage. This 'old-style' heavy attack allowed the *Nachtjagd* to penetrate the bomber stream, and 29 aircraft – 16 Halifaxes and 13 Lancasters – were lost. That same night, 238 Lancasters of 5 Group and 17 Mosquitoes and ten Lancaster 'active observers' of 1 Group attacked Brunswick, and another 181 aircraft bombed the locomotive sheds and marshalling yards at Laon in France. The Brunswick force was largely ignored by the night-fighters. Only four Lancasters failed to return. Nine bombers – four Lancasters, two Halifaxes and three Stirlings – were lost on Laon.

On 24/25 April, Karlsruhe was the target for 637 aircraft of all Groups, except 5 Group which followed up the raid on Brunswick with an experimental raid on Munich with a force of 234 Lancasters and 16 Mosquitoes and 617 Squadron using low-level marking. The raid cost nine Lancasters but the marking and controlling plan worked well and accurate bombing fell in the centre of the city. (While no award of the Victoria Cross was ever made for a Mosquito sortie, W/C Leonard Cheshire's contribution to the success of the Munich operation, in which he led four Mosquitoes of the Marking Force in 5 Group, was mentioned in his VC citation on 8 September.) Only the northern part of Karlsruhe was seriously damaged, most of the bombs dropped by the Main Force aircraft falling outside the target area, many in open countryside. Eleven Lancasters and eight Halifaxes were lost. The mainly 5 Group raid on Schweinfurt on Wednesday 26/27 April was inaccurate and unexpectedly strong headwinds delayed the Lancaster marker aircraft and the Main Force. German night-fighters carried out fierce attacks throughout the period of the raid, which resulted in the loss of 21 Lancasters. Another 493 Lancasters Halifaxes and Mosquitoes performed an accurate raid on Essen. It was an accurate attack thanks to good ground-marking by the Pathfinder Force. Six Lancasters and a Halifax failed to return. The Schweinfurt raid marked the last of the attacks on the Reich for some time and the start of the softening-up raids in readiness for the invasion.

On the moonlit night of 27/28 April, 322 Lancaster crews bombed Friedrichshafen and two other forces attacked

Photo reconnaissance picture of Mailly-le-Camp after the raid by 346 Lancasters and 14 Mosquitoes of 1 and 5 Groups on the night of 3/4 May 1944. (*Via 'Pat' Patfield*)

the railway yards at Montzen and Aulnoye again. Various diversions and other factors confused the German controllers, the Friedrichshafen force reaching the target largely without being intercepted and 1,234 tons of bombs being dropped in an outstandingly successful attack. While the raid was in progress 31 Bf 110s and three *Luftbeobachter* (air-situation observer) Ju 88s were successfully guided into the bomber stream and destroyed 18 Lancasters; one more Lancaster crashed on its return to England. At Montzen 14 Halifaxes and a Lancaster flown by the Canadian Deputy Master Bomber were shot down by German fighters.

On 1/2 May, Bomber Command sent six separate forces to bomb targets in France. At Chambly, near Paris, the main railway stores and repair depot for the northern French system was the target for 82 Lancasters and 16 'G-H'-equipped Stirlings. Eight Oboe Mosquitoes and 14 Lancasters of 8 Group marked the target, which was attacked in two waves. About 500 HE bombs fell inside the railway depot area causing serious damage to all departments, and the depot was completely out of action for ten days. Three Lancasters and two Stirlings

P/O (later S/L) T.N. Scholefield RAAF and his crew get kitted up in front of Lancaster B.I R5868 *S-Sugar* on 467 Squadron RAAF, which has 99 ops recorded on the nose. The Goering quotation, 'NO ENEMY PLANE WILL FLY OVER THE REICH TERRITORY', was added by LAC Willoughby, one of the engine fitters around the time that *S-Sugar* had completed 88 ops. Scholefield, who was from Cryon, New South Wales, and his crew, flew *S-Sugar* on four ops, including the 100th on 11/12 May 1944 when the target was Bourg-Leopold in Belgium.

south-east end of the barracks area, followed by another 173 Lancasters of 1 Group, most attacking the north-west end of the barracks while 30 aircraft were to concentrate on a special point near the workshops. There was nearly a full moon and the target was accurately marked, but an order on VHF by the 'Main Force Controller' to the Main Force to come in and bomb was drowned out by an American forces broadcast and the Lancasters were left circling the target for up to 15 minutes. When the order to bomb was finally given, the rush was like the starting gate at the Derby and approximately 1,500 tons of bombs were dropped on the target.

One of the Lancasters on the raid was *M-Midge* at Kelstern, named after the wife of the flight engineer, Percy Miller, who recalled:

> Mailly was supposed to be an easy raid. It was a French target and at the briefing we were told opposition would be light. It was being controlled by a master bomber and there was some kind of mix-up over the target and we were ordered to orbit on a certain course. We quickly realised the Germans were soon going to cotton on to this and we decided to shear off to one side and we watched as the fighters got in among the bombers and shot a great many down.

were lost. The Société Berliet motor vehicle works at Lyons, which was attacked by 75 Lancasters of 1 Group, was badly damaged, and railways and factories nearby were also hit. Forty-six Lancasters and four Mosquitoes of 5 Group flattened the Usine Lictard engineering works outside Tours, which was being used as aircraft repair workshops. At St-Ghislain 137 aircraft, including 40 Lancasters, attacked the railway yards with great accuracy. One Lancaster and a Halifax were shot down. Lancasters of 5 Group attacked the aircraft assembly factory at Blagnac airfield at Toulouse where the marking was successful and the factory was severely damaged.

On the night of Wednesday 3 May, Bomber Command attacked a Panzer depot and training centre at Mailly-le-Camp, about 50 miles south of Rheims, which was reported to house up to 10,000 Wehrmacht troops. Nineteen Airborne Cigar Lancasters on 101 Squadron at Ludford Magna jammed night-fighter communications, and a number of diversion operations were designed to disperse the night-fighter force in the area. The Main Force, provided by 1 and 5 Groups, was detailed to attack Mailly in two phases, 173 aircraft of 5 Group going in first and aiming their bombs at a point near the

In all, 42 Lancasters – 14 Lancasters of 5 Group and 28 of 1 Group, which was subjected to the greatest delay at Mailly – were lost.

After Mailly large numbers of aircraft no longer attacked a single target in a concentrated stream. Due to the risk to civilians, targets in France and Belgium now required greater accuracy from smaller formations of heavy bombers, but when on the night of 6/7 May, 149 aircraft attacked railway installations at Mantes-la-Jolie for the loss of two Lancasters and a Halifax, over 860 houses were destroyed and 54 civilians were killed. Lancasters and Mosquitoes of 5 Group raided an ammunition dump at Sablé-sur-Sarthe without loss, and another 52 Lancasters of 1 Group destroyed an ammunition dump at Aubigné about 120 miles inland of the Normandy coast. One Lancaster was lost. On 7/8 May, 471 sorties were flown to France to bomb five airfields and ammunition dumps and a coastal gun position. Twelve aircraft, including seven on the raid on an ammunition dump at Salbris by over 60 Lancasters and Mosquitoes of 5 Group, failed to return.

'The foray into France' on the night of 8/9 May, when raids were carried out on Haine-St-Pierre, Brest, Morsalines,

Berneval and Cap Griz Nez, recalled W/O1 Clayton Moore on 83 Squadron:

provided F/L Siddle's crew with some interesting entertainment.

We took off at 21.15 hours to attack the airfield at Lanvéoc-Poulmic just inside the coast at Brest, flying in *V-Victor.* On this occasion we were briefed to mess up the runways. We had no difficulty in finding the target and went straight in through moderate flak and a few searchlights to drop our load. The sky in the target area had a scattering of clouds in it, but we got a good view of the field, which was well lit up with flares, and I got a brief glimpse of the destruction wrought by our load as it splattered across the runways. Having got the photograph, we banked and went into a shallow dive to port on a course that would take us back out over the sea. The manoeuvre couldn't have been better timed if I had ordered it myself, because it was only then that I came to realise that we were under attack. As we turned I was presented with the full underbelly plan view of what was unmistakably a Focke-Wulf 190 as it soared skywards less than a hundred yards from my turret. Unbeknown to me, the fighter had been in the act of launching a deadly attack from beneath us as we had turned, thus robbing him of what would have been an almost certain victory.

'Dive port, GO!' I yelled into my microphone, watching the fighter which was now silhouetted against some searchlights. It was already banking to make a second attack. 'One-ninety, port beam, high. Stand by for attack,' I informed Gerry. 'I see him,' came from the mid-upper turret. I could feel my jaws droop under the pull of gravity as Bill tightened the turn and my arms felt like lead as I battled with the controls in an effort to bring my guns to bear. So far we were winning. The F/W was still unable to get a bead on us, despite its increased rate of turn, which had the effect of placing it on its back. But the fellow at the controls obviously knew a thing or two about flying an aeroplane, because he was gradually getting the upper hand and I could see that he would soon have us in his sights again. Although he was still beyond the range of our Brownings, I knew that his cannons could reach us even now, so decided that it was time for corrective measures. 'Climb port, GO!' I ordered, then realised that this wasn't such a good idea after all, because it was the next move in the standard corkscrew manoeuvre. The Jerry pilot had been expecting us to do this

S-Sugar with crew.

and soon recovered, lining us up in his sights. I remember thinking to myself, 'if he opens up now, he just can't miss!' The thought caused me some concern, so I decided it was time for drastic action.

'Flaps, GO!' I barked. It was my hope that this would cause the attacker to overshoot. At least, it would serve to shorten the duration of the attack. He had his aircraft righted and was probably at full throttle in an attempt to close on us once again. In the split second that it took Siddle to respond I sat almost transfixed by fear, waiting for the deadly volley of cannon fire that I knew must surely come. The fighter was bearing down on us and I had him firmly framed in my sights. I was just on the point of opening fire when things started to happen fast. Bill applied what must have been full flaps causing the Lanc to lift vertically and to slow dangerously near to stalling speed. The F/W grew large in front of me and then screamed above my turret, its pilot trying desperately to avoid a collision. In what couldn't have been much more than a millisecond, he was beyond the upward reach of my guns, which were still silent, but I heard the distinctive rattle of Gerry's twin Brownings as the fighter roared past, a few feet above us. OL-V was beginning to vibrate, indicating that she was nearing a stall so I gave the order for flaps to be raised and then enquired: 'Did you hit him, mid-upper?' 'Don't know for sure.'

'Where is he now?'

'Can't see him. He'll be up ahead somewhere.'

'Good show, gunners [this from Siddle]. Keep your eyes peeled, everybody.' Fortunately we didn't see him again.

We qualified for another aiming point certificate as a result of our night's work and Bill recommended Gerry and I for a DFM and DFC respectively. Although I considered such an award to be deserved in Gerry's case, I had serious doubts concerning my own entitlement. After all, I had so far completed 27 sorties without having fired a single shot in anger. In all, 452 sorties were flown on the night of 8/9 May and twelve aircraft including OL-T and F/O A.P. Whitford DFC RAAF and his highly experienced crew, failed to return.[13] I couldn't avoid wondering if they had been jumped by the same F/W that had given us such a rough time. Its pilot was nobody's fool and had come within an ace of downing us. The question as to why we hadn't been attacked again by the fighter remained open to speculation. Perhaps the pilot had lost us after the overshoot, or maybe he had decided to leave us alone and go in search of easier game. Or it could be that Gerry had damaged the aircraft. We would never know.

Attacks on coastal batteries continued on the night of the 9th/10th and heavy bomber raids were made by 5 Group Lancasters and Mosquitoes on the Gnome et Rhône Ateliers Industriels de l'Air factory and another factory nearby, at Gennevilliers, and a small ball-bearing factory at Annecy. Bomber Command claimed to have hit the Gnome et Rhône factories but 24 French people were killed and 107 were injured. Five Lancasters failed to return. The weather en route to Annecy was very bad but the target was bombed and all aircraft returned safely.

On the night of 10/11 May, just over 500 aircraft bombed rail targets at Courtrai, Dieppe, Ghent, Lens and Lille. At Ghent, 48 Belgian civilians were killed and 58 were injured. One Lancaster was lost on the raid on Dieppe. Twelve Lancasters were lost on the 5 Group attack by 89 Lancasters on Lille. These losses ended all talk of sorties to French targets counting as only one-third of an 'op' towards a tour. The crew of Lancaster ND898 Q-*Queenie* were part of the force attacking the marshalling yards at Lille. After releasing their bombs, skipper Reg Dear held *Queenie* straight and level for the aiming-point photograph. Suddenly, while rotating

his rear turret, Jim Johnson saw the black shape of another Lancaster loom out of the darkness on a collision course. Jim froze in horror expecting the worst as the whirling props of the other aircraft came nearer, but at the very last moment the other pilot took avoiding action by lifting his port wing and moving across to starboard. Unfortunately his efforts were too late to prevent his aircraft's port outer propeller slicing two feet off *Queenie*'s port tail fin and rudder. The rear turret gun barrels were also damaged. The impact made Reg Dear's rudder pedals oscillate wildly and this was shortly followed by a phlegmatic report over the aircraft's intercom from his shocked rear gunner: 'I think we have been hit by another Lanc, skipper.' Flight engineer Fred Charlton was then sent to the rear of the aircraft to make sure the rear gunner was unhurt and investigate the full extent of the damage to the tail fins. Meanwhile, Reg cautiously made a gentle turn away from the crowded target area and flew the less responsive *Queenie* back to Skellingthorpe.[14]

There was another lucky escape for a 550 Lancaster in May when the bomber was bounced twice by fighters on a return flight from a target in northern France. The pilot, Sgt T.A. Lloyd from Tunbridge Wells, put the aircraft into a steep dive to shake off the attackers and the Lancaster began to fill with flames and smoke. Lloyd ordered his crew to jump; four did before the pilot managed to regain control. The flight engineer and navigator had remained and set about tackling the fire in the aircraft, using the extinguishers and the contents of the crew's coffee flasks. Remarkably they succeeded and Lloyd was able to land at a fighter station on the south coast.

On 11/12 May, just over 420 bombers and 22 Mosquitoes of the Main Force carried out attacks on Bourg-Leopold in north-east Belgium, the railway yards at Hasselt, and at Louvain in Belgium, and just over 230 heavies and 20 Mosquitoes in three forces attacked the railway yards at Boulogne and Trouville and a gun position at Colline Beaumont. At Hasselt the target was marked and 39 aircraft bombed but all missed the railway yards because of thick haze, the Master Bomber ordering the bombing to stop. Five Lancasters were lost. Four others failed to return from the raid on Louvain where the railway workshops and nearby storage buildings were hit. The target for the 190 Lancasters and eight Mosquitoes of 5 Group with three Mosquitoes of 8 Group was the former Belgian *Gendarmerie* barracks at Leopoldsburg (Flemish)/Bourg-Leopold (French), which was being used to accommodate 10,000 SS Panzer troops who awaited the Allied invasion forces. The weather was bad with low cloud and poor visibility, and a serious error was made with the broadcast winds. Flare dropping was scattered and did not provide adequate illumination of the target. Haze and up to 3/10ths cloud conditions hampered the marking of the target and half the Main Force bombed before the 'Cease Bombing' instruction was announced by the Master Bomber. Five Lancasters were lost, taking the night's losses to 16 heavies. *S-Sugar* on 467 Squadron RAAF, flown by P/O T.N. Scholefield, who was from Cryon, New South Wales, returned to Waddington on this famous Lancaster's 100th completed operation.[15]

On the night of 19/20 May, operations resumed with raids by 900 aircraft on five separate rail targets, all of which lay within a radius of 150 miles of Caen, coastal gun positions at Le Clipon and Merville, and a radar station at Mont Couple, on the Picardy and Normandy coasts. At Longueau, south-east of Amiens, the Master Bomber cancelled the attack on the railway yards by over 120 Lancasters and Mosquitoes of 5 and 8 Groups because they were cloud covered. One Lancaster which failed to return is believed to have been shot down by a night-fighter north of Poix. A second aircraft crashed on returning to England. Of the other four rail targets that were attacked, three (Le Mans, Orléans-Les-Aubrais and Tours-St-Pierre-de-Corps) were severely damaged. Some 113 Lancasters and nine Mosquitoes of 5 and 8 Groups attacked Tours. The initial marking by 627 Squadron's Mosquitoes was accurate, and from start to finish the attack lasted 26 minutes, during which time 107 aircraft had dropped 477 tons of HE from an average height of 8,000ft. Boulogne's yards were damaged by 143 aircraft of 4 Group. At Orléans-Les-Aubrais 118 Lancasters of 1 Group dropped almost 1,500 bombs or 615 tons of HE, which left 200 craters in the marshalling sidings and destroyed a large number of goods wagons. One Lancaster failed to return. The raid on the marshalling yards and repair shops at Le Mans by 112 Lancasters and four Mosquitoes of 3 and 8 Groups claimed three Lancasters and included the 7 Squadron aircraft with W/C John Barron DSO DFC DFM, the Master Bomber, and that of his deputy, S/L John Mervyn Dennis DSO DFC, which collided over the target. 'Sitting in the flight-engineer's seat beside me,' recalls F/L Alex Campbell RCAF, a new pilot on 514 Squadron at Waterbeach:

was an experienced bomber pilot from another crew, F/Sgt Topham, who we called 'Toppie'. This was my first op and we were about to hit Le Mans. First off, I had joined

Lancaster B.I R5868 *S-Sugar* on 467 Squadron RAAF, 100 not out.

the Air Force to learn to fly aircraft. That was my desire and dream, as it was for many other young men at that time. I had no intention of being trapped underneath the sea in a submarine nor did I want to be inside a tank in the Army. I wanted to fly in the Air Force. Before I began operations I was really impressed by the might and power of Bomber Command as a whole. Also, I was very aware of the shaking and trembling of the earth as the bombers took off and assembled overhead to disappear into the cloud or darkened sky above. You could hear hundreds of engines increasing in sound as they reached altitude to set course, then gradually diminishing until they were no longer heard, as they flew eastward into enemy territory.

But there I was that night, around 17,000ft in a Lancaster Mark II and over enemy territory. I remember we were first shot at from heavy ground defences, then a short time later three enemy fighter attacks. The first two were Me 109s and the last one was a very sharp attack from a Focke-Wulf 190. Our many hours of practice in air-to-air firing and fighter affiliation paid off, however, as my gunners held their own on that first night. On our return 'Toppie' approached me and asked if I would consider trading tail-gunners. I couldn't believe it. I had no intention of doing so and break up our crew. A crew was considered something very sacred.[16]

S-Sugar at an 8th Air Force base with a B-17 and a Hurricane.

Lancasters and 16 Mosquitoes, and another, by 44 Lancasters and seven Mosquitoes on the Ford motor factory at Antwerp, were all without loss. Seven Lancasters and 18 Halifaxes failed to return from the Aachen operation.

On 26/27 May, Main Force crews of Bomber Command raided Aachen, Nantes, Rennes and coastal batteries on the French coast and Bourg-Leopold again. This time the force consisted of 150 Halifaxes of 4 Group and 32 Lancasters and 117 Halifaxes of 6 Group; 1 Group provided ten ABC-equipped Lancasters of 101 Squadron, and 8 Group, 14 Lancasters and eight Mosquitoes. A photo reconnaissance revealed that 150 personnel huts and all the main barrack blocks were destroyed or badly damaged. Shelter trenches had been partly obliterated by bomb craters. Considerable damage was also caused to the town of Bourg-Leopold, including a school and 20 houses. Civilian casualties were listed as 22 killed and a few wounded. Some 7,000 German soldiers were killed or missing and 218 German women were killed also. All the buildings 'including the big messes' had been destroyed. The 'Cavalry Camp', which held many Belgian political prisoners, survived the attack. The guards had closed the doors and run away, leaving the prisoners locked up unattended.

Nine Halifaxes and one Lancaster were lost over Bourg-Leopold. In all, 1,111 sorties in 17 separate operations were flown for the loss of 28 aircraft. Twelve Lancasters were lost on the raid on the Rothe Erde railway yards at Aachen, 45 miles south-east of Bourg-Leopold; a Lancaster was lost on the attack on a railway junction at Nantes; and one Lancaster and a Mosquito were lost on the raids on coastal batteries on the French coast.

Operations in support of the forthcoming invasion involved bombing batteries on the Channel coast of France and rail targets inland. On 31 May/1 June, 219 aircraft – including 125 Lancasters – bombed the railway marshalling yards at Trappes in two waves for the loss of four Lancasters. Another force attacked a coastal wireless transmitting station at Au Fèvre and a radio-jamming station at Mont Couple, and Lancasters attacked the railway junction at Saumur and a coastal gun battery at Maisy, where the Lancaster crews found their target covered by cloud and only six of the 68 aircraft bombed. Another 111 Lancasters and four Mosquitoes of 1 and 8 Groups were given the railway yards at Tergnier. F/O H.A. Vernon – 'an American from Maine or thereabouts' – and his crew, who were flying their first operation on 12 Squadron at Wickenby, went, as Campbell Muirhead, the bomb aimer, recalls:

On returning, 1 Group's squadrons found that mist was forming on the Lincolnshire Wolds and a number of Lancasters were diverted to airfields further to the south. One aircraft was so badly damaged by flak that it had to be written off. A further six aircraft received varying degrees of damage as a result of enemy action and five were damaged in minor incidents, including a collision over the target at Boulogne.

There were no Main Force operations on the night of 20/21 May, but the following night 532 aircraft attacked Duisburg. The target was covered by cloud but the Oboe sky-marking was accurate and much damage was caused in the southern areas of the city. Twenty-nine Lancasters were lost and three more went missing on mine-laying operations off enemy coasts. Tame Boar crews claimed 26 bombers shot down. Dortmund was attacked by over 370 Lancasters and Mosquitoes the next night. A second stream of 225 Lancasters and ten Mosquitoes meanwhile headed for Brunswick. Most of the bombs that were dropped on Dortmund fell in residential areas mainly in the south-eastern districts of the city. Eighteen Lancasters were lost. On the night of 24/25 May, four operations were flown by four different formations: Aachen, Eindhoven, coastal batteries and Antwerp. Antwerp was attacked by 44 Lancasters and seven Mosquitoes of 5 and 8 Groups; Aachen by 442 aircraft of all Groups. At Eindhoven bad weather prevented bombing of the Philips factory by a force of almost 60 Lancasters, who returned without loss. Raids on French coastal batteries by over 200 Halifaxes and

When we learned that *G-for-George* was to be our Lanc for our first op we all went out to her bay to go over her thoroughly, the fact that we'd flown in her twice before being completely beside the point – daft to take anything for granted. I had a quick look at the bombsight and then at the computing mechanism: after that, a glance, on the other side, at the bomb selector switches together with those governing the nose and tail fuse-settings; also at the bomb-distributor mechanism and its selector arm and the positions of the intercom socket and the oxygen connector. In the front turret I checked the Brownings and the loaded ammo trays.

All we knew was that Tergnier was a pretty late take-off (23.39 hours). My stomach tightened bringing with it quite a dull ache. Fear; an awesome trepidation, which makes your throat a bit on the dry side. God, why did I let myself in for all this? Nobody asked me to. I actually volunteered: and binded away at all sorts of authority to get here that much quicker. Must be round the twist. That knot was still in my stomach at Briefing; the fact that No.1 was to be an 'easy' op like the marshalling yards at Tergnier not lessening it at all. Stayed with me even as we were driven out to the bay where *G-for-George* was parked. Began to diminish as I inspected my bombs, checked my machine guns and so on. Diminished more as we climbed in and secured the hatch. Became no longer conscious of it as Vernon gave the thumbs-up sign to the two airmen operating the starter mechanism and the port-inner engine burst into life to be followed by the starboard inner then the starboard outer and, finally, the port-outer. After testing the Merlins and the magnetos, plus a crew check-round over the intercom, he eased the Lanc around until it was lined up directly with the exit from the bay. A short wait while another Lanc on the perimeter track rolled past the exit then we were out ourselves on to the track and round it. Held up at the chequered caravan while this Lanc before us took off. As it did, Vernon edged our Lanc forward into the position thus vacated. Next we got the green Aldis and Vernon swung her round on to the runway. Another green and he lined the Lanc up with the path of the runway. Maximum pressure on the brakes now, then, together with the engineer, Griggs, pleasant, maybe about a couple of years younger than we were, he opened all four throttles wide. The engines strained for their freedom, which was given when Vernon released the brakes. The Lanc raced forward along the runway rapidly gathering

speed. We were almost running out of the concrete when Vernon eased back on the control column and lifted her very smoothly into the air. Up came the undercart and we started climbing over base. Oxygen masks on at 5,000ft. At 8,000ft we set course for France. I made good pin-points over England, keeping us so dead on track that we hit the French coast exactly as briefed.

At the French coast some flak. Hard at first to realise it's aimed at you! But not much of it really and none of it actually hit us. Some more flak here and there across northern France. We took evasive action once or twice, but maybe we were over-reacting. Still, better to over-react a hundred times than not to react on that fatal one time. Then we were over Tergnier with its light and spasmodic flak. Down went the bombs. Target well plastered as far as I could make out. Then, with bomb doors closed, we turned for base. Some more light flak, then the French coast, then the English coast, then base. What an easy one for our first op. A very gentle blooding, one might observe. Still, it's the first over. Only another 29 to do …

On 19/20 May 1944, 118 Lancasters and four Mosquitoes of 1 and 8 Groups carried out a particularly accurate attack on the railway yards at Orléans. One Lancaster was lost. After further raids by Bomber Command, on 4/5 July, 282 Lancasters and five Mosquitoes of 1, 6 and 8 Groups accurately bombed Orléans and Villeneuve.

At Tergnier two Lancasters were shot down. No aircraft were lost on the raids at Au Fèvre, Mont Couple and at Saumur. Bomber Command returned to Trappes and Saumur in the coming nights and a German signals station and coastal batteries were hit up to and including 4/5 June. In Operation Flashlamp just over 940 aircraft dropped more than 5,000 tons of bombs on ten coastal batteries in the Bay of the Seine along the fringes of the Normandy landing beaches. It was the greatest tonnage in one night so far in the war. Twenty-four ABC-equipped Lancasters on 101 Squadron patrolled all known night-fighter approaches. Operation Taxable induced the German crews manning the radar installations on a part of the French coast, designedly left intact for the purpose, to believe that a large convoy was proceeding at 7 knots on a 14-mile front across the narrowest part of the Channel between Dover and Cap d'Antifer and heading straight for them. The necessary reaction on the radar screen was to be reproduced by the 16 Lancasters of 617 'Dam Busters' Squadron and by 18 small ships, some of which towed balloons fitted with reflectors to simulate echoes given off by a large ship. Operation Taxable began soon after dusk and 'went steadily and mercilessly on through the night'. At the same time the German radar was jammed but not too heavily. A similar operation (Glimmer) was carried out by eight Stirlings off Boulogne, while Halifaxes and Stirlings dropped dummy parachutists, rifle-fire simulators and other devices such as squibs and fireworks, which produced the sound of gunfire. Taxable successfully created the impression that an airborne landing near the village of Yvetot in north-west France was taking place. Then, in the early hours of 6 June, eight coastal batteries were attacked.

'Just after five pm on the evening of June 5th,' Clayton Moore recalled:

we were ordered into the air for a hurried night flying test on *S-Sugar*. Fifteen minutes after take-off, we were back at dispersal, surrounded by bomb trolleys, bowsers and ground crew bods. An hour later, we were sitting down to our pre-flight supper and the Mess was buzzing with speculation on whether or not this could be 'it' – the opening of the invasion. The briefing was very late, thus indicating that the trip must be a short one because of the limited period of darkness available to us. Then the curtain was pulled aside to reveal the target and our suspicions were confirmed. We were to attack some shore-based naval guns at a place called La Pernelle on the French coast. Strict orders were given concerning the need for pin-point accuracy and timing and it was stressed that we must not fly below a specified height and that bombs must not be jettisoned in the Channel under any circumstances. When asked if, in his opinion, it was the beginning of the invasion, the briefing officer refused to pass comment, other than to say that he didn't know any more than we did. But to us, all the indications were there. This was a maximum effort and, although we had carried out similar raids across the Channel before, this was the first time that we had been warned not to jettison any bombs in it and the reason for this was obvious to us.

We took off at 00.30 hours on June the 6th on a flight that was to take us just four and a quarter hours. Although the large numbers of aircraft with us was a good indication of the importance of the target, we saw nothing other than this to support our suspicions. The weather was good and we went straight to the target and dropped our bombs and markers with only moderate opposition, then turned for home. The return flight took us on a more direct route over the Channel to our landfall at The Needles, and it was during the Channel crossing that we began to detect the first significant signs of activity. As we neared mid-Channel, the navigator reported that he was picking something up on his H_2S set.

'Must be a convoy of ships, but it looks like a bloody great island and it shouldn't be there.'

Cheers came from the other members of the crew and Bill had to call for silence. I found myself thinking that brother Tom who was serving in the Royal Canadian Electrical and Mechanical Engineers was probably down there somewhere and I wished him luck. As we neared England, the dawn was beginning to show, enabling us to see hundreds of aircraft heading out in the opposite direction to us. Fighters, light and medium bombers and then, as we flew up England, the sky above us was filled with Fortresses. There was no doubt now; this was IT.

Above: Air Marshal Sir Arthur T. Harris and his staff at High Wycombe.
Right: Bomber Command booklet which appeared in October 1941, showing a plan view of a Whitley, one of several twin-engined types that preceded the Lancaster.

BOMBER
COMMAND

Left: *Admiral Prune* in early November 1942; the smiling face in the cockpit is that of P/O Jimmy Cooper. The white bomb symbol generally indicated a daylight raid. Because the squadron often dropped sea mines on 'Gardening' operations and at the time naval officers were attached to the unit, several of the aircraft displayed Admiral-prefixed nicknames.

Below: Lancaster B.I W4118/ZN-Y *Admiral Prune* on 106 Squadron at Syerston. The aircraft was lost while outbound on the Turin raid of 4/5 February 1943. Sgt D.L. Thompson RCAF and three of his crew were taken prisoner. Four other crew were killed.

Left: Rod-cleaning the front machine guns of Lancaster B.I R5666/KM-F on 44 'Rhodesia' Squadron. Another member of the ground crew is cleaning the cockpit windows. R5666 failed to return from Nienburg while being flown by F/O R.R. Michell and crew on 17/18 December 1942. Michell and four of his crew survived to be taken prisoner. Two men were killed in action. *(Imperial War Museum)*

Above: P/O A.S. Jess, the Canadian wireless operator on S/L Burnett's crew on 44 'Rhodesia' Squadron, carrying two pigeon boxes.

Lancaster B.I R5689 VN-N on 50 Squadron in flight on 28 August 1942.
On 19 September, Sgt E.J. Morley RAAF took R5689 off from Swinderby
on a 'Gardening' sortie only to crash at Thurlby, Lincolnshire, when both
port engines failed as the crew prepared to land. Sgt James Reginald
Gibbons RCAF, the tail gunner, died in the crash. *(Flight)*

S/L Hughie Everitt DSO DFC taxies Lancaster B.I R5689 VN-N on 50 Squadron out at Swinderby, Lincolnshire, on 28 August 1942.

A Lancaster on 50 Squadron at Swinderby in the summer of 1942.

Lancaster Is on 207 Squadron at Langar in formation on 20 June 1942. On 8/9 December 1942, EM-F R5570 was the only Lancaster of 108 aircraft that was lost on the raid on Turin by 133 aircraft of 5 Group when it failed to return with W/C Francis George Levett Bain and his crew. It crashed near Milan and all seven crew were killed. L7583/A survived the war and was scrapped in 1946; L7580/C was SOC in November 1945 after serving on 5 LFS (Lancaster Finishing School). (*RAF Museum*)

Left: Ground crew working on Lancaster B.I R5540 KM-O on 44 'Rhodesia' Squadron Conversion Flight.

Below: Ground crew working on Lancaster B.I R5540 KM-O on 44 'Rhodesia' Squadron Conversion Flight. This Lancaster I was later used by 1661 HCU (Heavy Conversion Unit) at Winthorpe and crashed on 28 January 1943.

Above: Lancaster B.I ED592 on 50 Squadron on an air test in January 1943. This aircraft was lost with P/O Francis Eric Townsend and crew on 1/2 March 1943 when it was shot down by a night-fighter. All seven crew were killed. *(Charles E. Brown)*

Right: It was rumoured that ED382 *J-Joe*, which has a portrait of Joe Stalin on the nose, may have been so named by a ground crew member who had a copy of the Communist *Daily Worker* delivered every day. *J-Joe* operated later on 625 and 300 Squadrons and other units and survived the war.

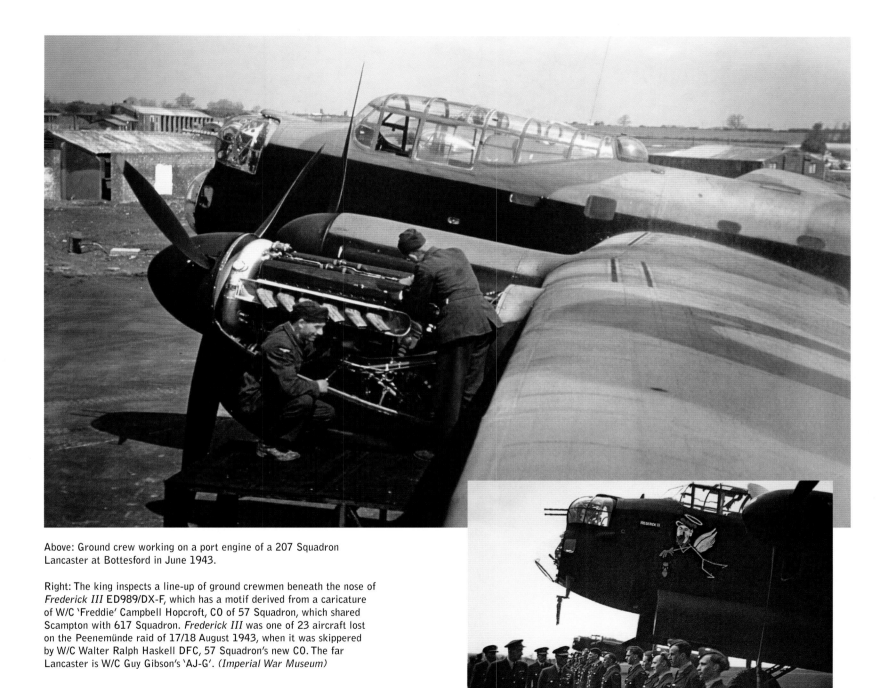

Above: Ground crew working on a port engine of a 207 Squadron Lancaster at Bottesford in June 1943.

Right: The king inspects a line-up of ground crewmen beneath the nose of *Frederick III* ED989/DX-F, which has a motif derived from a caricature of W/C 'Freddie' Campbell Hopcroft, CO of 57 Squadron, which shared Scampton with 617 Squadron. *Frederick III* was one of 23 aircraft lost on the Peenemünde raid of 17/18 August 1943, when it was skippered by W/C Walter Ralph Haskell DFC, 57 Squadron's new CO. The far Lancaster is W/C Guy Gibson's 'AJ-G'. *(Imperial War Museum)*

In August 1943, Woodford had three Lancaster assembly lines, two for Chadderton-manufactured components and a third for aircraft built by Metrovick at Trafford Park, Manchester. Bay One shown here contains Chadderton-built Lancasters. Lancaster III JB276 (left) was lost with W/O John Thomas Campbell Rhodes and his crew on 103 Squadron on 23 October 1943. Rhodes and five of the crew were killed in action and two were taken prisoner. (Avro)

Above: W/C Guy Gibson VC DSO DFC* with four of his crew who flew with him on *G-George* on the Dams raid: L–R: F/O Fred M. 'Spam' Spafford DFC DFM RAAF (bomb aimer); F/L R.E.G. 'Bob' Hutchison DFC* (WOp); P/O George Deering DFC (front gunner); F/L T.H. 'Torger' Taerum DFC RCAF (navigator); all of whom were killed in action on 15/16 September 1943 on *S-Sugar*, flown by S/L George W. Holden DSO DFC* MiD.

Above: S/L David Maltby DSO DFC with W/C Guy Gibson on 617 Squadron at Scampton. Maltby was killed in action on 14/15 September 1943 when he ditched in the North Sea off Cromer returning from an aborted sortie to Ladbergen.

Left: Lancaster fuselages at Chadderton. A.V. Roe produced the major components at Chadderton, with assembly and flight testing being undertaken at Woodford 20 miles away. Edwin Alliott Roe used his second forename and added his mother's maiden name, Verdon, in 1928 to become officially A.V. Roe. *(Avro)*

Above, left: S/L Dave Shannon DSO DFC RAAF, F/L Richard Algernon Trevor-Roper DFC DFM, Guy Gibson's tail gunner on the Dams raid in May 1943, and S/L George W. Holden DSO DFC* MiD, CO, 617 Squadron. Holden was killed in action on the 15/16 September 1943 Dortmund-Ems Canal raid. Trevor-Roper died on the Nuremberg raid, 30/31 March 1944.

Above, right: S/L Joseph Charles 'Big Joe' McCarthy DSO DFC* RCAF, a burly 23-year-old, 6ft 3in Irish-American from New York City on 617 Squadron, who flew on the Dams raid and the operation on the Dortmund-Ems Canal.

Left: Members of *R-Robert*'s crew on 460 Squadron RAAF at Binbrook in 1944, releasing a pigeon.

Left: Lancaster PB842 PG-Y on 619 Squadron, which F/O B.P. Curran RAAF crash-landed at Rinkaby in Sweden on 13/14 January 1945 when the target was Pölitz. (RAF Hendon)

K-King. (Via Eric Jones)

Above: Lancaster I L7583/EM-A on 207 Squadron which went on to serve on 1661 CU and 5 LFS before being scrapped in November 1946. (Charles E. Brown)

Right: Armourers manoeuvre a 4,000lb 'cookie' into position below the bomb bay of a Lancaster. The 'cookie' had a thin-walled steel casing filled with explosive for maximum blast effect with minimum fragmentation. Only one could be carried by Lancaster aircraft and none by Halifaxes.

Right: Bomb aimer's map of Nuremberg used on the catastrophic raid on the night of 30/31 March 1944 by Sgt Derek 'Pat' Patfield on P/O Denny Freeman's crew on *Q-Queenie* of 61 Squadron. *Q-Queenie* crashed at Foulsham in Norfolk on return. *(Via 'Pat' Patfield)*

Below: King George VI and Queen Elizabeth visit Warboys in 8 Group PFF on 10 February 1944. A 156 Squadron Lancaster is in the hangar.

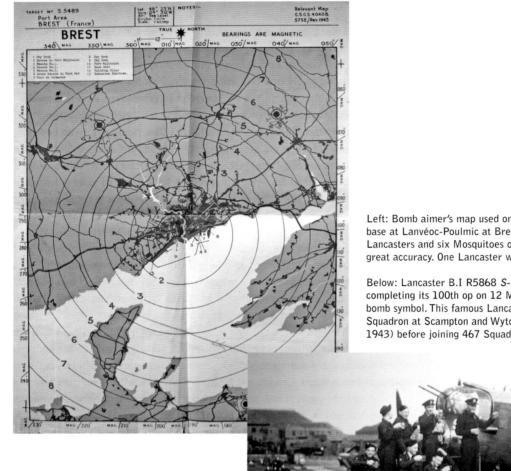

Left: Bomb aimer's map used on the raid on the airfield and seaplane base at Lanvéoc-Poulmic at Brest on the night of 8/9 May 1944 by 58 Lancasters and six Mosquitoes of 5 Group. The raid was carried out with great accuracy. One Lancaster was lost. *(Via 'Pat' Patfield)*

Below: Lancaster B.I R5868 *S-Sugar* on 467 Squadron RAAF after completing its 100th op on 12 May 1944; LAC Poone adds the 100th bomb symbol. This famous Lancaster had flown 68 operations on 83 Squadron at Scampton and Wyton as *Q-Queenie* (July 1942–August 1943) before joining 467 Squadron RAAF at Bottesford in September.

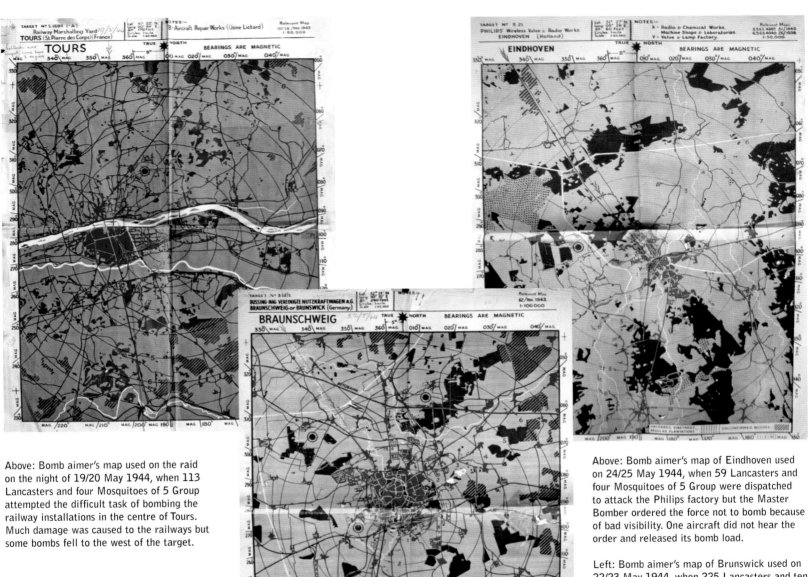

Above: Bomb aimer's map used on the raid on the night of 19/20 May 1944, when 113 Lancasters and four Mosquitoes of 5 Group attempted the difficult task of bombing the railway installations in the centre of Tours. Much damage was caused to the railways but some bombs fell to the west of the target.

Above: Bomb aimer's map of Eindhoven used on 24/25 May 1944, when 59 Lancasters and four Mosquitoes of 5 Group were dispatched to attack the Philips factory but the Master Bomber ordered the force not to bomb because of bad visibility. One aircraft did not hear the order and released its bomb load.

Left: Bomb aimer's map of Brunswick used on 22/23 May 1944, when 225 Lancasters and ten Mosquitoes of 1 and 5 Groups attacked the city. The raid was a failure and 13 Lancasters were lost. *(All via 'Pat' Patfield)*

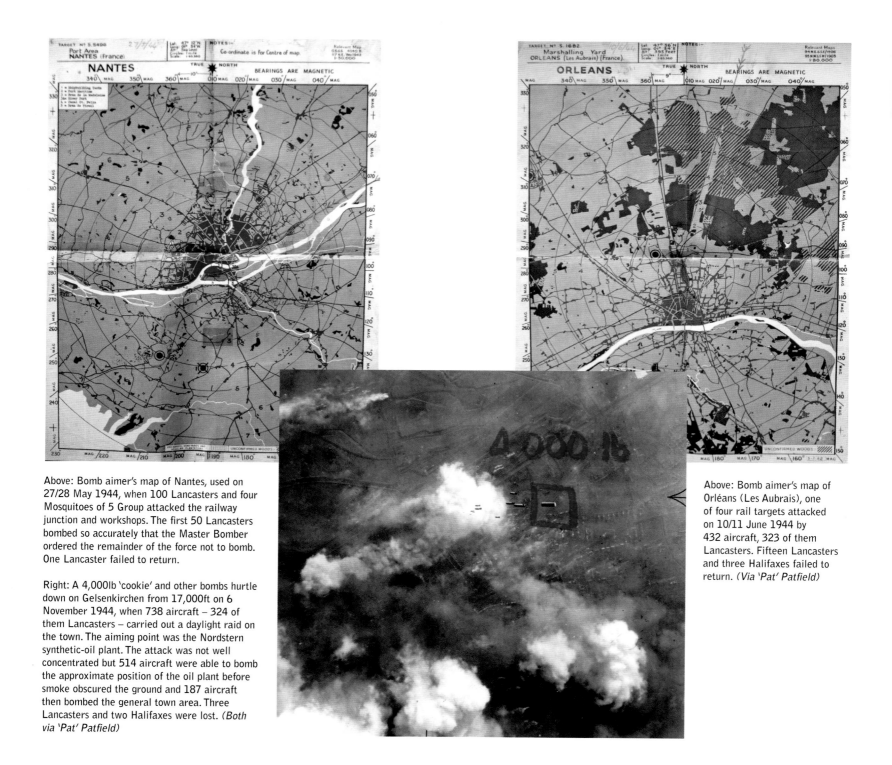

Above: Bomb aimer's map of Nantes, used on 27/28 May 1944, when 100 Lancasters and four Mosquitoes of 5 Group attacked the railway junction and workshops. The first 50 Lancasters bombed so accurately that the Master Bomber ordered the remainder of the force not to bomb. One Lancaster failed to return.

Right: A 4,000lb 'cookie' and other bombs hurtle down on Gelsenkirchen from 17,000ft on 6 November 1944, when 738 aircraft – 324 of them Lancasters – carried out a daylight raid on the town. The aiming point was the Nordstern synthetic-oil plant. The attack was not well concentrated but 514 aircraft were able to bomb the approximate position of the oil plant before smoke obscured the ground and 187 aircraft then bombed the general town area. Three Lancasters and two Halifaxes were lost. (Both via 'Pat' Patfield)

Above: Bomb aimer's map of Orléans (Les Aubrais), one of four rail targets attacked on 10/11 June 1944 by 432 aircraft, 323 of them Lancasters. Fifteen Lancasters and three Halifaxes failed to return. (Via 'Pat' Patfield)

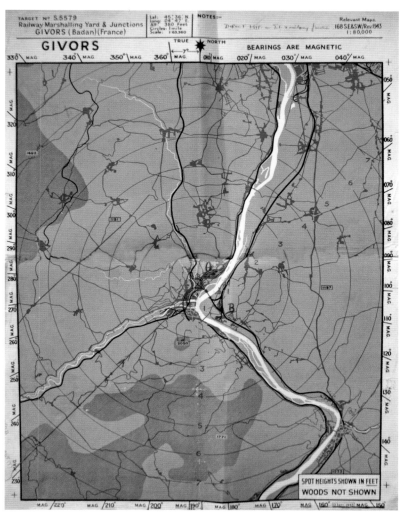

Bomb Aimer's map of Creil (Thiverny) in July 1944, where there was a flying-bomb storage site which was attacked by several Groups. (*Via 'Pat' Patfield*)

Bomb aimer's map of Givors (Bedan) marshalling yard (A) and junctions (B) used on 26/27 July 1944, when the railway yards were accurately attacked by 178 Lancasters and nine Mosquitoes of 5 Group. Four Lancasters and two Mosquitoes failed to return. (*Via 'Pat' Patfield*)

Above: Lancasters en route to their target in daylight. On 18 July 1944, 1,032 bombers – including 667 Lancasters – attacked five fortified targets south and east of Caen in support of Operation Goodwood, the imminent armoured attack by the British Second Army.

Right: The name and the ghoulish figure were the creation of Sgt Harold 'Ben' Bennett, *Phantom of the Ruhr*'s first flight engineer who had been a ground engineer in Fighter Command in the early part of the war.

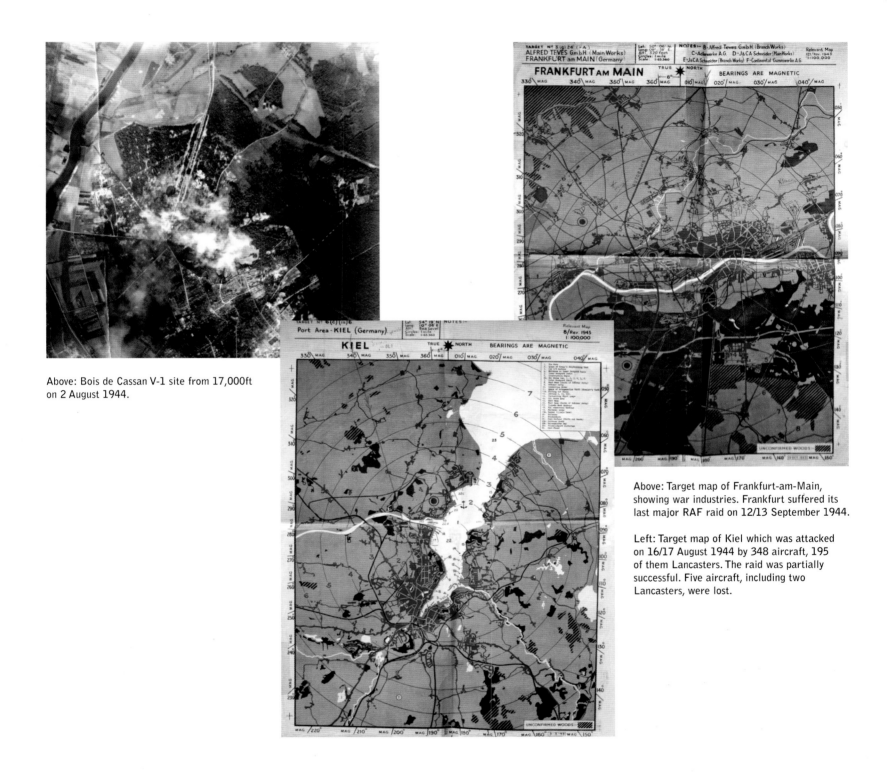

Above: Bois de Cassan V-1 site from 17,000ft on 2 August 1944.

Above: Target map of Frankfurt-am-Main, showing war industries. Frankfurt suffered its last major RAF raid on 12/13 September 1944.

Left: Target map of Kiel which was attacked on 16/17 August 1944 by 348 aircraft, 195 of them Lancasters. The raid was partially successful. Five aircraft, including two Lancasters, were lost.

TIME

THE WEEKLY NEWSMAGAZINE

HARRIS OF THE R.A.F. BOMBER COMMAND
His bombs are hammering the prelude to invasion.

Above, left: Lancaster III QR-M **EE176** *Mickey The Moocher* on 61 Squadron at Skellingthorpe in 1944 had received a Walt Disney cartoon of Mickey Mouse pulling a bomb-trolley on which sat a bomb.

Above: Arthur Harris on the cover of *Time* magazine.

Left: W/C Guy Gibson VC DSO DFC*, who was killed on the operation on Rheydt on the night of 19/20 September 1944 when flying a Mosquito. He was controller for the raid.

Left: On the night of 21/22 November 1944, 123 Lancasters and five Mosquitoes of 5 Group breached the Dortmund-Ems Canal near Ladbergen.

Below: The same night about 15 miles north-west of Osnabrück an aqueduct on the Mittelland Canal which carried the Ems-Weser Canal over the River Glane was destroyed by another 5 Group force of 138 Lancasters and six Mosquitoes. Next day, photo reconnaissance revealed 59 coke barges stranded on a 30-mile section completely drained of water and breaches in the embankments on both sides.

Left: The Ruhr or 'Happy Valley' as it was known by RAF bomber crews.

Below: A 4,000lb 'cookie' going down on Cologne on 2 March 1945, when two raids were mounted against the city, the first by 703 aircraft – including 376 Lancasters – and the second by 155 Lancasters of 3 Group. *(Via 'Pat' Patfield)*

Below, left: Lancaster crew prepare for an 'op'.

'THIS IS WAR AND SOMEBODY'S GOT TO DIE!'

I saw a Lanc buy it; he just seemed to explode in mid air. The flak was light to medium before and after target and very heavy and concentrated over the target. Searchlights were very concentrated; three main cones, about 60 over target itself; 30 to the north and 40 to the south. Saw two aircraft coned. Concentrated bombing of target: heavy and widespread fires raging. Flak, searchlights and fighter flares lit up the sky to such an extent that you could almost have read a newspaper in the aircraft. Flak bursting all around the aircraft. Rear gunner says he saw one of ours go down in flames just after target.

Campbell Muirhead, Gelsenkirchen raid, 12/13 June 1944

In all, RAF Bomber Command flew 1,211 sorties on D-Day; a new record. On 6/7 June, 1,065 RAF four-engined bombers and Mosquitoes dropped 3,488 tons of bombs on nine choke points, including bridges and road and rail centres behind the Normandy battle area. Four Lancasters were lost on the 5 Group attack on Caen, where the Main Force of bombers had to wait for the target to be properly marked and then fly over an area teeming with German units and guns at bombing heights below 3,000ft. One of these was *B-Baker* on 83 Squadron at Coningsby. F/L Bill Siddle's crew on the squadron had volunteered for a second tour and when asked if he was coming with them, Clayton Moore could see little point in refusing. His second tour began at 30 minutes past midnight on Wednesday 7 June, when the squadron had lifted off for the hastily called attack on the marshalling yards at Caen:

Because of the considerable confusion caused by the swift and unpredictable arrival of the Allied armada, we encountered only moderate opposition over the target area. This was to be expected in the circumstances. Because of the vast area of coastline to be defended against possible invasion, the German resources in guns and troops were scattered and sparse and would remain so until these could be concentrated in the area of the attack. In the initial stages of the invasion we could expect to be called upon frequently to attack important road and rail centres so that the transportation of vital enemy troops and equipment to the front could be delayed. The attack on Caen turned out to be a fairly easy one for us and it was the first operation I could recall during which I had not seen a single aircraft being shot down.

The centre of Caen was left in flames, the river barrage over the Orne was destroyed, four other bridges were destroyed or had their approaches blocked and the main roads from the town to Falaise and Bayeux were badly cratered. At Vire two of the three Lancasters lost were shot down by enemy fighters, the other to flak. The raid, by 1 Group, was over in about five minutes, during which time the bombers hit all the choke points and partly destroyed the railway station. The centre of the town was in ruins, with rubble blocking the roads. Lancasters of 5 Group attacked Argentan and another force bombed Lisieux and Condé. At Achères, near Paris, about half of the 97 Lancasters of 1 Group did not bomb, as Campbell Muirhead recalled:

Duff weather, heavy cloud and rain. Was map-reading my way quite easily up to the target, had even selected and fused my bombs, when the Master Bomber of the Pathfinder Force came on the air and ordered us to take our bombs back to base. Having real finger trouble, the PFF, not being able to identify the railway junction. (Yet, to be fair, perhaps the cloud beneath the Master Bomber was much thicker than the stuff below us.) Was annoying, really, to be able to identify that rail junction so clearly, even to see it sliding so steadily up my bombsight towards the graticule (as you get it on to the graticule you press the tit) then to be told not to bomb it. Felt inclined to press the tit despite the order, but thought better of it. Not that I'm becoming 'bomb-happy' or anything like that, but for all I know, because we didn't bomb, a German troop train, maybe even carrying Tiger tanks which can knock hell, so we understand, out of any tanks we or the Yanks possess, bound for Normandy might, as a result, be able to pass through that junction before dawn.

News of the D-Day landings is announced to 166 Squadron air crew at Kirmington by F/L Frank Tighe, the station adjutant. On the left is the squadron CO, W/C Don Garner DSO. Among the group is F/O Donald Pleasence, later to become one of England's best-known actors. He was the WOp on F/O E.B. Tutty RAAF's crew on 31 August when they were shot down over Agenville and he was taken into captivity. Tutty and Sgt R.D. Butcher evaded capture. Two of the crew were killed.

So we returned to base with our bomb load. Our turn over Achères brought us over the outskirts of Paris. They don't have a black-out there; it's a form of 'blue-out' and you can see, dimly, the outlines of certain streets. Now and then torches could be seen flashing 'V'. Suppose the French would have to stand on roofs to do that, otherwise the Germans would shoot them. No flak or searchlights over Paris. In fact, this evening we haven't had flak or searchlights anywhere. Wonder if I'll ever visit Paris. Before the war only the comparatively rich could afford to visit France (when everything was very cheap, a bottle of wine, I'm told, costing about a tenner). Maybe my turn will come after the war.

Seven aircraft missing from this operation, though none from Wickenby, which shows that there must have been quite a few night-fighters knocking around. Surprising they made so many kills what with all that heavy cloud and without searchlights to guide them. Reckon their victims must have been Lancs who got out of the bomber stream and were then sitting ducks, easily picked up by the radar sets the night-fighters are supposed to be able to use with great accuracy once the German ground control has guided them to the vicinity of the wandering Lanc. It's the old, old story; allow yourselves to get out of the bomber stream and almost certainly you have had it; even on a five hour round trip into northern France.

A Lancaster which failed to return crashed at Éragny with the loss of all seven crew. More than 1,060 aircraft had attacked and dropped 11,500 tons of bombs. F/L J.S.A. Marshall RAAF recorded in his log book a tragic moment on returning from the trip to Argentan: 'On crossing the Channel the Navy (ours) shot down a Lanc about 200 yards on our port. Trigger-happy bastards.'

Raids on the communication targets continued on the night of 7/8 June, when 337 aircraft were dispatched to bomb railway targets near Paris at Achères, Juvisy, Chevreuse and Versailles-Matelet and an important road and rail junction at Massy-Palaiseau about 14 miles south of Paris. This target was very well marked with red and green TIs, and bomb bursts were concentrated amongst them. The railway track could be seen in the light of explosions. Crews bombed from 6,000ft and at this height they encountered intense light flak. On the leg into the target they also met considerable fighter opposition and eight bombers failed to return. Twenty-eight Lancasters and Halifaxes were lost on the raids on the road and rail targets. On the raid on Chevreuse, 115 Squadron at Witchford lost six Lancasters, one of which exploded over Paris. In another attack, which was requested by the US First Army, 112 Lancasters and ten Mosquitoes of 1, 5 and 8 Groups carried out a raid on an important six-way road junction in the Forêt de Cerisy halfway between Bayeux and St-Lô. The area was believed to contain fuel dumps and German tank units preparing to attack First Army units. The bombing was not successful, however, as the Main Force bombed on a stray marker, which had been dropped in error 6 miles from the target. A Lancaster on 101 Squadron failed to return.

The night following, 483 aircraft attacked rail targets at Alençon, Fougères, Mayenne, Pontaubault and Rennes to prevent German reinforcements from the south reaching the Normandy battle area. Three Lancasters and a Mosquito failed to return. On the night of 8/9 June, the first 12,030lb 'Tallboy' bombs developed by Barnes Wallis were used operationally for the first time when 25 Lancasters on 617 Squadron attacked the Saumur railway tunnel near Tours, which had to be destroyed to prevent the 17th (SS) Panzer Grenadier Division's move from south of the River Loire to the Normandy front by train. The bridge was demolished along much of its length and was not brought into full use until four years after the end of the war.

On the night of 10/11 June, 432 aircraft – including 323 Lancasters – attacked rail targets in France, at Achères,

Dreux, Orléans and Versailles. Clayton Moore recalled: 'We were called upon to carry out yet another attack on a transportation target. This time it was on Orléans, about 70 miles south-south-west of Paris. Again we had an easy ride. We took ND933 for the trip and the squadron came through the attack unscathed.'

Fifteen Lancasters and three Halifaxes were lost. One of the two missing Lancasters on 100 Squadron at Grimsby was *K-King* flown by P/O H.W.L. Skinner, which was badly shot up by a night-fighter returning from the attack on Achères. Sgt D.R. Foggo, the rear gunner, was badly wounded. After the bale-out order F/O Dickie Carroll, the 34-year-old Irish bomb aimer from County Waterford, regardless of his own safety, delayed his exit to go back along the fuselage and help the wireless operator drag Foggo out of his turret, which had jammed. They moved the barely conscious rear gunner to the main door, put his hand on his ripcord for him and pushed Foggo out. Then Carroll went forward again to collect his own parachute, but found that it was missing, presumably fallen out through the open hatch in the nose. Skinner suggested that they jump together. Since a parachute is designed to take a strain of more than a ton during opening, the prospect of a safe dual descent was considered to be reasonably healthy. The big obstacle, however, was that the escape hatch on the Lancaster was too narrow for two men to squeeze through side by side and so it was that on the few occasions when parachutes were shared, if they could not reach the main exit door, the 'passenger' usually rode on the parachutist's back. Unfortunately, in this position it was almost impossible for the 'passenger' to get a firm enough grip that would withstand the opening shock – it has been described as the shock experienced in violently braking a car travelling at 100mph. With great difficulty, as *K-King* was now diving steeply, Carroll climbed on Skinner's back and after a long struggle they squeezed together through the narrow escape hatch and dropped out. As soon as they had fallen clear the pilot pulled his ripcord. He looked round. The parachute was open but Carroll was no longer with him.[1]

On the night of 11/12 June, 329 aircraft – 225 of them Lancasters – attacked railway targets in northern France. Three Lancasters and a Halifax failed to return. Campbell Muirhead on 12 Squadron flew his seventh op, to the marshalling yards at Évreux:

Take off, in *K-for-King*, was at 22.15 hours. Our bomb load was 11 x 1,000lb, 4 x 500lb. Flak was very occasional

F/Sgt Jack Goulevitch sports a top hat in front of his crew's Lancaster on 460 Squadron RAAF. On civvy street he was an undertaker!

indeed along the route, hardly enough to comment on. But over target the light flak (from 88s, I should think) was very accurate, bursting all around the a/c, virtually cocooning it. Heard it rattling around on the outside of the fuselage. Nine-tenths cloud over target, but was able flying as low as 2,500ft to spot the marshalling yards surprisingly clearly. There seemed to be quite a few trains in the yards; you'd think they'd be able to damp down the engines when informed of the likelihood of an air attack. I could see the steam rising quite clearly from at least a couple of locomotives: Would be troop or ammo trains, of course: doubt if any civilian traffic would be permitted in these parts. I am so fond of railways and engines that I get no satisfaction from bombing such installations. Yet it's got to be done accurately: absolutely no point in taking off with a bomb load of 12,000lb, exposing yourself to danger getting here (and getting back again) and then, when over the target, being slip-shod and careless. I have a good run-in, get a good aiming point and drop them in series right across the marshalling yards. Some mighty explosions which suggest ammo train coming skywards. But some

O for Oomph Gal.

bomb aimer has been careless (maybe more than one bomb aimer) because, before we left, we saw fires raging in the SE part of the adjoining town. It's this 'drop-back' that they keep on and on about at briefing: bomb aimers lying there, thumb poised over the tit all ready to press and the target sliding slowly, oh, so slowly, up the graticule: and they can't wait, their thumb drops just that fraction of a second before the target is fully on the cross graticule and, gradually, if one or two bomb aimers act similarly, the entire bombing effort more or less creeps backwards, until it's the area before the target which is getting it, not the target itself. They're quite extensive fires and I wonder how many French people die as a result. Also, how many French railway workers are killed in our attack on the yards.

In the event, we now also get eggs and bacon when we get back – if we do. These fresh eggs, which are in very short supply all over the country, are for operational air crew only. A bit embarrassing at times for you to sit down and be served with a delightful fresh, fried egg while beside you some ground-crew officer is trying to force down that dreadful powdered egg which I for my part find absolutely inedible. Wouldn't give it to the cat. The cat wouldn't eat it anyway.

Campbell Muirhead's eighth op was on the night following, when F/O Vernon's *K-for King* carrying a 4,000-pounder and 18 500-pounders was one of 671 aircraft in a stream of 286 Lancasters with 17 Mosquitoes headed for the synthetic-oil refinery at Gelsenkirchen while 671 aircraft, including 285 Lancasters, attacked communication targets in France. Muirhead wrote:

First one on the Reich itself. And so bloody different: it made one feel we'd simply been playing at the game up until now. Indeed, compared to this, all those previous efforts over Occupied Europe seemed like pleasant little picnics, events you looked back upon with something akin to pleasure (if you can equate dropping bombs with that word). I was, as per usual in the front turret: I always climbed into it shortly after we were airborne. And there was no need to be dramatic and shout, 'Enemy coast ahead,' because all could see it except, of course, the wireless operator and the navigator, both of whom were in their little blacked-out compartments. Yes, you could see it: maybe not the actual coast, but the flak, seemingly solid walls of it hosing up into the sky. We were in the second wave, so it was the first one which was getting it.

Crossing the coast, the flak (which, I'd think, was from those extra-heavy 10.5 and 12.8cm guns we'd heard about) seemed to reach out for you alone as if it was something personal. The searchlights too: again one received the feeling that it was your Lanc only that they were trying to fasten onto. Over the coast and inland, leaving that wall of flak behind us now occupied with the third wave of Lancs. Still some flak and searchlights of course and all predicted stuff. But now the flares were bursting all around us, dropped, I should imagine, by Junkers 88s trying to illuminate us so that the night-fighters could come in. Now and then I saw flurries of tracer as interceptions were made. Occasionally I saw the silhouette of other Lancs: comforting, because it confirmed that we were still in the bomber stream. Get out of the stream and you've absolutely had it: you're a sitting duck and can be picked off easily.

Next, Cartwright the rear gunner, a quiet, dependable chap who, for the most part, kept himself to himself, was yelling, 'Corkscrew, corkscrew,' like a man demented and Vernon had her in a stomach-wrenching dive to port followed by a steep climb to starboard then another almost vertical dive, but to starboard this time. Cartwright then announced that he was certain we'd lost whatever was after us and we levelled off. Was glad to note the odd Lanc around us and that therefore we were still in the stream.

The rear turret was the least enviable position in the Lanc. An attack almost inevitably came from behind and the rear gunner received the full cannon blast. Sometimes he was the only one who got it, the Lanc having started corkscrewing immediately on the attack and thereby escaping but, in so many cases, too late to save the gunner. Cartwright made light of it saying that if he has to be hosed out of the turret on return to base – as has happened on some occasions – at least he'd know nothing about it. Some rear gunners received dreadful injuries in the attacks and even if the aircraft managed to make it back to base, were often almost impossible to extricate from the turret. It was difficult enough to get out of in any event, the exit being so small. Certainly the most dangerous position in the aircraft. The mid-upper turret wasn't all that easy to get out of either, particularly with P/O Horsfall being so big, but was a piece of cake compared to that rear effort. Horsfall, a regular who volunteered for aircrew, was perhaps a couple of years older than we were; a Yorkshireman with rather an abrasive personality; a health fiend; rugger and all sorts of dreadful exercises. My front turret I could get out of easily, no difficulty at all there. And if I was in my bomb aimer's compartment I could get out of that Lanc quicker than anybody because I lay on top of the escape hatch: one pull and it's up and all I had to do was to fall out. Out and down. Had one great fear which was that, in the panic which would be enveloping me if we had to bale out, I'd forget to clip on my parachute pack first. Told it had happened.

Approaching Gelsenkirchen at last. More flak than ever, it seemed, and more searchlights than ever. Fires caused by first wave intensive. Green Target Indicators for us; the second wave. I got them clearly into my sight and then down went the 4,000-pounder and the eighteen 500-pounders. Shouted for bomb doors to be closed: flew on to turning point. As we banked for our next course I could not help staring at the fires burning below which, even from 19,000ft, one could see clearly. Stupid doing that really; it must have destroyed my night vision.

That was really widespread bombing. What is now referred to as an 'area bash'. Some of my load must have fallen on houses; maybe even on air raid shelters as well. A faint niggling at the back of my mind; how many women

and children had I killed simply by pressing that little tit? No point in deluding oneself over that one had killed people. But I suppose it's daft to permit your mind to think along these lines. Trouble is I'm too tired. It's now almost 08.00 and honestly I don't know why I stay up writing this when I could be in bed and sleeping, or trying to sleep, anyway. One more thing, I didn't feel scared during it; but after it felt something akin to a tremble coming on. And yet, surely that was easy stuff compared to what the bomber crews suffered in 1942. Even last year was worse when you look back on Bomber Command's losses over it, both in air crews and aircraft. Should consider ourselves lucky, I suppose.

Lancaster pilot P/O Peter Johnson with his sisters Margaret and Pat while on leave from 619 Squadron at Dunholme Lodge. Johnson was killed in action on the night of 21/22 June 1944 when *D-Dog* crashed off the Dutch coast on the operation on Wesseling. *D-Dog* was one of six Lancasters on the squadron that failed to return.

Seventeen Lancasters were brought down on the Gelsenkirchen raid by the German defences, and 17 Halifaxes and six Lancasters failed to return from the communication raids, one of which was on the railway yards at Cambrai. *A-Apple* on 419 'Moose' Squadron RCAF was flown by P/O Arthur de Breyne RCAF. A Ju 88 attacked from below and astern and set both port engines on fire. The flames soon became fierce and 'Art' de Breyne ordered the crew to abandon the aircraft. Sgt Jack Friday RCAF, the bomb aimer, was knocked unconscious as he opened the front escape hatch and was bundled through the hatch by others of his crew as one of them pulled the ripcord of his parachute. P/O Andrew Mynarski, the mid-upper gunner, left his turret and went towards the escape hatch. He then saw that F/O George P. Brophy RCAF, his friend and rear gunner, was still in his turret and apparently unable to leave it. The turret was, in fact, immovable, since the hydraulic gear had been put out of action when the port engines failed and the manual gear had been broken by Brophy in his attempts to escape. Without hesitation, Mynarski made his way through the flames in an endeavour to reach the rear turret and release Brophy. Whilst so doing, his parachute and his clothing, up to the waist, were set on fire. All his efforts to move the turret and free the gunner were in vain. Eventually Brophy clearly indicated to him that there was nothing more he could do and that he should try to save his own life. Mynarski reluctantly went back through the flames to the escape hatch. There, as a last gesture to the trapped gunner, he turned towards him, stood to attention in his flaming clothing and saluted, before he jumped out of the aircraft. Mynarski's descent was seen by French people on the ground. Both his parachute and clothing were on fire. He was found eventually by the French but was so severely burnt that he died from his injuries. Brophy, thrown clear of the wreckage when the Lancaster crashed, had a miraculous escape. Had Mynarski not attempted to save his life, he could have safely left the aircraft. Mynarski was awarded a posthumous VC for his most conspicuous act of heroism.[2]

On 14 June, more than 220 Lancasters and 13 Mosquitoes carried out two separate raids on Le Havre, where 15 E-boats and light naval forces were a threat to Allied shipping off the Normandy beachheads just 30 miles away. The raid took place in two waves, one during the evening and the second at dusk. Most of the aircraft in the first wave were from 1 Group and the second from 3 Group, both waves being escorted by 123 Spitfires of 11 Group. In all, 1,230 tons of bombs were dropped on the pens. Just before the first wave bombed, 22 Lancasters on 617 Squadron released their 'Tallboys' on red spot fires dropped by three Mosquitoes, including one flown by W/C Cheshire. They marked the E-boat pens in the eastern area of the port and 11 Lancasters attacked this aiming point. One 'Tallboy' blasted a 16ft-diameter hole in the north-west corner of the blast pens. The total shipping destroyed was 53 boats of various kinds. When Barnes Wallis saw the PR photos that confirmed that the pens had collapsed along much of their length he commented that the best use of 'Tallboys' against such structures was a deliberate near miss. 'Their

earthquake effect undermines the foundations and thus their massive roof weight then becomes a tremendous liability!' The light flak was 'tremendous' but only one Lancaster failed to return and one on 617 Squadron had to be written off after returning.[3] That night over 330 aircraft bombed the vital road centre at Aunay-sur-Odon and German troop positions at Évrecy near Caen. At Aunay, where at least 31 civilians had died two days earlier in the daylight bombing, another 165 civilians were killed. While no Germans were present, the town's six medium-class and four minor roads were considered by Allied commanders as being vital for German reinforcements arriving in the area. All the bombers returned without loss. Another 330 aircraft took off for railway targets at Cambrai, Douai and St-Pol, but all were either partially cloud covered or affected by haze, so the bombing was not completely concentrated or accurate. A Halifax and a Lancaster carrying the Master Bomber failed to return from the Douai operation.

On the night of the 15th/16th, railway yards at Lens and Valenciennes, an ammunition dump at Fouillard and a fuel dump at Châtellerault were bombed by just over 450 aircraft for the loss of 11 Lancasters. On 15 June, the French Resistance reported that all rail traffic in south Brittany had come to a halt except for one single-track line running from Nantes to Rennes. The following night, 405 aircraft of Bomber Command began a new campaign against V-1 flying-bomb launching sites in the Pas-de-Calais, with raids on four targets accurately marked by Oboe Mosquitoes. No aircraft were lost on the raids in northern France but 309 bombers – 147 of them Lancasters – that set out to bomb the synthetic-oil plant at Sterkrade/Holten were hit hard. The route the bomber stream took passed near a German night-fighter beacon at Bocholt just 30 miles from Sterkrade and the *Jägerleitoffizier* had chosen this beacon as the holding point for his night-fighters. Nine Lancasters and 12 Halifaxes were shot down.

On 19 June, after standing by for three days waiting for cloud over the Pas-de-Calais to clear, 18 Lancasters and two Mosquitoes of 617 Squadron, with nine Mosquitoes of 8 Group providing preliminary marking, set out to attack the Blockhaus flying-bomb store in the Forêt d'Éperlecques, a mile from the village of Watten. The conditions proved too difficult for accurate marking and the nearest 'Tallboys' to the main building were 75ft and 100ft away from it. The next day, 17 Lancasters and three Mosquitoes of 617 Squadron, with escort, were dispatched to the V-2 launching and storage bunker in a quarry near Wizernes with 'Tallboys' but they were recalled just before they reached the French coast because of 10/10ths cloud at the target. Watten was bombed again on 25 July by 16 Lancasters of 617 Squadron with a Mosquito and a Mustang marking the target. The Germans ordered that Watten be abandoned.

'In the early afternoon of June 21st,' recalls W/O Clayton Moore on 83 Squadron at Coningsby:

the entire squadron took part in a two-hour formation exercise involving a couple of hundred Lancasters. The news was circulating that the RAF was to take part in daylight attacks on the enemy, in addition to the customary method of bombing during darkness. The prospect was one that I found daunting, particularly in view of the loss rate being suffered by the Americans despite the recent introduction of long-range fighter escorts. It became obvious to me that a lot of training would be needed before our pilots would become proficient in the execution of the new technique.

The weather, as with previous days, remained dull and the slight northerly wind kept temperatures a little chilly. At Coningsby on the shortest night of the year, a Tannoy announcement informed a number of crews, including F/L Siddle's, that ops were on that night and that the designated crews would be taking part. 'Briefing was scheduled to take place at 7 p.m.,' continues Clayton Moore, 'so we guessed that the trip would most likely be a short one, probably to the invasion area once more. Imagine our surprise when at the briefing it was learned that we were to be sent to the heart of the Ruhr to attack an oil target at Wesseling, nine miles south of Cologne.' One hundred and thirty Lancasters would be accompanied by five crews in 1 Group, and six Mosquitoes would mark the target using the 'Newhaven' method. Another 123 Lancasters and nine Mosquitoes of 1, 5 and 8 Groups would attack a synthetic-oil plant at Scholven-Buer. Plans were to bomb both targets simultaneously. Clear weather conditions were expected at both target areas but as the midnight hour approached the weather deteriorated.

'We lifted off in *Sugar* forty minutes before midnight,' continues Clayton Moore:

and we managed to dodge much of the action that was taking place around us during the approach. It wasn't until we began our run-up to the target that things started

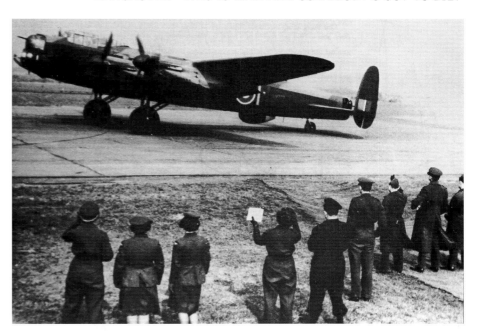

to get hectic. To begin with, there was a lot of heavy flak coming up at us and there was no doubting its accuracy. Then, just as we were lining up for the bombing run, we got coned in the searchlights and the Jerries started firing up the beams at us. Having decided that discretion made more sense than valour, the skipper ordered the postponement of the straight and level bit and proceeded to throw us all over the sky in a manner that threatened to do more damage to our Lancaster than the Cologne defences could inflict. After about a minute of this, during which time I had banged my head against the top of the turret a couple of times, Bill decided that the tactic wasn't working, so put us into the screaming dive manoeuvre. This produced the desired effect and we soon found ourselves free of the flak and searchlights, but only because we were by then well out of the target area. This made it necessary for us to go around again in the hope that Lady Luck would be more kind towards us this time. To this end we succeeded and the run-up proved uneventful. Not so for some of the others, however, and I was again made to witness the horrendous sight of a couple of Lancaster being blown apart and falling into the inferno beneath.

Once rid of the bomb load we headed west, soon to be hidden from sight by the darkness that I found so

Lancaster B.I LM227 *I-Item* on 576 Squadron, fitted with the H₂S airborne radar transmitter and with P/O J.R. 'Mike' Stedman at the controls, is waved off by WAAFs and RAF men at the end of the runway at Elsham Wolds for its first bombing raid on 4/5 July 1944, to Orléans. By the end of April 1945 *I-Item* had flown 93 sorties and one 'Exodus' and six 'Manna' trips to make the 100.

Above: A Royal visit to Mildenhall on 5 July 1944, when crews in 15 and 622 Squadrons operated from the station. Far right is Sgt Bill Matthews and next left, 37-year-old Sgt C.A. Canday, both of whom were killed in action that night when F/O Matthew Michael Golub RCAF and his crew on 15 Squadron were shot down and lost without trace. *(Via Fred Coney)*

Right: Sgt Bill Matthews' last letter. *(Via Terry Owen)*

comforting. The tracking of a bomber by electronic means had not yet advanced to the stage which would enable a fighter pilot to launch an attack without first establishing visual contact with his quarry. First he must find us and this was one advantage that we had over the poor bloody Yanks on their daylight raids. They were there for all to see.

Shortly after leaving the target I detected a malfunction in my gun turret. Bill was in the habit of executing a gentle weave whenever we were over enemy territory and I now felt that the turret was wobbling each time the aircraft banked. My suspicion was confirmed when I positioned the turret on the beam and sighted one of the canopy frames against the nearby rudder fin. The turret was undoubtedly loose and I reasoned that the mounting must have suffered damage during the violent evasive action that we had been forced to take over the target. I at once reported the fault to the skipper.

'Is it serious?' he asked.

'I'm afraid it might easily part company, Skipper.'

'Then you'd better get out of there and come forward.'

'I'd like to stay here while all these fighters are around, Skipper,' I countered. 'I've got a seat pack parachute, so I'll be able to bale out if the turret does come off and I'll be

Sgt W MATTHEWS 1446208
As Usual. 6-7-44

Dear Folks,

Yes its me again and I have just been talking to some of our very old friends. Namely, King George, Queen Mary (sic) & Princess Elizabeth. The king asked me quite a few questions, how many times I had been over etc. If I liked daylight raids and so on. Did I find time for much sleep. (The answer to that is 4 hours in the last 47 hours). He came to convey Genl Montgomery's regards for the great show we put up for the boys at Villers-Bocage and said the praise the troops gave the RAF, in France now has to be heard to be believed. I wasn't a bit dressed up for the effort, I had just my flying boots on, Battle dress, no shirt, no Cap, hadn't shaved for 2 days. There I was with the crew talking and chatting as if at the local boozer. Elizabeth seems a nice girl, but she doesn't talk a lot. Incidentally, George (Mark W) talks perfectly. No trouble at all. I guess it will be in the 'news' next week or in the paper tomorrow. I may not be in the picture.

By the way the photo we had taken last week I will be able to get pretty soon as a presentation of Bomber Command. Say! We do see life in this game. Wouldn't change my job for the World now.

Well folks, its Weds today (or is it Thursday) and I start leave next Tuesday all being well. You should find the gen in my log book quite interesting. Will also give you some idea of the action I've seen just lately. Ha! I'm the kid. I'll tell you about the night when only 27 came back out of 60. Boy! We have great fun.

Well folks, I guess thats all for now. I'm hoping to get an hours sleep now. Hope you don't think I've changed too much when you see me.

Love

Bomber Bill

alright in that respect. But you'll know what's happened if you don't hear from me.'

'OK but I want you out of there as soon as we cross the enemy coast.'

I stayed in the turret until we were out over the water on the way back. Apart from not relishing the idea of going down in 'the drink' without a dinghy (particularly since I was a non-swimmer), I didn't want to be in the turret during the landing, just in case. In the event, Bill sat her down to a landing which was so smooth as to be uncharacteristic of his standards and I was thus able to present the ground crew waiting at dispersal with the proof of my complaint. On the following afternoon I visited the armoury and was told that all but two of the bolts which secured the turret to the rotating service joint had sheared off and that it was improbable that the turret would have remained in place on the airframe during any further violent evasive action. I also learned that the squadron had lost two aircraft on the raid.

Both targets were covered by 10/10ths cloud, so at Wesseling H$_2$S was used and at Scholven-Buer Pathfinder aircraft provided Oboe-based sky-marking instead. Only slight damage was reported at Wesseling, though the plant suffered a 40 per cent loss in production for a while after this attack and Scholven-Buer suffered a 20 per cent loss in production following the raid. Eight Lancasters were shot down and a 61 Squadron aircraft which landed at Woodbridge was wrecked when it was hit by a Flying Fortress on 214 Squadron that was badly damaged on a special operation. More than 50 kills were claimed by *Nachtjagd* as the result of an effective Tame Boar operation against the Wesseling force, I./NJG1 being credited with 16 victories by eight He 219 *Uhu* ('Owl') crews. In all, 37 Lancasters were lost on the Wesseling operation. At Dunholme Lodge, 44 and 619 Squadrons each lost six Lancasters. Five crews on 207 Squadron at Spilsby and three on the two Lancaster squadrons at Waddington were also lost. At East Kirkby, six Lancasters on 57 Squadron and four on 630 Squadron were shot down. At the target, *G-George* on 630 Squadron, flown by 22-year-old P/O Lionel N. 'Blue' Rackley RAAF, a red-headed Australian, was shot up by a Ju 88 night-fighter and its control surfaces badly damaged. On 24/25 April, on the raid on Munich, Rackley had got his crippled Lancaster to Borgo on the island of Corsica, though his rear gunner had died in the resulting crash-landing. His replacement was P/O William Alfred Cyril

Davies, who had been the sole survivor from a take-off crash on 20/21 February.

'We set off for home,' says Rackley:

and as we went three things became very clear. First, with luck we would reach England – our engines were still intact. Second, there could be no question of controlling the aircraft for a landing – we would have to bale out. Third, 'Taffy' Davies the rear gunner's parachute pack had been hit by a shell and was useless. And there was no spare chute for him.

Back over England in the vicinity of Henlow *G-George*, a map of Corsica resplendent on its nose, became uncontrollable and Rackley gave the order to bale out. The bomb aimer, F/Sgt Doug Morgan RAAF, offered to attempt a jump taking 'Taffy' Davies with him. While the two men stood face to face the crew roped them together. Davies clung to Morgan with arms and legs and together they dropped out of the forward hatch. However, the moment the parachute snapped open the sudden deceleration broke the ropes, the two men lost their grip on each other and the gunner fell away. Subsequently Morgan was awarded the BEM for his brave attempt.

Meanwhile, the other crew members jumped and Rackley followed them out into the night. From 10,000ft he could see a layer of cloud far below. 'For some time I didn't seem to be getting any nearer the cloud. The illusion was so strong I convinced myself I was being carried upwards not downwards.' Rackley dropped through a shallow cloud layer in seconds and landed on a moving passenger train travelling at some speed from Luton to London. His parachute was entangled on one of the carriages. 'All I knew was that I was being dragged along at a terrific rate and rattling beside my ears was the distinctive sound of train wheels. I was bumped about violently, but for some odd reason it didn't seem to hurt; nor was I aware of any fear.' Then Rackley blacked out. How far he was dragged he will never know but at some point he was saved by the quick-release box on his parachute harness. It was smashed, allowing the parachute to be ripped off his body. 'The next I recall was wandering along a railway line. It was still dark. I hadn't the least idea where I was. I had lost my flying boots, my parachute had disappeared, my head was bleeding and I couldn't move my right arm.' Rackley waved down a slow-moving goods train and a suspicious train crew, not wholly convinced he was not an enemy airman, put him in the guard's van and took him to a signalbox from where an ambulance collected him. Rackley

Lancaster I LL842 NF-F *Phyllis Dixey* which went missing in action with F/O G.H. Parker RCAF (killed in action) and crew on 24/25 July 1944.

Lots of Spits, plus some Lightnings, swarming about, spoiling for a fight. Would be a brave Fw 190 or Me 109 to venture into the air. Sometimes I think these fighter boys, in the absence of the Luftwaffe, would be more than prepared to take each other on … Was able to spot quite easily the sort of ramp-like things from which these robot bombs are launched. Camouflaged, I know, but even from about 15,000ft could be identified. Odd, but you could make out the outlines of paths to them from the edges of the surrounding fields. Would have credited the Germans with more thoroughness than that. Paths across grass made by feet stick out a mile when viewed from the air, especially on a clear sunny day such as this. With a straight bombing run, not bothered too much by the flak, was able to straddle the site: think I got two of the ramps. There were only three or four Lancs on this one and all approached the site from different directions pretty well, as far as I could make out, blanketing it.

I then suggested to Vernon that we go down to strafe the flak posts. He thought this might be a good idea: so did the others, except Dunn the wireless operator and Norman the navigator, both of whom declared we'd gone round the bend. Dunn had nothing to do for most of the time except to maintain a listening watch. During our operational flights he sat in his blacked-out compartment and read Westerns. Someone asked him if he can't read anything better than these wretched Westerns. He said he'd tried murder mysteries, but that they're not exciting enough. Imagine, all that going on and he sits there reading Westerns! Norman came from somewhere about the Midlands, I think. Right from the beginning he maintained his firm intention of remaining behind the curtains in his little navigation compartment getting on with his work, that he didn't want to see anything that was, in his own words, 'Going to blast up my arse'. Vernon was a very good pilot. When we went into the bombing run he responded rather immaculately to my 'left-lefts', 'right a wee bit' and so on. Then held her as steady as a rock. Does not comment much on the proceedings except to observe, on occasions when the flak is intense, 'Christ, look at all that shit coming our way.'

Vernon put the Lanc into a steep dive, pulling up at about 1,000ft. We headed for the nearest flak post which, very surprisingly, was barely camouflaged. I held off as long as possible then opened up with my twin Brownings, keeping the trigger pressed and taking a chance on their jamming. As we passed the post, Vernon banked slightly giving

later heard that the parachute had travelled all the way to London and it caused quite a commotion on arrival at the terminal. A search party was sent back up the line but by then Rackley was in bed in the Luton and Dunstable hospital.

On 22 June, 234 aircraft – including 119 Lancasters – attacked V-weapon sites and stores in northern France again. F/O Vernon's crew on 12 Squadron went to Wissant, as Campbell Muirhead recalls:

Another daylight op. Target sighted visually: followed my bombs all the way down and saw them explode. Hate to admit it, but I found that a most fascinating sight. Cover of Spitfires and Lightnings. A beautiful summer's afternoon with France below us seemingly stretching on and on and looking so delightful, so peaceful, so welcoming. I must visit that country (but not, I hope, by courtesy of the Germans!) Yes, if I make it, I'm going there after the war. God, what we're all going to do after the war! Chaps in the Mess sometimes talk about after it's over: but only half-heartedly – got to survive the damn thing first.

Horsfall the opportunity to bring his two Brownings to bear. Then, once over it, Cartwright, from his rear turret, let fly with his four guns. Vernon next pulled her round and we did the same from the opposite direction. As we did so, we noticed another Lanc similarly on the strafe nearby. Must give the German gunners their due: no running away – they kept on firing all the time, even when we were bearing down on them with guns blazing.

I don't know for sure whether we knocked out those flak posts or not, our guns firing, of course, only .303 bullets. And, naturally, we didn't mention this little bit of what might be termed 'free enterprise' at debriefing. Certainly not, they view that kind of thing somewhat sourly; you might lose them one of their precious Lancasters. (And just think what 'Butch' [Harris] would say were he to learn that he'd lost one of his heavies, grudgingly diverted from the bombing of Germany, as a result of unauthorised activities over France. God, he'd have had a flaming fit!)

On the night of 23/24 June, 390 heavies and 22 Mosquitoes carried out attacks on four V-1 sites in the Pas-de-Calais, and 203 Lancasters and four Mosquitoes of 1 and 5 Groups set out for rail yards at Limoges and Saintes. All four flying-bomb sites were hit and the rail yards were bombed with great accuracy. A Lancaster was lost on the l'Hey operation; two Lancasters were missing from the raid on Saintes and four failed to return from the raid on the flying-bomb site at Coubronne. One of them was *H-Harry* on 35 Squadron, flown by S/L George Francis Henry Ingram DFC. F/Sgt Robert M.J. Gill DFM, the mid-upper gunner, recalled:

We were on our way back from the target the second time about ten minutes later; we were hit from underneath by a fighter with *Schräge Musik*. I heard the thump, thump, thump of cannon shells and then the aircraft was alight in the wing and fuselage. The ammunition was going off and I lost my eyebrows and eyelashes and some hair and skin was burned … I went down to the fuselage door and kicked it but I couldn't get it open so I went forward over the main spar and through the open hatch at the front. Just after I left I saw the flash as the aircraft exploded and the wreckage came down at Nieuport, just across the river from Dunkirk. The pilot and rear gunner [P/O Douglas Alfred Weatherill DFC] were still onboard.

A veteran Lancaster on 300 'Masovian' Squadron wrecked at Faldingworth after a belly landing. In April 1944, 300 Squadron at last converted from the Wellington and became the only Polish Lancaster squadron in Bomber Command.

Gill and the five other members of the crew were soon captured and spent the rest of the war in captivity.

One of the missing Lancasters on Saintes was ND528 *B-Baker*, which was recently given to P/O Vernon's crew on 12 Squadron as their 'own' Lancaster, but as the crew were stood down this night *B-Baker* was allocated to a F/O Frank Armstrong Jeffrey RCAF and crew for their second op. 'Don't know what the target was,' recalled Campbell Muirhead:

being stood down, we didn't bother to find out. Looks as if that second op was also their last: *B for Baker* reported missing with no word of crew. Next day still no word so that's definitely curtains for *B-for-Baker* and seven inexperienced air crew members. One can only hope they've ended up as PoW.[4]

On 24 June, 321 aircraft – including 200 Halifaxes and 106 Lancasters – attacked three V-1 sites in the clear in the Pas-de-Calais. All targets were accurately bombed and no aircraft were lost. Sixteen Lancasters and two Mosquitoes of 617 Squadron returned to the flying-bomb site at Wizernes escorted by 11 squadrons of Spitfires. This time the target was clear and Leonard Cheshire dived to mark the huge dome with spot fires, but his stores hung up and he called in Gerry Fawke, who dropped his smoke bombs as an area marker instead. His Mosquito was hit several times by flak, as were four Lancasters, but he accurately marked on the lip of the dome. The bombing was excellent and well grouped. A hit at the top had caused a big landslide, while others bombs had destroyed secondary installations and work at Wizernes continued. The heavy anti-aircraft defence was very accurate and the long, steady and sustained run-up that the Stabilized Automatic Bombsight demanded for its superb performance resulted in more damage than usual being sustained by the Lancasters; one was shot down. A few hours later, 739 aircraft of all groups attacked seven V-1 sites. It was a clear, moonlit night and 22 Lancasters were lost, mostly to night-fighters (617 Squadron returned to Wizernes on 17 July and again on the night of 20/21 July with 1 Group, after which the site was abandoned by the Germans).

On Sunday 25 June, 323 aircraft – including 106 Lancasters – carried out what is believed to be accurate raids on three V-1 sites. Early in the morning 17 Lancasters, each carrying a 'Tallboy', and two Mosquitoes were dispatched to a flying-bomb launching bunker under construction at Siracourt in the Pas-de-Calais. W/C Cheshire accompanied them in a P-51 Mustang with the purpose of using it as a low-level marker aircraft in the early morning attack. Cheshire, who had not flown a single-engined aircraft for three years, took off an hour after the rest of the aircraft and caught them up on the way. He dived the Mustang from 7,000ft to 500ft through accurate heavy flak to drop his two under-wing red spot fires and was followed by S/L Dave Shannon in one of the two Mosquito marker aircraft carrying four smoke bombs and two red spot fires each. They were so accurate that the second Mosquito did not need to mark. The aiming point was easily identified and 12 'Tallboys' burst close to it, four of them very close. When the smoke cleared there appeared to have been one direct hit on the concrete bunker. The 'Tallboy' had penetrated the 16ft-thick supposedly bombproof roof and had caused a massive explosion. Other bombs had fallen on the east and west sides of the building. Cheshire landed safely back at Woodhall Spa at 1018hrs.

On the night of 27/28 June, over 720 aircraft attacked railway targets at Vaires and Vitry. Campbell Muirhead on 12 Squadron flew his 13th op, on Vaires:

> Good concentrated bombing. Boring, this effort. Funny, but before we took off, there was everybody saying well, it's our 13th, let's hope it's not the unlucky one; why can't we call it number 12A? Almost jittery simply because of that '13'. Then during it no flak, no night-fighters and just the odd searchlight. We had a feeling of anti-climax, I suppose. Yet, one a/c (P/O M.A. Guilfoyle) from here missing. Couldn't have been flak so practically the only conclusion can be that

he must have strayed out of the stream and that, although we didn't sight any night-fighters, they were around somewhere, maybe not in the vicinity of Paris, but perhaps spaced out along the stream. Hope that he and his crew are in the bag.[5] One thing of beauty: dawn breaking as we crossed back over the French coast. The colours were magnificent. A feeling of serenity flying there at 13,000ft with another Lanc or two for company and there is this wonderfully beautiful dawn ahead of us. Almost welcoming, it seemed. A less attractive welcome as we dropped down through the cloud to find it bucketing down over Lincolnshire.

While the yards at Vitry were hit only at the western end, the attack on Vaires was particularly accurate. Four Lancasters failed to return from the raids.

On 28/29 June, 230 bombers hit the railway yards at Blainville and Metz for the loss of 18 Halifaxes and two Lancasters. On 30 June, two German Panzer divisions, the 2nd SS 'Das Reich' in Toulouse and the 9th *SS*, were believed to be en route to Normandy and would have to pass through a road junction at Villers-Bocage in order to carry out a planned attack on the Allied armies in the battle area. Some 258 aircraft dropped 1,176 tons of bombs on the junction. The Master Bomber ordered the bombing force to drop down to 4,000ft in order to be sure of seeing the markers in the smoke and dust of the exploding bombs. Two aircraft were lost. One eyewitness reported that he had seen the remains of a German tank strewing the top of a two-storey building. The planned German attack was called off.

July began with a series of heavy raids on V-1 flying-bomb sites in France save for an attack by 467 aircraft on 7 July when they were dispatched to help relieve the situation at Caen. On 2 July, Campbell Muirhead was on the operation on the V-1 supply site at Domléger:

> Another of these robot-plane sites. Those things must be doing an awful lot of damage judging by the number of times we have to bomb them. Sky cloudy, but site well plastered. And, of course, so much fighter cover we're in danger of being crowded out of the bloody sky. What a beauty that Spitfire is! She's so graceful. (But then, so is our Lanc.) Yet, despite the beauty of the Spit and the fact that they're swarming all around us, I still don't like these daylight efforts. And never will. I announce, after I've dropped my bombs, closed the bomb doors and had a quick look up the bay for

hang-ups that today is my 23rd birthday. To which Horsfall retorts that surely I don't expect a present from him? I reply that from a Yorkshireman you don't expect anything other than brass neck. Vernon cuts into this cross-talk by mildly observing that he wouldn't like to bet on my making my 24th, that we are having it easy on those French robot-plane sites etc, but that when they get back to putting us on to devastating the Third Reich my chances of making that 24th will be reduced by, he reckons, about 75 per cent. He doesn't mention that that pessimistic percentage applies to him also. All this cross-talk is very unprofessional. Shouldn't be indulging in it. You're not 100 per cent on the ball for eventualities when you're 'yacking' away over the intercom. Sign of over-confidence: you could pay for it.

On 4 July, 328 bomber crews – mostly Halifaxes – attacked V-1 flying-bomb sites in northern France without loss. That night, 282 Lancasters and five Mosquitoes of 1, 6 and 8 Groups attacked the railway yards at Orléans and Villeneuve. At Orléans the bombing height was 18,000ft but at North Killingholme, P/O E.S. Vaughan's crew on 550 Squadron, who normally flew *P-Peter*, were given *S-Sugar* because after an air test it was found that *Peter* would not fly above 16,000ft. *S-Sugar* was brand new and had only arrived at the crew's dispersal number 13 that morning. Vaughan's crew had arrived on the squadron soon after the Normandy invasion and had been alarmed to find that that their dispersal number was 13 but Sgt Tom Elliott, a pre-war regular from Berwick-on-Tweed and the WOp/AG on the crew, seemed to lead a charmed life. 'At OTU everyone was sent into a hangar to "sort themselves into crews". I "crewed up" with six other sergeants but my best mate said that he "didn't fancy my chances with that lot" and he joined a crew which included some very experienced officers. He was lost after 14 ops.' Vaughan's crew would complete their tour but when the pilot took *S-Sugar* off for Orléans the Lancaster began to swing alarmingly. Instead of shutting off the power and aborting, Vaughan opted to carry on and just as he hauled the bomber clear of the ground *S-Sugar* went out of control and crashed at the end of the runway. A wheel came off, an engine flew into the air and finally the Lancaster came to rest with its back broken. As the crew walked away, *S-Sugar* burst into flames.

S/L Noel 'Paddy' Corry, a veteran of well over 1,000 flying hours (including 300 hours on night-fighting operations on 25 Squadron during the Battle of Britain) who joined

Minnie the Moocher, a name derived from 'Cab' Calloway's popular slow blues song, and ground crew on 12 Squadron at Wickenby in July 1944.

12 Squadron as 'A' Flight commander on 31 May, flew his eighth raid on 4 July (a night of the full moon):

It was a six-hour-ten-minute trip to bomb the marshalling yards at Orléans. I and 'Wally' Waldron (having a break from his W/Op duties and on the look-out while stretching his legs), spotted a Fw 190 a little below us in the moonlight, but we immediately altered course to slip away from him and there was no combat. Others were not so fortunate that night for we saw three aircraft shot down over the target.

Orléans and Villeneuve were accurately bombed but 11 Lancasters were lost on Villeneuve and three on Orléans. Another 231 Lancasters and five Mosquitoes of 5 Group were given a special target: a flying-bomb site in limestone caves at St-Leu-d'Esserent near Creil, 30 miles north of the centre of Paris. The French originally used these caves for mushroom growing but the Germans had commandeered them for the storage of their V-1 weapons. The roof was about 25ft thick, consisting of a 10ft layer of soft clay surmounting 15ft of limestone. It was to be hit first by 17 Lancasters of 617 Squadron supported by a Mustang and a Mosquito and then immediately after by the rest of 5 Group carrying full loads of 1,000lb bombs, with some Pathfinders. The markers were accurately laid and the bombing was very concentrated, although six of the 617 Lancasters brought their 'Tallboys' back as smoke obliterated the red spot fires before they were able to bomb. One of the 'Tallboys' scored a direct hit on a cave entrance, causing a large subsidence. A near miss had also helped to render this entrance useless and the railway line had been destroyed.[6] Thirteen Lancasters were lost.

On Wednesday 5 July, HM King George VI, Queen Elizabeth and the Princess Elizabeth toured RAF Mildenhall before holding an investiture. The occasion made a big impression on 22-year-old Sgt 'Bill' Matthews, flight engineer on P/O Matthew Michael Golub RCAF's crew on 15 Squadron. The young man from Bow, Mile End, had been an enthusiastic cyclist and often cycled to Newmarket to take part in 25-mile cycle races and then cycled home to Mile End again afterwards. At the age of 15, he cycled on his own to the Isle of Wight for a week's camping holiday! After leaving school with a glowing report he had joined a company in Finsbury Square in the City as a trainee bookkeeper and accountant. He was called up by the RAF and served initially as a flight mechanic, but though he had promised his mother that he would not volunteer for air crew, after a time he did just that! Late that Wednesday evening Bill wrote a letter to his parents about the royal visit and that, all being well, he would be home on leave on the following Tuesday. At 2259hrs Bill joined the other members of Golub's crew and took off in *T-Tommy*. The night was clear and there was a bright moon when the Main Force left England. Golub's crew were part of a force of 542 aircraft of four groups – 321 of them Lancasters – detailed to bomb two flying-bomb sites and two storage sites at Wizernes in northern France. Another 154 Lancasters of 1 Group hit the main railway area at Dijon and all aircraft returned safely. The V-1 sites were hit and at Wizernes 81 Lancasters dropped mixed loads of 11 1,000lb American-made semi-armour-piercing bombs and four 500-pounders from 8,000ft after marking by five Mosquitoes, but three Lancasters were lost without trace and another was

brought down by flak. At Mildenhall there was no word from *T-Tommy*. Another crew saw LL890 get caught in searchlights as it went out across the Channel near Dunkirk after dropping its bombs on the site at Wizernes. All seven of the crew, including Bill Matthews, 'who would not change his job for the world', were dead.

On Thursday 6 July, 551 aircraft attacked five V-Weapon sites. W/C Leonard Cheshire accompanied 17 Lancasters and two Mosquitoes of 617 Squadron in an afternoon raid on the uncompleted V-3 long-range gun site at Marquise/Mimoyecques, 9 miles north-east of Boulogne and 95 miles from London, when he piloted a Mustang marker aircraft. The damage was later assessed as one direct hit against a corner of an 18ft-thick slab of concrete that had been cast to cover three of the five gun shafts on the hilltop, blasting a crater 35ft across and 100ft in diameter. It was later found that this bomb had also blown a cavity beneath the slab, causing part of it to collapse inwards. Up to 3,000 Germans and other workers were entombed when the water table was breached and flooded the lower workings. Another four 'Tallboys' were within 60yd of the aiming point, one being a near miss at the north end of the slab. All this led to the collapse of one of the gun shafts and some of the upper level of tunnels. Upon his return Cheshire was ordered to leave the squadron and rest. He had completed four tours and had flown at least 100 operations. On 12 July he was replaced as CO of 617 Squadron by W/C James 'Willie' Tait DSO* DFC.[7]

Meanwhile, on 7 July at dusk, 467 Lancasters and Halifaxes of 1, 4 and 6 Groups and 14 Mosquitoes of 8 (PFF) Group were dispatched to help the Canadian First and British Second Armies held up by a series of fortified village strong points north of the Normandy city of Caen. It was Campbell Muirhead's 19th op:

Another daylight op. We were told that very accurate bombing was essential: 'Bring the things back if you can't be sure of your aiming point.' This because we were bombing only 1½ miles ahead of our troops (Canadians, evidently). My sighting was not right so I didn't press the tit. I told Vernon to go round a second time. God, the language which came over that intercom! Interspersed with references not only to my complete inadequacy as a bomb aimer but also to my parentage. Can't exactly blame them; there we were, the only Lanc left over Caen and what flak there was beginning to concentrate on us. But there was no way I was going to drop 13,000lb of high explosive when there was the slightest possibility of the dreadful stuff killing or wounding our own troops: would have taken the load back to Wickenby first. However, I got a perfect sighting on that second run (despite what was still being said over the intercom) and placed my bombs exactly where I wanted them. Despite that second time round, on the way back we caught up with the stragglers (*B for Baker* being brand-new and with that extra few knots more than most). Flying almost level with another Lanc, who was limping. Just crossing the French coast when up came quite a scything of flak. It got him (not, by the grace of God, us). He started diving straight down. We counted four parachutes opening, prayed for more but to no avail. The Lanc struck the water, burst into a terrific sheet of flame. We all fell silent until over base, everybody thinking of the three men inside.

Watched by cheering troops, in clear weather conditions the Main Force dropped 2,500 tons of bombs on to well-laid markers. Caution had dictated that the bomb line should be well ahead of the British forward positions so the aim points were situated more in the city centre rather than on the German positions on the northern outskirts. Smoke and dust soon obscured the markers but the Master Bomber, W/C 'Pat' Daniels of 35 (PFF) Squadron, returned the bombing to the target area. 'OK, fellows, find a nice spot and lay them down gently.' There was little German defence in Caen itself and because of a six-hour delay between the end of the bombing and the ground attack, only newly arrived troops of the 16th Luftwaffe Field Division were affected by the bombing (the 12th SS Panzer Division had recovered before the Allied troops attacked). After some fierce fighting, the Allied troops finally gained the centre of the city, only to be delayed by rubble.

During the air attack, intensive flak from the south of the city was encountered, as F/Sgt Eric Haslett, a flight engineer on 626 Squadron, recalled:

This should have been an easy one. There was a vast concentration of bombers and as we approached the coast the Germans let up a huge amount of flak. I had never seen so much before. One could almost step out and walk on it. We were hit by the flak many times and a small piece came through the fuselage on my side, passed by me, went through the navigator's legs, through the radio operator's legs and was found afterwards in the radio operator's upholstery.

Despite the flak we bombed on target and made it back to Wickenby.

Minnie the Moocher on 12 Squadron over a V-1 site in July–August 1944.

Only two Lancasters were shot down. A 550 Squadron Lancaster crash-landed at Manston on return. That night, 123 Lancasters and five Mosquitoes of 1 and 8 Groups accurately bombed the railway yards at Vaires without loss, and 208 Lancasters and 13 Mosquitoes, mainly from 5 Group but with some Pathfinder aircraft, attacked the flying-bomb site at St-Leu-d'Esserent again. Bombs destroyed the mouths of the tunnels and the approach roads so that the enemy was no longer able to use the flying-bomb site. Its stores and the weapons were transferred to the Nucourt caves, the entrances to which were attacked on Monday 10 July and again on Saturday afternoon, 15 July. The closure of St-Leu-d'Esserent came at a price: 32 Lancasters were shot down by night-fighters, though one was believed to have been shot down by another Lancaster. Five out of 16 Lancasters dispatched by 106 Squadron at Metheringham were lost.

By the middle of July, when they could be identified and marked, railway targets in France were being bombed on a regular basis. On the 12th/13th, 378 Lancasters and seven Mosquitoes of 1, 5 and 8 Groups sought marshalling yards at Culmont, Revigny near Nancy, and Tours. Culmont and Tours were accurately bombed but cloud interfered with the

Opposite: Jack Ryan
RAAF with air and
ground crew on
576 Squadron.

all-1 Group raid at Revigny. Half of the force brought their bombs home and those that had dropped them did little damage to the railways. Two nights later, 242 Lancasters and 11 Mosquitoes returned to Revigny and also to the railway yards at Villeneuve. At Villeneuve the railway was hit, though much of the bombing fell to the east of the target. The raid on Revigny was abandoned because the railway yards could not be identified. Seven Lancasters failed to return from the raid. On the night of 15/16 July, 222 Lancasters and seven Mosquitoes attacked the railway yards at Châlons-sur-Marne and Nevers, neither of which was successful.

On 18 July, 1,032 bombers attacked five fortified targets south and east of Caen in support of Operation Goodwood, the imminent armoured attack by the British Second Army. Donald Falgate, a bomb aimer on 463 Squadron RAAF, was on one of the 126 Lancasters of 5 Group[8] that attacked Colombelles at altitudes of 6,000ft–9,000ft, dropping mostly 1,000lb bombs on to prepared positions of the 16th Luftwaffe Field Division:

It was the first attack I made in daylight conditions, so I could see the target, which was a very large steelworks where the Germans had dug themselves in. We were the first aircraft to bomb. It was quite something to see the bombs going down and actually hitting the target. I had not actually seen that on any of the night raids. Because the raid was low-level and the steelworks were so big, there was no way you could miss. I could see all the German tanks entrenched. It was very satisfying to see the stick hitting them. The opposition was insignificant.

Bomber Command dropped more than 5,000 tons and American heavy and medium bombers, 6,800 tons. Despite heavy flak around Caen, only six aircraft – five Halifaxes and a Lancaster – were lost and four more bombers were lost on their return over England.

Crews returned to be told that they were 'on' again that night. The 5 Group target was a railway junction at Revigny. In all, 972 sorties were flown that night, the two biggest raids being oil plants in Germany by 1, 6 and 8 Groups. Five heavies were lost on the raids on the oil plants. Two Lancasters were lost on the raid on the Aulnoye railway junction, which formed part of the attack on Revigny by a total of 253 Lancasters and ten Mosquitoes of 1, 3, 5 and 8 Groups. At Revigny the illuminating flares went down over the 'brilliantly lighted target' but no target indicators followed. The crews could

hear the Master Bomber and his deputy talking but a German night-fighter hit the Master Bomber's Lancaster and the marker flares 'went down all right'. A few crews bombed the flares in the burning aircraft but they were not in the railway yards. Fighter flares were then dropped, air battles broke out and 24 Lancasters went down in flames. At Dunholme Lodge 619 Squadron lost five of its 13 aircraft taking part in the raid. The Deputy Master Bomber called to abandon the attack. By the end of the night 'trains were still running through Revigny'.

Lancaster III *J-Jig* on 300 (Polish) Squadron at Faldingworth in Lincolnshire, piloted by F/Sgt Z. Stepian, was one of just over 940 aircraft bombing fortified villages at Émiéville near Caen on 18 July when it was badly damaged by flak. The rear gunner, F/Sgt M. Zentar, was rotating his turret searching for fighters at the time and the blast swung the turret beyond its usual position, ripped open the door and sucked him out of his seat. He fell out, but his left foot jammed in the doorway and there he hung head downwards. F/Sgt Derewienko, the mid-upper gunner, and Sgt Jozef Pialucha, the flight engineer, went to his aid but could not pull him in. His foot began to slip out of the shoe, so one of them grabbed his trousers, which, however, began to tear. Pialucha then clambered out (the aircraft was now over the sea), precariously held in place and looped a length of rope round the rear gunner, which he then made fast to the seat. He then returned to his task of nursing the damaged Lancaster back to England. The Lancaster limped home with the rear gunner hanging head down from the tail and those watching at Tangmere airfield saw him swing his head to one side to avoid hitting the ground as the bomber touched down. He was bleeding from ears and mouth, but was not badly hurt. He was subsequently able to boast of being the only man to have flown upside down from Caen to Great Britain. Sgt Pialucha was given the immediate award of the Conspicuous Gallantry Medal for his outstanding courage and initiative.[9]

That night, 194 aircraft – including 77 Lancasters – attacked the synthetic-oil plant at Wesseling, and 157 Lancasters and 13 Mosquitoes bombed another oil plant, at Scholven-Buer near Gelsenkirchen. One of the Lancasters that attacked Scholven-Buer was *B-Baker*, carrying a 4,000-pounder and 18 500-pounders, piloted by F/O Vernon, as Campbell Muirhead, who flew his 20th op, recalls:

Got it again: that knot in the stomach, the instant when, at the bomb aimers' pre-briefing, the target was revealed as Gelsenkirchen. Suppose it was brought about by the

D-Dog Mk.II on
460 Squadron RAAF.

recollection of our earlier raid on that city plus the acceptance that all the easy stuff over France was now finished and that this was the resumption of the big ones on Germany. Stayed with me during the main briefing right until we took off. Then, true to form, it disappeared. Must have a thing about this knot in the guts, the number of times I refer to it. The main briefing, I must record, was just as intense and detailed as the earlier one for the same target. Routes were slightly different, but not much more, and about the only humour around was when one pilot observed plaintively that he had bombed that bloody oil refinery only about five weeks ago and was told that if he and his 'oppos' had bombed the bloody place properly that time we wouldn't all be having to go back there tonight. He accompanied his nodded acceptance of this with a request for a promise that if we knocked out the oil refinery completely tonight we wouldn't have to go back a third time. His request was ignored.

Quite a lot of light flak along the route, but very heavy stuff flung up over Gelsenkirchen: but having said that, it didn't appear to me to be nearly as heavy as it was when we visited there on 13 June (God! that was all of five weeks ago; our eighth one and now we're on to our twentieth). It could be that I only thought it wasn't as heavy, that earlier effort over the place being our first German raid and in any case now being more used to having flak bursting all around us. Anyway, it was heavy enough to tear a large number of holes

out of the Lanc. (Was illogical enough to feel anger when I examined them. Imagine, actually trying to wreck beautiful B-for-Baker!)

That target got well and truly plastered. Those fires! Like some Dante-visioned Hades. As we approached Gelsenkirchen we saw perhaps half a dozen heavy fires raging in different parts of the city (maybe where the oil refinery is, maybe not) which had been started by the first wave. Fires which were all converging on each other to such an extent that, by the time we were over the target, it was simply a sea of leaping flames. I dropped the load right in the middle of it.

Can understand now these stories we hear from time to time about aircrew baling out over a city which they have destroyed like this and being strung up by civilians from the nearest lamp post. Could be true enough. I suppressed the faint niggling at the back of my mind re women and children. No faint niggling this time; didn't even give it a thought. Wonder, as I write this, if I'm becoming hardened or am just not allowing myself even to contemplate the number of civilians these bombs of mine must have killed. And it is 'must have'; you can't drop a 4,000-pounder and eighteen 500s and fool yourself that the only harm you did was to destroy either part of an oil refinery or some uninhabited buildings. Anyway, the Germans are doing the same thing to us.

I had thought that the German night-fighter controller's job was only to guide his flock on to us, but this character over the Gelsenkirchen area extended his remit to include quite a harangue (in a most beautifully modulated English accent, probably acquired at Cambridge or Oxford). He told us that there were no military installations in Gelsenkirchen and that we were killing innocent women and children, that when we were shot down by the valiant Luftwaffe they (the Luftwaffe presumably) couldn't be responsible for our safety at the hands of those we were bombing so indiscriminately. He then went on to call us, among other things, 'English Terror Flyers'. I asked WOp to put me on the air for a quick second. Vernon said no, to cut it out, but I persisted. When Dunn said OK, I was on, I pointed out to the controller that there were Scots as well as English attending the proceedings. A pause, then a somewhat petulant, 'Oh, very well, Scots Terror Flyers also.' I felt much happier at this inclusion; one doesn't like all the credit to go to the Sassenach.

But am hellishly tired: which is why I haven't recorded any losses we might have had, can't be bothered going down to the interrogation room to find out. Went to bed but couldn't sleep because of my mind dwelling on that horrific 4,000-pounder I'd dropped on Gelsenkirchen. The bomb had a delayed-action device on it. When I fused it in the Lanc I really primed it so that it didn't go off when it hit the deck, neither when moved would it detonate, but the moving of it, however slight, activated the delayed-action mechanism so that about 20 minutes later it exploded. Nice, pleasant, friendly world we live in, eh? (One of the Intelligence types told us that the Germans get concentration camp prisoners to move those monstrous things, promise them food for doing so. Yes, a nice, pleasant, friendly world all right). Back to bed again now. Still got ten to do. Hope no more like this. No, still can't sleep. So might as well keep on writing until the Mess starts serving breakfast. Well, that's two-thirds of the tour over (always was quite good at arithmetic!) and we've been very lucky having had only two of that lot over Germany. Looks, however, as if all the rest might be over the Reich, a reflection which keeps one rather subdued.

On 20/21 July, 971 sorties were flown for the loss of 30 Lancasters and seven Halifaxes. Eighty-seven heavies bombed V-weapon sites at Ardouval and Wizernes, and 153 heavies and 13 Mosquitoes of 4 and 8 Groups attacked the synthetic-oil refinery at Bottrop, while another 147 Lancasters and 11 Mosquitoes of 1, 3 and 8 Groups attacked an oil plant at Homberg. A diversionary sweep was made over the North Sea by 106 aircraft from training units, while Lancasters and Mosquitoes made a 'spoof' raid on Alost. The marshalling yards at Courtrai in Belgium were bombed by 302 Lancasters and 15 Mosquitoes of 1, 5 and 8 Groups. One of the Lancaster crews that took part in the Courtrai raid was F/O Vernon's, as Campbell Muirhead recalls:

Target area appeared very quiet when first wave was over it. But the second wave, of which we were part, encountered quite a bit of fighter opposition. Many fighter flares dropped. We did some corkscrewing and were not attacked. But we don't get off scot free from those comparatively quiet efforts on marshalling yards: Nine missing from this raid including three from Wickenby. The entire night's ops, which included raids on the Ruhr, buzz-bombs sites, etc, cost the RAF 31 a/c missing. That's over 200 blokes. Not all killed, of course, but for the majority it's the chop. A sad, sad, note: our old friend H for Harry who saw us through four of our earlier ops bought

Goofy carrying a bomb and operational symbols on a 460 Squadron RAAF Lancaster.

it on this one. Truly, feel like weeping. Also for the crew – their very first operational flight. This old game, I suppose: the older, almost clapped out, kites go to the 'sprog' crews, the powers-that-be reckoning that green, inexperienced aircrew are more likely than anybody else to get themselves shot down and that, this being the case, it's going to be an old Lanc that buys it, not a brand-new effort equipped with H_2S and such-like. Brutal, but I suppose it has to be. After all, that applied to us also at the end of May/early June.

A different atmosphere on this one. There we were, after Gelsenkirchen, absolutely convinced that from now on it was the Reich for us and that we had kissed goodbye to those easier French jobs. And all becoming unwound on discovering that we were back on the marshalling yards efforts (in this case a Belgian rather than a French target, but there is no difference, really).

Too much chat when heading for the enemy coast all because of this unwinding. I contributed to it by adding my piece that chat on the intercom meant less vigilance and that it was high time everybody dried up. Vernon took the hint and ordered no talking unless necessary. It was about thirty seconds later that Horsfall, obviously having spotted what he thought was an enemy fighter, bawled out: 'Corkscrew, corkscrew' and added: 'For f***s sake!' His extra three-word

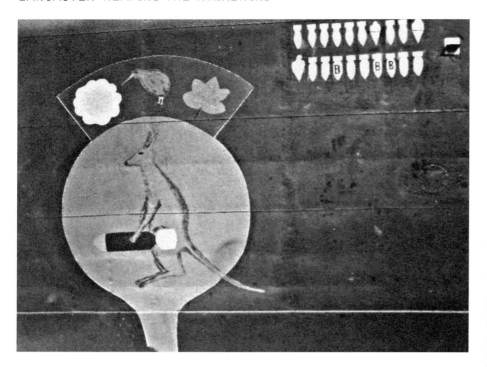

Kangaroo holding a bomb, a Kiwi and a Canadian maple leaf, denoting a multi-national Lancaster crew on 460 Squadron RAAF.

As the blaze intensified, the skipper gave the order to bale out. Blood was pouring from my wounds but I made my way towards the escape hatch at the front of the Lancaster where I was greeted by the strange sight of the bomb aimer crouched over the hole in the floor where the escape hatch should have been. When he saw me, he screamed at me to stay back and not come any closer. I ignored this remark and proceeded forward, but then I stopped when I saw he had drawn the six-inch knife that we all carried as part of our escape equipment. With his face creased in anger, the bomb aimer explained that in his panic he had accidentally dropped his own parachute out the hatch. He then warned that 'if he couldn't escape no one else would'.

West had no option but to retreat and inform Gaston of the bomb aimer's actions. By this time, the engineer had managed to put out the fire, enduring severe burns in the process. In the circumstances Gaston decided that, though the Lancaster was losing height, he would try to keep flying as long as he could. Eventually they reached the English coast, with the Lancaster almost touching the water of the North Sea. As the aircraft struggled towards Woodbridge, the two remaining engines stuttered, while the undercarriage would not lock down because of the lack of hydraulic fluid. The pilot then made a crash-landing only for the undercarriage to collapse and the bomb bay doors to swing open. Once the aircraft had come to a halt after falling over to the port side, the injured crew was taken to hospital.[10]

Nine Lancasters were lost on the Courtrai raid, and 21 Lancasters and seven Halifaxes went missing on the Bottrop-Homberg raid. At Mepal 75 Squadron RNZAF lost seven of its 25 aircraft. On 23/24 July, over 620 bombers were detailed to hit Kiel in the first major raid on the city in two months. Attacks by other forces on an oil refinery and storage depot at Donges, near the mouth of the River Loire, and flying-bomb sites in France were also ordered. At the target no contact could be made with the Master Bomber and his deputy took over. Marking was by H_2S and considered accurate, although bombing was rather scattered in the early stages of the attack. Both marking and bombing appeared to become more concentrated towards the end of the raid and many explosions and fires were seen. In a raid lasting only 25 minutes, 612 aircraft got their bombs away on the glow of green and red TIs and sky-marking flares, green with red stars, as seen through the 10/10ths cloud that covered the target.

exhortation was quite unnecessary because even before it Vernon had put her into a headlong dive to starboard. But, although it maybe sounds self-satisfied to be writing so, it was complete justification of my earlier bit about less vigilance if chatting. Better to be self-satisfied than dead.

Forgot to moan that we nearly got bombed by a 'friendly' aircraft on that one. Not quite on the run-up when Horsfall yells out that there's a Lanc right above us with its bomb doors open. (It must have been almost on top of us if he could see that.) We alter course slightly to take us out from beneath it so that when he drops his load none of it will hit us on its way down. In fact, quite a few of our aircraft have been struck by bombs from above. Not surprising considering the numbers which are usually milling around over the target. Don't think the bombs actually explode, but can cause problems.

On Homberg, Sgt Jack West, navigator on F/Sgt J.M. Gaston RCAF's crew on 115 Squadron at Witchford, recalled that his Lancaster was fired on by flak and attacked by a fighter which destroyed most of the instruments, stopped two of the engines, started a fire in the cockpit and killed two of the crew:

Severe damage was caused to the north-east portion of the Deutsche Werke shipyards, and hangars at Holtenau airfield were partially destroyed. A considerable amount of damage was done to the facilities and barrack area near the torpedo boat harbour and moderate damage to a torpedo components and electrical signalling works. Sixteen medium-sized buildings in the Marine Artillery Depot were partially destroyed. Four Lancasters were shot down on the operation without a single survivor among them.

S/L 'Paddy' Corry on 12 Squadron was on his 11th Lancaster raid. He recalls:

We took off at 22.30 from Wickenby and we were airborne for a total of five hours. I have no record of the actual time of the combat, but we had a determined attack from a Ju 88 night-fighter shortly before reaching the 'enemy' coast. There was a new moon and Sgt Frank 'Piwi' Faulkner the rear gunner saw the Ju 88 coming in from about 800 yards on the port quarter at our level. 'Piwi' opened fire and alerted the crew and Sgt 'Ozzy' Osborne our mid-upper gunner also opened fire. 'Piwi' called immediately for 'Corkscrew port'. I threw the Lanc into a corkscrew to port – it was my practice to yell, 'Brace, brace, corkscrew!' as I rolled the aircraft into the vertical to start the corkscrew – but the Ju 88 persisted with his attack, despite the ferocity of our evasive action, through two complete corkscrews and had closed up with us considerably when suddenly all our rear turret guns went u/s and ceased firing. Fortunately, just then the Ju 88 broke away to starboard down into dark sky, the enemy pilot probably deciding that it would be prudent to seek another target with less alert gunners. There was no indication to our gunners that they had hit the Ju 88 severely; during the combat 'Piwi' fired 400 rounds and 'Ozzy' 200 rounds.

We only had one other actual combat with a night-fighter; on the night of 23 September, while we were on a five-hour-15-minute trip to bomb Neuss, near Düsseldorf. My gunners saw the twin-engined night-fighter as he swung in to attack us and gave him a long burst as he scampered away after we had taken violent evasive action. Again, I think the night-fighter pilot broke away into the darkness to seek easier prey.

The oil refinery and storage depot at Donges was attacked again on Monday night, 24 July, by just over 100 Lancasters of 5 Group. Another 461 Lancasters and more than 150 Halifaxes

L-Leader on 460 Squadron RAAF.

were dispatched to Stuttgart in the first of three heavy raids on the German city in five nights. Three aircraft were lost on the raid on Donges and 21 aircraft – 17 of them Lancasters – failed to return from the attack on Stuttgart. Campbell Muirhead recalled the raid:

E/A were all over the place. Saw interceptions take place even before we'd hit the enemy coast, they seemed to be coming out to sea to meet us: indeed, we were stalked while still over the water but by violent corkscrewing were able to ditch him. Another interception when only a few miles inland: saw a flurry of tracer just slightly to starboard, but it looked like two-way stuff (that Lanc would have been much better corkscrewing rather than trying to fight it out). Someone else coming up behind us shortly after that – could have been the same fighter, of course – so off corkscrewing again. Didn't see anybody buying it, though, but that could have been because of the heavy cloud all around us. Bombing of Stuttgart not as concentrated as it might have been, again owing to the heavy cloud. Purely an area bash, this: nothing military about it. Saw what looked like a Me 110 when leaving the target: was flying on the same course as

MacNamara's Band on **101** Squadron in a hangar at Ludford Magna. Note the 'crochet' bomb symbols.

There followed a raid next day on aircraft assembly depots and wireless/telegraphy stores at St-Cyr on the outskirts of Paris by 100 Lancasters and Mosquitoes of 5 Group. The only Lancaster lost on the entire operation was on 463 Squadron. Another Main Force raid took place that night on Stuttgart again, with two other forces going to the Krupp benzol plant at Wanne-Eickel and to three flying-bomb sites in France. The eight Lancasters and four Halifaxes that failed to return were all lost on the raid on Stuttgart. Campbell Muirhead recalled the raid:

Quite a muttering when, at Briefing, the cover over the Order of Battle was whipped off to reveal the name of the target. 'But we were there only last bloody night,' complained a pilot in the front row. He was congratulated on such a feat of memory and informed that if Stuttgart wasn't flattened on this effort he might find himself over the place for even a third night. But Germany two nights running. I think all of us in the crew were now almost reconciled to Germany for our remaining seven. Lucky seven? Could be.

As opposed to the previous night, there were very few night-fighters. No doubt German Control didn't reckon on us bombing Stuttgart two nights running and vectored their night-fighters elsewhere. There were quite a few fighter flares along the route and all sorts of coloured things dropped on us, but that, I think, was about all – the Luftwaffe was certainly not out in force. There were massive fires this time. The fires multiplied as we approached the target and each wave of Lancs deposited its 2,000-pounder and its clusters of incendiaries. An out and out area bash, of course. We could see the fires, despite the amount of cloud, for well over 100 miles after leaving the target.

Age 23 and your sole purpose in life, apart from trying to keep it, is to convey that bastard 2,000-pounder of High Explosive plus the J-type canisters crammed with their dreadful incendiaries to Stuttgart. The HE strikes first, tearing buildings apart; the incendiaries come next, setting aflame the resulting masses of rubble. That's what's awaiting them at Stuttgart: brought there by me and hundreds of others like me. Also wonder whether the sirens have already sounded there – once you're well into Germany their Controllers are able to calculate what your likely target is – and whether people are already cowering in air raid shelters waiting for the fearful onslaught. People, that's who you're after as well as industry, houses, schools, anything that will burn: men, probably mostly elderly, the fit being

ourselves and at about the same height. Some heavy cloud nearby so we slid into it.

Odd how, despite the heavy cloud over Stuttgart, the TIs seemed to be brighter than ever before. These were usually dropped by the Pathfinders and are, in effect, bombs which explode at a set height above the target, releasing ignited pyrotechnic candles which, in turn, slowly cascade earthwards spreading as they do so. When they land they form wide circles of bright fire of varying colours, easily seen from the air. If there's a Master of Ceremonies present he will instruct by wireless which colours to bomb. There may be circles of reds and you're told to drop your loads on that colour. Then this almost inevitable 'creep-back' could occur as a result of bomb aimers releasing their bombs prematurely and causing the approach to the target to bear the brunt of the attack rather than the target proper. In such cases the MC might drop a fresh set of TIs – say coloured green – and intimate that the reds were now to be ignored and that the greens only were to be used as aiming points. But they were certainly brighter than I've ever seen them before. Probably new models with more powerful candles. They're always 'improving' on any items which will hasten the destruction of the Reich that was, according to Adolf, going to last for a thousand years.

in the Forces, women maybe some of them factory workers but probably most of them older. And children. Children of all ages. I know the Germans brought it upon themselves! I know nobody can stop it until they surrender. But, God, it's a terrible killing time. Almost glad when some event around us jerks you away from such a gloomy reflecting. You really can't afford such brooding.

Late in the afternoon of Wednesday 26 July, 178 Lancaster and nine Mosquito crews of 5 Group were alerted for an attack on the rail yards at Givors near Lyons. At Coningsby newly promoted F/L Bill Siddle and his crew found themselves on the battle order for the third night out of four, as W/O Clayton Moore recalls:

We lifted off just after 9 p.m. on what was to prove one of the most tiring and exacting eight-hour flights we'd ever undertaken. The Met office had got it wrong this time and we found ourselves flying in thick cloud for most of the outward flight. Then, just as we neared a point about half an hour from the target, we started to ice up badly, so Bill decided with reluctance that we should abort the mission and head back to base. Despite the gloom and murk that was all around us, I continued to carry out a search for any fighter that might have managed to get off the ground in the prevailing conditions. I had my turret on the starboard beam when I saw what looked like the lighted windows of a farmhouse drifting past beneath us. I at once called Bill and asked, 'What's our height, Skipper?'

'Fifteen thousand. Why?'

I reported what I had seen, then added, 'There goes another one and it's above us this time.'

Realising the full significance of the mysterious lights was immediate. Because we had turned around we were now flying directly into the path of the advancing bomber stream and the 'lighted windows' I was seeing were the exhaust flames of other Lancasters as they roared past us in the opposite direction and it was with some alarm that I envisaged the fireworks display that would result from a head-on collision between two bomb-laden Lancasters.

'Hang on, everybody,' Bill announced. 'We're going down,' and I at once felt my turret shoot skywards into the pitch darkness that surrounded us.

The manoeuvre served a dual purpose. Not only did it remove the danger of collision, but the lower altitude alleviated the icing problem. Furthermore, now that we were free agents, untroubled by a preconceived operational flight plan, we could plot our own course and altitude for the return flight. But the situation had some disadvantages. For one thing, we no longer had the collective security that the main bomber stream would provide. We were alone and our trace would stick out like a sore thumb on the German radar screens. But there was plenty of cloud cover around and this would make it difficult for an enemy fighter to locate us. There was one other problem: the fact that we carried a full bomb load, so it was decided that we would jettison this at the earliest. So it was that an unfortunate German village was treated to a somewhat noisy awakening as we made our way back home. Because our duties had not been carried out, I was led to understand that the flight could not be counted as an operational sortie, but I drew consolation from the fact that a number of other crews had also aborted.

Four Lancasters and a Mosquito failed to return.

Raids on V-1 flying-bomb sites followed and then, on the night of the 28th/29th, 494 Lancaster crews were given Stuttgart again. Another 307 aircraft – 106 Lancasters, 187 Halifaxes and 104 Mosquitoes – were to bomb Hamburg for the first time since the Battle of Hamburg in July 1943. At Waterbeach, crews on 514 Squadron gathered at the briefing to learn what their target was. F/L Alex Campbell RCAF, who was flying his 25th trip, recalled:[11]

All eyes were on the platform. Although small in stature and build, our squadron CO held the respect and admiration of us all – W/C Mike Wyatt had a flashing smile and an impressive display of gongs. A hush fell over the assembly. 'Good evening, chaps,' he began. 'You will be further tested tonight and it's with the greatest confidence in you that I shall outline the details of tonight's operation against the enemy.' Then the heavy black drapes were drawn aside to show a detailed map of Europe. On it was a bold zig-zag ribbon which traced a route from our base to the target – Stuttgart. The CO allowed the murmur and comments to subside before going on. 'I can't over-emphasize the importance of tonight's offensive.' How often had we heard that statement? We often heard other sayings like, 'the war effort … shortening the war … dedication to duty … maximum effort'. My mind, meanwhile, was busy re-enacting the trip to the same target only two nights before – there were the cloud formation, the fighters and the flak bursts. We saw railcars floating skyward like tiny matchboxes as our bombs dropped down. Coming home from every operation, there was always the grave possibility of ditching at sea as we were running out of fuel. We all wondered what was in store for us tonight.

My attention was jarred back to the briefing room by such phrases as 'bright moonlight … expected fighter activity … mobile anti-aircraft guns'. The nav leader then told us that the track and turning points would be identical to the previous three raids on Stuttgart. Surely the enemy would be waiting for us all along the route. Adding to this glum news, we were then informed that the cloud cover would be with us until the last fifty miles, when that would disappear abruptly and we'd have bright moonlight all the way to the target. This would expose us to the fighters there and back.

Once we were dropped off at dispersal, in front of our Hercules-powered Lancaster Mark II called Charlie Squared we checked the exterior of it over. We had certain rituals and nervous jokes etc. But once aboard, it was strictly business for everyone as more checks and inspections were done at each crew member's station. After starting up and taxying around the perimeter track; we arrived at the marshalling point. It was now time for Jock Donaldson, the flight engineer, to help me with the pre-take-off check: H-T-M-P-F-F-G-S. That stood for hydraulic, trim, mixture, pitch, flaps, fuel, gyro and switches. All was in order. Shortly after 22.00 hours the tower gave us clearance for take-off and a moment later I lifted the heavily laden Lancaster off runway 23.

Once we were in the air, it went like this:

Navigator to skipper, 'Set course 117 degrees at 7,200ft. 22.19 hours. Nav lights out.'

'ETA enemy coast 22.43 hours.'

'Roger, nav,' I said.

Navigator 'Judy' Garland was kept quite busy on most trips and depended on commentaries from the rest of us, who had visual contact with the outside surroundings. As a welcome cloud cover now gently engulfed us, I talked briefly with the crew members. 'Chappy' Chasinger had things under control, as usual, in the bomb aimer's compartment and he enquired as to the well-being of Sam Harvey in the tail gunner position. Sam gave us assurance that, if the rest of us did our jobs, we'd be OK from stem to stern. 'Jonesy' in the mid-upper turret cheerfully replied, 'Everything is copasetic [US slang for 'excellent'] skipper, thank you.'

Our cockney wireless-operator, 'Ben' Lyons, made his usual derogatory remark about Jock and his Scottish ancestry, as well as his ability to perform as our flight engineer. Occupying the engineer's usual seat was Bob Giffin, a new bomber pilot. He had re-mustered to aircrew and was on his first operational trip as a second dickey. We had just crossed the enemy coast when Bob remarked, 'If this is ops, we should have brought along our knitting.' I heard the click of more than one microphone as crew members prepared to respond. However, I quickly assured Bob that he wouldn't be bored for too much longer.

I was already becoming uneasy at the increasing amount of light filtering through the cloud cover. In order to establish the thickness of it, I quickly eased the Lancaster upwards until we found ourselves in brilliant moonlight. It was a beautiful but sinister sight, this huge white blanket glistening below us. I immediately shoved the column forward and continued our descent until I broke through the overcast. We could now see the rectangular light and dark patches of summer cropland in France, with the occasional silvery glint of a river or railway line. Back up into the cloud we went, with the knowledge that the cloud area was barely 500ft thick and was soon likely to disappear completely. The chilling prospect of being exposed for four hours in bright moonlight over enemy territory was most unwelcome.

The following events happened in quick succession and in much less time than it will take to explain them:

The time was now approaching midnight. Ben suddenly reported an unidentified blip which his visual screen had picked up – an aircraft low on the starboard side. It ceased to close in and remained at about 300 yards. The gunners couldn't get any visuals, as we were still hanging onto a diminishing cloud layer. Ben reported another 'bogey' which was dead astern and below, pulling up at a similar range. Then yet another blip pulled up at the same range and altitude, this time to port. Then Ben informed me that he had to leave his screen and tune in for the hourly broadcast from base in case there were any crucial changes or information affecting our mission.

Suddenly we burst into startling, brilliant moonlight! I immediately put the Lancaster into an evasive-action corkscrew manoeuvre. But it was too late. Our gunners could already see the tracers firing from an airborne radar-equipped Ju 88. The 20mm shells were thumping and tearing at our aircraft. Meanwhile, our gunners retaliated by firing back with their rapid-firing Browning .303s in an attempt for survival. The German then broke off momentarily, probably to assess the damage.

Our entire port wing, which contained about a thousand gallons of high-octane gasoline, was fiercely ablaze and spewed a brilliant flaming trail behind us. Although the situation was desperate, we weren't completely disabled. Then the Junkers took a steep turn and came around for another attack, with the end result being that only one of the four Lancaster engines was left running. As well as rupturing and setting the port fuel tanks ablaze, the first well-aimed shells tore into the main fuselage, destroying instrument panels and setting the two port engines on fire. The extinguishers momentarily subdued the roaring flames, only to have the fire rekindle when the chemicals ran out. I'd already feathered the props and shut off the fuel supply to the engines, but the flames persisted. I had no choice but to give the order to hit the parachutes and abandon ship. All the stations reported back except the mid-upper. There was no time to dispatch another crew member to the mid-upper position.

Departing in order, 'Chappy' was the first to jump out the forward escape hatch, followed by the second dickey, Bob. They both disappeared forward, down the hatchway. But as Jock moved forward to follow, I was horrified to see Bob struggling back up in an attempt to move to the rear. We talked about this scene a little later. We were sure Bob was trying to clear the exit for the rest of us as he somehow pulled his ripcord too soon. The cockpit and passageway were filled with billowing silk and entangled straps as the slipstream whistled through the now shattered windscreen. Continuing my struggle with the controls, I could see Jock and Bob in a tussle, until they finally both disappeared down and out of my view. 'Judy' now moved forward and hesitated for a moment to give me one of those famous Winston Churchill 'V for Victory' signs. Then he too vanished forward and down.

As for the remaining crew members – Ben, Jonesy and Sam – I heard only from Sam. He kept insisting on more time before jumping. He always did have a stubborn nature. With three ammo tracks out of commission, his solitary Browning sounded more like a pea-shooter compared to the customary sound of all four machine-guns. When he finally quit firing, I gave Sam a few seconds before attempting my own departure. I then removed my helmet and intercom cord.

Lancaster III EE139 *Phantom Of The Ruhr* on 550 Squadron at North Killingholme with F/O Joe C. Hutcheson's crew just before its 100th op on the 5 September 1944 daylight raid on Le Havre, France. (Note the mustard-coloured circular gas detection panel, which appeared on aircraft of 1 Group Bomber Command, and the four ice-cream cones in the first two rows of the bomb log denoting raids on Italian targets in 1943.) EE139 began its career in 100 Squadron at Waltham near Grimsby in May 1943 and flew at least 29 sorties with this unit until, in November 1943, it joined 550 Squadron. During 11/12 June 1943–21 November 1944 *Phantom Of The Ruhr* completed 121 trips, surviving five night-fighter attacks and returning with severe flak scars on five occasions. (Len Browning)

I released my tight grip on the steering column, whereupon I immediately sailed weightlessly into the air. In no more than a second I was dumped down into the aisle with my seat pack, parachute and dinghy firmly wedged between the throttle quadrant and the pilot seat. I had to scramble back into the seat and, hugging the equipment tightly behind me, manoeuvred back to starboard and then dived almost headlong down towards the forward escape hatch.

With a sudden stabbing blow to my chest, I virtually became impaled or hung up! I had struck the flight engineer's telescoping tubular foot rest, which I had failed to retract completely. This would normally be done by the flight engineer, but on this trip his station was occupied by Bob, the second dickey, who was unfamiliar with our ditching and bale-out drill.

With the Lancaster in a deep diving attitude, I was suspended in mid-air by my battle dress gear. Then a violent explosion threw me against the starboard side of the fuselage. Quickly, I doubled my knees and, with a strenuous effort, I catapulted myself straight towards the escape hatch. What a demoralising shock to find myself coming to an abrupt stop with my head protruding outside. The tremendous roar of the slipstream almost drowned out all my thoughts. At the time I did not realise that the hatch cover had become diagonally wedged across the opening. I was now securely jammed with my shoulders and arms pinned inside the aircraft. Although it took only a second

Lancaster III EE139
Phantom of the Ruhr at
North Killingholme, with
six air-crew members on
550 Squadron.

of my harness had slipped almost completely off and were now looped around my feet. My fingers quickly analysed the situation as hand-over-hand I pulled the harness toward me until I could firmly grab the D-ring and give it a good strong pull. I now dug my fingers desperately into the harness. Almost right away there was a flutter, then a snap as my fingers were torn loose from their grasp. My heels cracked together and my head snapped back as I hung suspended upside down under a beautiful canopy of white.

I could see a rooftop and I knew I had to prepare for an immediate landing. I had time only to bend my head forward with the intention of grasping my ankles to absorb some of the shock of hitting the earth. Then a rather subdued thud momentarily put my mind out of commission. After what was only a minute or so, I became aware of a prickly sensation on my wrists as my fingers explored a soft, crumbling material. I then noticed a slight vibration, accompanied by the receding drone of thousands of engines. I was flat on my back in a wheat field behind enemy lines, in France and alive. It appeared now that I would be able to see my twenty-first birthday.

or two, a series of visions and events floated through my mind. I could see bacon and eggs, our debriefing breakfast on our return. I could see my mother. Also, a small wading pool that my father and I dug on the farm. And many more visions. Then I suddenly became struck with anger and I jolted myself into immediate action. With extra strength, I hollered, prayed and swore, all the while twisting and turning, until I felt both flying boots whisked off by the slipstream. Then a black and orange hulk slipped quickly down and passed me. It was the Lancaster and I was free from it! I immediately grabbed at my chest for the D-ring of the chute, not waiting for the customary '10-count'. It was not there, or any other part of my shoulder or thigh straps! As I recall, I seemed almost amused that I could escape the burning aircraft only to end up dying without a parachute.

As I plummeted headfirst toward the earth, my eye caught sight of a silvery glint above and I could also make out a dark shape fish-tailing downwards after me – my parachute pack! Grasping upwards for it I instinctively bent my knees and as I did so I felt a distinct tug at my heels. The straps

Aided by the bright moonlight, *Nachtjagd* units destroyed 39 Lancasters on Stuttgart. The force of 187 Halifaxes, 106 Lancasters and 14 Mosquitoes that raided Hamburg was attacked by Tame Boars on the bombers' homeward flight and 18 Halifaxes and four Lancasters were shot down.

On Sunday 30 July, almost 700 aircraft – 462 of them Lancasters – were dispatched to bomb six German tank concentrations in front of a mainly American ground attack in the Villers-Bocage/Caumont area. Only 377 aircraft were able to bomb, on to Oboe markers because of cloud, and only two of the targets were effectively hit. Four Lancasters failed to return. Just over 120 Lancasters bombed the railway yards at Joigny-la-Roche on Monday 31 July. The raid was accurate and one Lancaster was lost. At Le Havre just over 50 Lancasters of 1 Group bombed the port area. One of the Lancasters failed to return. Sixteen Lancasters and two Mosquitoes on 617 Squadron and 97 Lancasters and six Mosquitoes of 5 and 8 Groups attacked the railway tunnel at Rilly-la-Montagne 7 miles south of Rheims, which was being used as a V-1 store. They were escorted by 102 Spitfires. W/C Willy Tait and a second Mosquito marked for 617 Squadron, flying at 12,000ft despite the Main Force being detailed to arrive at 18,000ft. The marking and aiming points were the north and south entrances to it, and 617 Squadron would bomb both of them with 'Tallboys'. Six Lancasters on 83 Squadron were detailed to destroy enemy communications and storage installations. The formation met its escort at the French coast and flew to the target without incident. Mosquitoes marked the target, so well in fact that the two 617 Squadron Mosquitoes were not required. The formation achieved a good bombing concentration, attacking both ends of the tunnel with what was considered equal success. Main Force bombs cut the track on both sides of the tunnel while those 'Tallboys' that fell apparently caved in the entrances.

F/L Bill Reid VC dropped his 'Tallboy' and was about to turn away when his bomb aimer, P/O 'Les' Rolton DFC, asked him to hold his course for another 28 seconds until their aiming-point photo had been taken. Rolton had been Reid's bomb aimer on the Düsseldorf operation when their Lancaster had been shot to pieces by fighter attacks and for which Reid was awarded the VC. Rolton had accompanied his skipper when Reid had recovered from his wounds and went back on ops in January 1944. The Scot had also invited Cyril Baldwin to join him on 617 Squadron, but the 5ft 4½in mid-upper gunner had declined, saying he had only a few more ops to complete his tour. And also, his brother-in-law, an engine fitter on 617 Squadron, had warned him that the Dam Busters' squadron was full of 'bloody lunatics'.

Now, as Reid held his course a 1,000-pounder dropped from above hit *S-Sugar*. Reid wore a harness with chest clips. As his flight engineer handed his parachute to him, another bomb struck them, the two hits causing damage to one of the port engines as well as the fuselage. *S-Sugar* began to spin down in two pieces. As Reid unstrapped himself and struggled to open the crash exit overhead, the nose seemed to break away, leaving him to tumble through empty air. He pulled his ripcord but kept a tight grip on the lines, being uncertain as to whether he had properly locked the parachute to his harness. Landing in a tree, he slid down it, with injuries to his face and one hand. Reid applied his field dressing to his cut face and then headed south towards Paris, 30 miles away. He carried a silk escape map and was confident of his ability to speak French, but his attempt at evasion was cut short when three German soldiers challenged him. Reid was taken to a flak site and attended to by a doctor, where he met up with F/O Dave Luker, his wireless operator, who had also been thrown out of the Lancaster when the nose of the plane broke off as it spun down. The other five crew members were killed.[12] Reid and Luker spent ten

months in *Stalag Luft III* Sagan and *Stalag Luft IV* Bellaria. A 9 Squadron Lancaster crashed at Puisieulx near Rheims with the loss of all the crew.

The Main Force now reverted to tactical targets, assisting the army against the Wehrmacht and operations against the 'V' bomb weapon sites. In consequence, operations, both by day and by night, were carried out against such places as Thiverny, Bois de Cassan near Paris, Forêt de Nieppe, Forêt de Mormal and Hazebrouck. On 2 August, 394 aircraft – 234 of them Lancasters – attacked flying-bomb sites in northern France. Two Lancasters were lost on the raid on Bois de Cassan. Next day, when 1,114 aircraft, including 601 Lancasters, attacked V-1 storage sites in northern France, five Lancasters were lost on the Trossy-St-Maximin operation and a Lancaster on 514 Squadron at Waterbeach failed to return from the Bois de Cassan raid. Attacks on the flying-bomb sites continued on the next two days. On the 4th, oil-storage depots too were hit, and 27 Lancasters on 617 Squadron attacked a railway bridge at Étaples. Some hits were scored but the 1,000lb bombs used failed to destroy the bridge. The next day, 742 aircraft – including 257 Lancasters – attacked flying-bomb storage sites in northern France once more. Small formations of Lancasters attacked Brest and Étaples and 306 Lancasters attacked oil-storage depots at Bordeaux, Paulliac and Blaye on the River Gironde near Bordeaux. Campbell Muirhead, who flew his 27th operation on Blaye, recorded his thoughts in his diary:

Another daylight op. And still I don't like them. Flew at sea-level due west and then turned into Bay of Biscay still

Lancaster B.I ME758 PH-N on 12 Squadron at Wickenby being awarded a DSO and a DFC by W/C Mike Stockdale on completion of *N-Nan*'s 106th operational sortie. *N-Nan* completed 108 trips and was SOC in October 1945.

at nought feet (all this so as to escape German radar). Passed over a couple of fishing boats and exchanged waves. A Lanc flying on our starboard side suddenly dived into the water. I could hardly believe what I was seeing: one second he was there, next second he was under. Climbed to 8,500 to bomb. Unfortunately, electrics for the bomb doors failed (probably all that sea spray) so couldn't release my bombs. Brought them back. Weather clamped at base so diverted to Church Broughton. Returned to base at 15.00 hours the following day.

The entire effort one long bore: especially going all that way and not being able to bomb these oil storage tanks. There they were and we couldn't get at them. I hope *B-for-Baker* all repaired now: don't like flying in any other kite. We have been told (orally!) to take the French letters off our guns. Owing to shortage of armourers an order had come out saying all gunners and bomb aimers had to service their own guns after an op. OK if you've fired your guns, but not if you haven't. So, in order to keep the moisture from the barrels when flying, we pulled French letters over the muzzles and thus saved ourselves some effort. Next, about half the Squadron was taking off with eight condoms hanging from the .303s much to the amusement of the WAAFs. Not to the amusement of a visiting Group Captain,

however, so off they came. (They say we are to be issued with round Perspex caps to fit over the muzzles: I think this is purely coincidental and in no way connected with our 'illegal' use of contraceptives …)

Shortly before noon on 6 August every German fighter unit in the Paris area was scrambled to intercept a formation of American bombers with little or no escort, but the 220 aircraft were Lancasters and Halifaxes ordered to attack the flying-bomb supply sites at Bois de Cassan and Forêt de Nieppe once more. Bombing was scattered and was no more successful at Hazebrouck, where just over 60 aircraft, mostly Halifaxes, attempted to bomb the railway centre. Four Lancasters failed to return from the Bois de Cassan operation. That evening a dozen Lancasters and three Mosquitoes of 617 Squadron and 16 Lancasters of 106 Squadron, escorted by 17 Spitfires, set out to bomb the U-boat pens at Lorient, which was due to be attacked by the US 4th Armoured Division the next day. Heavy flak hit four of 617 Squadron's Lancasters and several of 106 Squadron but there were no losses. Three 'Tallboys' hit the roofs of the pens without penetrating them. The bombs dropped by 106 Squadron, which were timed to explode between 6 and 36 hours after the attack, appeared to have fallen east of the pens. Despite being surrounded by the Americans, the German defenders in Lorient refused to capitulate and only finally surrendered shortly after VE Day.[13]

On 7 August, at Wickenby all operational air crew were ordered on the Tannoy to report for briefing at 1400hrs. Campbell Muirhead wrote:

At our special bomb aimers' pre-briefing at 13.00 the pain that shot into my guts was so fierce I could hardly breathe. Had thought myself pretty well experienced by now, virtually well able to take all that could come. But not when I saw 'Berlin'. Guy next to me at the pre-briefing mumbles 'we've had it now'; that our chances are virtually nil. And the short walk to the main Briefing Room is more of a despondent shuffle, hardly anybody speaking. Can read what you like in the newspapers about the high morale of our aircrews, but now it had gone through the floor. Mine certainly had and similarly so had that of, I think, every single one of us. Rest of the crews, as we entered, could tell from our faces that it was a nasty one. Confirmed when some of us whispered the target as we sat down. But even though, by then, the target was generally known in the Briefing Room, when the cover of Order of Battle was removed to reveal Berlin, an anguished groaning arose all around. 'Winco' Nelson sensibly waited for it to subside. I looked yet again at the long route taking us out over the North Sea, across the Dutch coast then into northern Germany towards Berlin, the Luftwaffe night-fighter radio and visual beacons clearly marked on it, and was just about to say something to Vernon when 'Winco' Nelson called for silence. He had just started off on the importance of the Battle of Berlin when a WAAF officer came on to the platform holding what appeared to be a teleprint. She gave it to him. He read it carefully and then looked at us. 'It's scrubbed,' he announced. Silence. But only a brief one as the message sank in. Then cheering: not wild, exciting cheering, more a subdued expression of extreme relief. In war films you see characters displaying stiff upper lips as they go to what, in the film, is certain death. Reckon there had been very few stiff uppers in that Briefing Room until the cancellation teleprint. Certainly mine hadn't been so although I would not have said it was actually quivering. Like everybody else, when 'Winco' Nelson announced the scrubbing, I felt the relief sweep over me and my stomach was back to normal even before I was aware of it.

An hour before midnight on 7 August, 1,019 heavy bombers set out to blast the Normandy battle area again. Five aiming points on German defences protecting the approaches to Falaise were marked for air bombardment, which was joined by 720 artillery pieces delivering high explosive and flares. The targets were in front of Allied ground troops 16 miles from Falaise, so bombing was carefully controlled and only 660 aircraft bombed before 600 Canadian tanks attacked, followed by infantry in armoured personnel carriers. By dawn on 8 August, the Canadians had penetrated the German defences for 3 miles. Ten heavies were victims of *Nachtjäger* and one to flak. Oberleutnant Heinz Rökker claimed three Lancasters north-east of Le Havre for his 39th–41st victories.

The night following, 170 Lancasters and ten Mosquitoes attacked oil depots and storage facilities at Aire-sur-la-Lys and the Forêt de Lucheux in France for one Lancaster lost. The raid on Aire-sur-la-Lys was Campbell Muirhead's 28th trip of his tour:

I was glad to be back in good old *B-for-Baker* once more: nothing went wrong this time. No enemy fighters were sighted. A few believed to have been knocking around, but I for

my part didn't witness any interceptions. Bombing was very concentrated. Oil tanks left blazing behind us. The way we and the Yanks are going for oil targets suggests Germans must be getting short of it. Why, despite this easy and comparatively 'gentle' operation, did I find myself full of fear and foreboding? It reminded me of our first ops. Is the reason that, in the beginning, one resigned oneself to the seemingly inescapable fact that this was it, that one would never complete more than a few before buying it? Could be. And now, coming to the end of the tour, one senses just the chance that we might make it? I think that's it. But all I know is that, right now, I'm getting the jitters. So are some of the others.

F/L H. Topliss of British Columbia on 100 Squadron at Waltham shakes hands with Sgt H. Williams from Cheltenham, the NCO in charge of the ground crew, after Lancaster ND644 HW-N *N-Nan* had completed its 112th operation. *N-Nan* was lost with F/O George Alfred Osborn Dauphinee RCAF and crew on its 15th operation on 16/17 March 1945.

On 9 August, 12 Lancasters of 617 Squadron and 17 Lancasters of 9 Squadron escorted by a dozen Mustangs and 44 Spitfires bombed the U-boat pens at La Pallice. Six 'Tallboys' hit the 20ft-thick pen roof, one blasting a crater 5ft deep and another penetrating one of the pens to 9ft before exploding. (Damage was not decisive and 617 Squadron returned to La Pallice on Friday 11 August with 2,000lb armour-piercing bombs because of the shortage of 'Tallboys'.) On the night of 9/10 August, 176 Lancasters and 14 Mosquitoes of 1 and 5 Groups successfully attacked an oil-storage dump at Forêt de Châtellerault. Two Lancasters were lost. Another 311 aircraft – 171 of them Lancasters – attacked four V-1 launching

Lancaster III ED888 *Mike Squared* on 576 Squadron at Elsham Wolds receives a bar to the DFC. ED888 completed the highest number of trips of any Lancaster in Bomber Command – 140 on 103 and 576 Squadrons (the latter as PM-M2) and 103 Squadron again, May 1943– December 1944.

There was a pregnant silence. And it was true, tonight of all nights they hardly needed reminding of their duty.

We crossed the French coast at 12,000ft. Bill gave me the approach course for Lille. A few searchlights fingered the sky aimlessly. The flak was light and sporadic. In the near distance we saw a shower of brilliant red TIs floating down. A moment later we heard the Master Bomber calling in 5 Group to drop their 'Tallboys'. Dead on time and track we commenced the run-in to the target with bomb doors open. Mac was sizing up his first drop. Everything seemed fine. But when the Master Bomber's voice sounded again it was to specify an undershoot by two seconds. This was a tall order even for an experienced bomb aimer. 'Mac' [F/Sgt Inia Maaka, the Maori bomb aimer] began some furious mental arithmetic. He did not want to kick off his tour with a dummy run. But then the Master Bomber thought better of it and ordered a fresh marking. A second shower of TIs floated down, this time bright green.

'Hit the greens, please. Hit the greens,' he said, very matter-of-factly and quite devoid of the tension which gripped me.

Lying over his bombsight 'Mac' was forced into more last-second adjustments. The next words were his as our bombs and incendiaries thumped out of the racks and *S-Sugar* wrestled to gain height. I closed the bomb doors and held the aircraft straight and level for the photo-flash. As I dived *S-Sugar* away we saw massive fires below us. From his rear gun turret Norrie Close gave us a blunt, Yorkshire assessment of prospects for Fort d'Englos.

'Well alight, is that. That's had it. I bet it's warm down there. I wouldn't fancy being a fireman in that lot.' He was still going on about it as we flew over the Channel. I don't think he'd talked so much in the whole of the previous six months.

We landed at Mepal after two hours and forty minutes of flying. All seventeen aircraft returned safely. The raid was somewhat marred by the marking fiasco but there was no doubt in my mind that the target had burnt. As regards my crew, with the exception of Mac we'd had an undemanding opener. They all performed well. The tension was gone. We were happy, excited and ravenously hungry. After de-briefing we bolted to our respective messes for a king-size breakfast. It was only egg, bacon and chips but it tasted like prime sirloin. I climbed into bed at two o'clock and this time, went out like a light.[14]

sites and the Fort-d'Englos storage site. F/O Harry Yates on 75 Squadron RNZAF at Mepal was on his first op with his own crew after having flown as a 'second dickey' on the raid of 8/9 August on Lucheux:

We tagged on to the queue of Lancasters, dramatic in the fading light, engines throbbing, moving slowly towards the take-off runway. The moment came near. The pilot in front of us opened his throttles and thundered away. I turned on to the strip and ran up the engines on full brakes, checking things one last time. *S-Sugar* shook and strained with anticipation. The seconds dragged out until the control caravan gave us our green. I released the brakes and we were away. We were going to war.

It was ten minutes past ten as I banked at 300ft to begin the slow climb up through the circuit. Half an hour later at 10,000ft Bill Birnie the navigator gave me a course and we left Mepal behind. All around in the darkening sky were dozens of Lancasters and Halifaxes bound, like us, for Fort d'Englos.

Out over the Channel, the gunners requested a test firing. 'Yes,' I said, 'but don't forget Jerry can see your night trace. Keep your eyes skinned.'

On the morning of 10 August, ten Lancasters on 617 Squadron, led by Willy Tait, and 68 Lancasters of 1 Group were dispatched to bomb the U-boat pens at Brest. Tait's 'Tallboy' hit close to the centre of the pens and all others were on target except for one which fell about 100yd from the front of the pens. The tidal wave sank a tanker and damaged a U-boat, while a second tanker capsized and blocked the harbour. (Lancasters of 617 and 9 Squadrons and others carried out further attacks on the U-boat pens at Brest, La Pallice and IJmuiden during August.)[15] Later the same day 117 bombers attacked fuel dumps in the Forêt de Montrichard and blasted U-boat pens at La Pallice and Bordeaux. Eight Mosquitoes of 100 Group provided a fighter escort for the bombers attacking the French Atlantic coast and no bombers were lost during the day's operations. The same evening 12 Lancasters sowed mines off the Biscay ports and Campbell Muirhead went to the Gironde area at Cordovan for his 29th operation:

Only eight a/c on this boring effort. Not much excitement in flying 3 hours there and 3 hours back just to drop six bloody sea mines. The code name for this kind of operating is 'Gardening'. One missing (F/L G.C. Owens), believed hit by flak and crashed in England.[16] Saw a couple of fighters. Looked like Ju 88s. But no attacks made on us: just as well as it was a clear, cloudless sky. On reflection, I shouldn't have moaned about this mining effort because I reckon all of us would give a hell of a lot to have our final op as easy as that. Just one more to go and it's all over. We must have *B-for-Baker* for our next and last op (we flew *A-Apple* because *B-for-Baker* was on a 50-hour inspection): Must have her, oh, Christ, we must!

Three railway yards and a bridge were attacked by 1, 3, 4 and 8 Groups on 11 August. At Mepal, Harry Yates DFC on 75 Squadron RNZAF lifted *U-Uncle* off the runway about seven from the end of the order. His target was the marshalling yards at Lens:

As we climbed in the warm, afternoon sunlight the full justice of the term 'bomber country' was revealed. Above neighbouring Witchford, beyond to Waterbeach and into the distant haze the sky was aswirl with the dark shapes of heavy bombers, hundreds of them, rising inexorably in their circuits. We had climbed to little more than 7,000ft before the early ones detached themselves to fly south. Bill gave

me a course and we joined an extenuated gaggle above the peaceful fields and towns of southern England.

Out over the Channel our escort appeared as if from nowhere, high above us. Looking down on the disorder below, those boys must have felt like cowboys on the range trying to control an unruly herd of cattle. In the event, it made no difference how unruly we were. Lens was not important enough for the Luftwaffe to risk a confrontation. A few light AA guns offered some half-hearted resistance. But the German aces relaxed, no doubt, over a quiet – and safe – game of chemmy in the mess.

The marshalling yards came up on track, faintly visible through the ground haze. Mac unloaded his bombs and watched them down into the centre of the target. It was 16.33 hours as the photo-flash burst. We banked away with no regrets about the lack of drama. Our escort turned north with us, still weaving white trails of condensation, probably still hopeful of a scrap. If so, they would return to base disappointed. Over mid-Channel I looked up through the Perspex to find that we were alone. We entered the Mepal circuit at just before six o'clock. Control brought us down in a matter of minutes. Number three was in the bag. If only the next twenty-seven could be so easy.[17]

That night 179 Lancasters and ten Mosquitoes of 1 and 8 Groups attacked the railway yards at Givors near Lyons again. No aircraft were lost. At Coningsby there was a period of enforced inactivity for F/L Siddle's crew because of the shortage of members, as W/O Clayton Moore recalls:

We were without a mid-upper gunner and gunners were in great demand due to the commencement of the daylight attacks, on which the front turret was usually manned. The daylight raids began on July 30th and the squadron took part in seven of these between that date and August the 6th. These were followed by night raids on Secqueville and Bordeaux. On the August 6th raid *Queenie* was lost with F/L A. Drinkall and crew. The target had been Bois de Cassan and – because it was a daylight raid – the aircraft had carried an extra gunner. Because of my aircrew grade (and because I was still on the list of volunteer gunners) I was also in danger of being nobbled, but had somehow managed to miss out on the daylight flights, not that I was unduly perturbed about that. My name came to the top of the list on Friday, August 11th when I was listed as replacement tail gunner with F/L Eric Young's crew. Young, a New Zealander, was an experienced pilot with an experienced crew, so I didn't mind having to dice with them. A night attack, the target called for my return to Givors and I suspected that this was because of the undisputed failure of the Command's visit some two weeks previously.

The weather report was better this time and I got the impression that the crews were keen to make up for the earlier disappointment. During the briefing, I had a word with the mid-upper, P/O Craig, on the crew's evasion tactics and was relieved to find that these were similar to our own. Prior to closing the briefing, W/C Pat Burnett DFC (a recent replacement for Dixie Deane) rose to give the customary summing-up. This usually consisted of words intended to instil confidence and invariably ended with the CO wishing the men a good trip and good luck. Not this time, however. He began his tirade by pointing out the failure of those involved in the previous attack on the target and pressed home his insistence that a better show was to be expected this time. He then polished off the pep talk by saying, 'This is war and somebody's got to die!' and then strode from the room, leaving behind a deathly silence.

We lifted off with PB240 (OL-J) at 21.50 and climbed into the darkening sky in the direction of the continent. As we crossed the English coast at our operational height I listened to the business-like exchanges between the various members of the crew and drew comfort from the impression of quiet efficiency and discipline that I got. I hadn't flown with the crew before, but I felt confident that it was every bit as good as our own and this was evident from their previous trouble-free operational record. The flight, although met with little enemy opposition, was a lengthy one lasting eight gruelling hours and we were all feeling shattered when we finally touched down at Coningsby. After the de-briefing, we trooped into the mess for breakfast before. I retired to my room and locked the door before falling into bed. I couldn't recall ever having felt so totally exhausted and I had a blinding headache again, so didn't want to be disturbed for some hours. Within a few minutes of hitting the pillow, I was out for the count. I remained in a blissful state of unconsciousness until well after 5 p.m. at which time I got up and had a refreshing shower before going for a meal. Down in the mess room, I met Dick Lodge.

'Where the hell have you been?' he wanted to know. I answered the question and then asked him to explain his concern.

'You're on with Young's crew again tonight. They've been trying to raise you all afternoon, so you'd better get down to the flights right away.'

Although feeling half starved, I immediately left the mess and hurried over to the gunnery office for instructions. Over at the flight offices, I was told that the Gunnery Officer was already attending the briefing and that this had begun ten minutes earlier. I was also told that, since I couldn't be found, W/O Dennis Beck DFC had been detailed to fly in my place. Since the last-minute swap was out of the question, I returned to the mess and collected a meal from the serving hatch. After eating, I went back to the flights, intent on finding W/O Beck. After a while the operational crews began to drift in, having had their pre-flight meal of two eggs, etcetera. On finding Beck amongst the throng, I proceeded to apologise to him.

'That's all right mate,' he assured me. 'Anyway, I want to get some ops in.' Beck was a quiet but jovial Londoner with good looks and an ever-ready smile. He had been with the squadron for some time and I had heard that he was married, with children.

'I went to Givors with Young last night,' I told him. 'They're a damned good crew, so you should have a safe enough trip. What's the target?'

'Brunswick. Looks like a maximum effort. Take-off is set for 9 o'clock.'

I wished him luck and set off over the field for the long walk to the caravan, where I joined the assembled 'press on gang' of well-wishers waiting to watch the aircraft taking off. *J-Johnnie* was in the leading line-up of Lancasters waiting near the caravan and I got a wave from Young before she began to roll. When the rear turret bounced past, I gave Beck the thumbs-up and got a broad grin in reply. Once the remaining aircraft had taken off, I walked back to the mess for a few drinks and a singsong around the old piano before I retired to my room with a mug of cocoa and a plateful of Leicester cheese chunks and bread buns. Once settled in, I proceeded to write a few letters as I disposed of the snack. I awoke the following morning feeling refreshed as a result of the lengthy rest I had enjoyed during the previous 24 hours. As I walked into the mess for breakfast, I was struck by the unusual silence of the assembled diners. Gone was the familiar babble of conversation and laughter and the resulting sombre atmosphere caused me to suspect something was seriously wrong. I collected my breakfast from the serving hatch and joined Jock Wilson at his table. He was staring disconsolately at his plate and didn't see me as I sat down across from him.

'Good morning, Jock.' He looked up to greet me, but with only the ghost of a smile.

'It's no' such a good yin,' he answered. 'Last night was the biggest chop night the squadron's had for months.'

'How many?'

'Three.'

'Good God! Who?'

'Keeling, Erritt and Young.'

'Young!' I echoed with disbelief. I didn't know F/L Keeling's crew very well because they were on 'A' Flight, but P/O 'Tubby' Erritt had shared my room in the Sergeant's mess until a few weeks ago, when he had been granted his commission. It was the news about Young that really stunned me. I should have been with that crew. I had flown with them in 'J' the night before last and now they were gone and W/O Beck had taken my place because I had failed to report in time for the flight. I tried to imagine what had gone wrong with such an efficient body of men. No doubt they had been feeling tired. Most crews on the squadron, with the exception of our own, had been pressed almost to the limit since the invasion started and there was a point beyond which even young men could not progress before their vigilance and efficiency suffered.

The news brought home to me the realisation that I had once more managed to cheat death. Such were the stakes in the game of aerial warfare. Once the cards were dealt, you picked up your hand and you played it out to the best of your ability. If luck was on your side, you might be able to cut your losses, or you might even win. It all depended on how many aces you held. Luck had been with me last night, but not with Beck, and I felt nothing but shame because his going had been because of me. Maybe, I thought, things would have been different had I been with them and, then again, maybe not. There were so many imponderables. You weaved to port instead of starboard and this brought

F/O Hal Ollis RCAF and his crew on *Ollie's Bus* on 625 Squadron at Scampton.

you within view of an enterprising Jerry fighter pilot, or it placed you at the exact spot at which the next flak shell was going to burst. Maybe some members of the crew had been feeling the strain of too many nights over Germany. Being tired played tricks with your ability to remain alert and it affected your eyesight. Under this condition, your imagination began to dupe you into seeing things that didn't exist and you sometimes failed to notice those that did. The month of August proved disastrous for the squadron. Six more aircraft and crews were to be lost during the remaining seventeen days, bringing the total for the period to nine. Despite all this, plus our enforced state of inactivity, the month had its compensations for the crew and for me. Bill Siddle was awarded the DFC and some of the others received the DFM. I had to be content with nothing more prestigious than the Canadian Operational Wings, which served to signify that I had completed a tour of thirty trips.[18]

THE MEANS OF VICTORY

Fighters are our salvation, but the bombers alone provide the means of victory.

Winston Churchill

On Saturday 12 August 1944, the Tannoy at Wickenby ordered all operational air crews to report to the briefing room at 1600hrs. The navigators' and bomb aimers' pre-briefing was slightly in advance. Campbell Muirhead began writing his diary before he went:

Afraid my hand is trembling slightly. Dreading the target simply because it's the last target of the tour. And yet, maybe I shouldn't be panicking as I am; this last, this final, target could well turn out to be some piece of cake over France. Still, quite a few of the French raids have been dicey enough. Well, I'll know quite soon now. It's 15.30 so in less than 30 minutes we'll learn the target. But wish it was over. The operation, I mean. Now, if it only could be about 8 o'clock tomorrow morning then I would have it all behind me one way or another.

Twenty-nine ops, only four of them being over the Reich and we end up with a bastard target like Brunswick. The seven of us sat there in the Briefing Room looking glum, not speaking, trying to take in what was being said: maybe the others, like me, reflecting just our hellish luck not to finish with a comparatively easy French effort. The silence was broken when briefing was over. All started snarling away. At the RAF; at 'Butch' [Harris] for being so dedicated to destroying German cities; at each other. Particularly the latter, our language worse than ever before. None of us exactly twitching, but not very far off it. Some frantic smoking on the way out to the bay and before we climb aboard. Plus more snarling and cursing.

Brunswick was to be an experimental raid by 242 Lancasters and 137 Halifaxes. No Pathfinder aircraft would take part and there would be no marking. The intention was to discover how successfully a force of aircraft could carry out a raid with each crew bombing on the indications of its own H_2S set. Crews had

been told at briefing that the bombers would fly past Brunswick and then wheel in a slow deliberate formation and start the time run from east of the city. On the approach, the bomb aimers would pick up on the H_2S the huge factory owned, according to Intelligence, by Reichmarschall Hermann Goering. From the factory, they would begin the bomb run on Brunswick. Another 297 aircraft meanwhile, 191 of them Lancasters, would attack the Opel factory at Rüsselsheim. Another force of 144 bombers, including 91 Lancasters, would attack enemy troop concentrations and a road junction north of Falaise.

'Take-off,' continues Campbell Muirhead:

was not as clever as usual: so much petrol and such a heavy bomb load, Vernon has to fight like hell to get her airborne. A slow climb for height then on course for the Dutch coast. And there the flak's coming up giving the appearance of coloured, twinkling, confetti in reverse. Bursting all around us, the shrapnel sounding like hailstones as it hits the fuselage. And it's not a case of getting through it and into the clear; there's flak inland too; heavy, concentrated stuff, well predicted. Then cone after cone of searchlights. Are trapped ourselves by one. Christ, it's like suddenly finding yourself naked in front of thousands of people. Violent corkscrewing, really headlong stuff, which brings the sour taste of vomit to your mouth before we manage to get free. During the manoeuvre catch a glimpse of another Lanc being held by a cone. Why isn't he corkscrewing? Then lose sight.

Now time for me to start slinging out the 'Window'. This to upset German radar. Which it does: all the searchlights become stationary for a while before evidently going on to manual and the predicted flak tails off. One or two of the searchlights remain pointing almost vertically upwards and I catch sight of a Me 110 below us flying through the beam of one. Maybe only momentarily but long enough for me to see the German crosses on the wings. So that's the night-fighters on the scene, probably vectored here from every available base. A sudden flurry of tracer underneath us and to port signals an interception. Shortly after, another eruption of tracer – much nearer us this time, but can't see whether or not the fighter's made a kill because Cartwright is yelling, 'Corkscrew, corkscrew' and we're in an acute dive to starboard.

Flying level again and my windowing over, I climb into the front turret. Quite a view! From above us a constant flow of flares to guide the fighters is being dropped very accurately all along the bomber stream (no doubt by older aircraft). As I cock the Brownings one of the flares bursts ahead slightly to port and in its light I see a Me 110 almost level with us. I yell, 'Corkscrew, corkscrew, corkscrew' hysterically I think and as Vernon dives headlong to starboard I fire my guns at the Me 110. I'm convinced this is a night-fighter come-hither, the tactics being for No.1 to fly level with the bomber, just near enough to be spotted, then, while the crew of the latter are straining away to identify him, No.2 creeps up on the Lanc's tail and that's the end of the matter as far as that particular bomber is concerned. Remain convinced of this tactic despite our experts saying a co-ordinated effort like that between two fighters is simply not possible. Level again and back at 19,000ft, I see more interceptions and two of our kites going down. Shouldn't stop my own searching, but find my eyes drawn to the dreadful sight of two bombers, flames streaking aft, describe graceful curves earthwards at a seemingly slow pace. My eyes keep on them, but I don't see them hit the deck. Hope the crews managed to bale out.

F/O George Jarratt RAAF and the crew of *K-King* on 460 Squadron at Binbrook were on their 13th trip. On an earlier raid they had lost the bomb aimer's hatch on take-off and Jarratt had circled until all the Binbrook aircraft were airborne before landing back on the runway with a full bomb load and fuel load. The hatch was found and fitted and Jarratt took off again. Halfway to Brunswick *K-King* was hit by flak which cut through the hydraulic pipes to the rear turret. They also had oxygen trouble and *K-King* dropped from 18,000ft to 8,000ft. Only with the help of Sgt Andy Andrews, the English flight engineer, did the pilot, who was from Sydney, manage to pull the Lancaster out of the dive. They bombed from 7,000ft. Coming out of the target area *K-King* was attacked by night-fighters and, seeing a burning rocket coming towards him, Jarratt managed to climb out of its way. A Me 110 followed

up the attack and Bob Compton, the mid-upper gunner, claimed the fighter as 'destroyed'.

Campbell Muirhead continues:

Nearing Brunswick now, so time I got out of the turret and down into my bombing compartment. Tell Vernon I'm doing so and have just unplugged my intercom when there's a stream of tracer flying past us. Another violent corkscrewing while I'm trying to make my compartment.

Everything ready and Brunswick coming up. A fearsome barrage of flak set to our height on the approach. So thick you wonder if there's possibly one bit of the sky ahead that could be free of it. You know you've got to fly into it and through it and (sounds so bloody dramatic as I sit here writing!) you say to yourself that this is the chop. On your very last one and where has your effing luck gone?

Can smell the cordite from the bursts all around us. There's also the rattle of the shrapnel hitting us; it just seems to go on and on. Like being in a greenhouse during a prolonged hail-storm. Louder noises suggest we've been hit more severely, but the bomb doors are open and we're on the bombing run now. This is what we came for and I must get it right. Have to give Vernon only two slight corrections and I have the aiming point sliding up smoothly. It hits the graticule and I press the tit. Down go the heavy boy and the clusters of incendiaries. Into what, even from 19,000ft, seems a furnace of swirling, angry flame. Order the bomb-bay doors to be closed and at the same time wonder if anybody can be alive down there. Stop wondering when, on Cartwright's yell, we're off corkscrewing again. Have to wait

Lancaster ND465 OL-L on 83 Squadron which crashed at Coningsby returning from the raid on Givors railway yards on 11/12 August 1944. There were no injuries to F/L A.J. Saunders RAAF and his crew. The crew were shot down in the target area on 18 August 1944 on the operation on L'Isle-Adam. Saunders and the two gunners evaded capture. The four other crew members were captured.

155

Canadian-built Lancaster X KB745 VR-V of 419 'Moose' Squadron RCAF, flown by F/O Rokeby, photographed over Normandy in the summer of 1944 by First Lieutenant Joseph H. Hartshorn DFC, an American pilot on the squadron who flew 34 ops. The grey streaks on the wing were caused by the exhaust gases from the leaded petrol. KB745 and F/O George Ross Duncan RCAF and crew flew into a hillside in Scotland setting course for Norway and an attack on the U-boat pens at Bergen on 4 October 1944. All the crew died.

roll must be massive. But Hitler should have thought about retaliation when he started the war.

Yes, over now. Only got to land safely and that's it.

But it isn't quite. Making our approach to the runway and, suddenly, a Lanc ahead of us erupting. We get the code word. It's 'Intruders' (Ju 88s: no bombs, just loaded with cannon.) They slip into the returning bomber stream, accompany the Lancs returning to base and into the circuit then wait until one of them is landing and thus a sitting duck. Climb to a certain height, fly northwards on a certain course, the theory being that any aircraft not doing so is enemy and can therefore be shot down. (But what about any of our own who, through damage, don't receive the secret signal?) Not so sure it was Intruders anyway – there haven't been any around for ages. Maybe simply a badly damaged Lanc exploding on landing and Control understandably playing it safe. In any event, we are recalled after about 15 minutes and land safely.

Winco Nelson's there when we climb out of *B-for-Baker*. Has beer for us in his car. Never did beer taste as good as that bottle I had. Wonder if the Winco greets with a bottle of beer all his crews who complete a tour. Asks us what it was like. Vernon says 'hellish' which is about the mildest expletive I think we've used since yesterday's briefing. We all thereafter, clutching our bottles of beer, go round the Lanc examining the holes. Not as much damage as I had expected: a host of smallish holes plus two or three rather big tears. Winco says she'll be operational again by tomorrow. He then tells us that a special outfit of Lancasters called 'Tiger Force' is to be formed to go to the Far East to bomb Japan and would we like, as a crew, to volunteer? He takes no offence to the rude reply to this: especially as the word 'sir' is affixed to it. He grins and says, 'The Japs cut these off only for starters.' After he goes off and while we're awaiting the transport, I pat *B-for-Baker* and whisper thanks to her for being so perfect, so faithful and so wonderful. But I'm careful not to let any of the others see or hear me doing this otherwise they'd think I really have gone round the bend.

until we've ditched the fighter before opening the panel to check no bombs still hanging up. If there were, the Germans would get them, not the Dutch. But all OK.

Onto turning point beyond the target then starting on homeward journey. Flak still tearing at us but now not quite so heavy. But more fighter flares. I see from the front turret quite a number of interceptions. At least one of them is a kill because a bomber explodes in mid-air. More flak and searchlights on way towards the Dutch coast. Flak just as heavy there as on our way in. Night-fighters still after us and more interceptions; they even follow us out to sea. Eventually, after what seems ages, the English coast. Some humour around now: suddenly everybody loves everybody. Jesus Christ! We've made it. We've done the thirty even though the last was a bastard like Brunswick. No thought of the number of people we must have killed. God, with the sky over Brunswick virtually crowded with bombers the death

George Jarratt got *K-King* back to Binbrook. On his last trip, when Jarratt would have been first back to Binbrook, he and the crew flew a few more miles to the west to pay their 'last flying respects to Lincoln Cathedral'.

The raid on Brunswick was not successful and there was no concentration of bombing, with bombs falling in the central and Stadpark areas while other towns up to 20 miles away were bombed by mistake. The German defences destroyed 17 Lancasters and ten Halifaxes. At Rüsselsheim most of the bombs fell in the surrounding countryside. Thirteen Lancasters and seven Halifaxes failed to return.

On Monday 14 August, over 800 aircraft – 411 Lancasters, 352 Halifaxes and 42 Mosquitoes – attacked German troop positions facing the 3rd Canadian Division, which was advancing on Falaise. A careful plan was prepared with Oboe and visual marking and with a Master Bomber and a deputy at each of seven targets. Harry Yates on 75 Squadron recalled:

To maximise concentration our bombing height would be just 7,500ft. A Spitfire escort would provide cover en route and counteract any spoiling tactics by enemy fighters over the target.

At 1.30 p.m. the Bedford crew truck slowed to a halt.

'R-Roger,' announced the WAAF driver.

R-Roger the samba dancer. The newest art form at this time was undoubtedly the painting of aircraft insignia. Everyone admired the American talent for adorning B-17s with improbably pneumatic young ladies in diaphanous robes. Not to be outdone – well, yes actually, to be completely outdone – some Mepal fitter more familiar with a monkey wrench than an artist's brush had let loose his carnal fantasies on poor R-Roger. The result was … unusual. Rio Rita we called her, after the song of that name. She certainly didn't stimulate my imagination. But I suppose there was a chance that an enemy pilot coming in to attack might be laughing too much to draw a decent bead on us.[1]

At 2.05 p.m. I returned the waves from the band of onlookers gathered as always by the caravan and we steamed off down the runway. At 90 knots R-Roger was ready to bear us and her bomb load of eleven 1,000-pounders and four 500s aloft. We reached our required height quickly and joined the gaggle on the first leg to the south coast. Our escort arrived thousands of feet above us, weaving their sinuous, mesmeric patterns and shining in the afternoon sunlight like diamonds. I couldn't comment upon their opinion of us. We were distributed all over the sky again at first. But some suggestion of orderliness had developed by the French coast. Not formation flying exactly, but it was progress.

Visibility was indeed good. In the countryside south of the shattered city of Caen I saw afternoon shadows among the mottled browns and greens and the linear grey of the long, straight Falaise road. Small groups of charred buildings, mostly farms, dotted the landscape. The farmers had survived four years under the Nazi heel only to be burnt out by the fires of liberation. Others trod through their broken homes, not people at all in the ordinary sense but soldiers busy at their anti-tank and machine guns, artillery pieces and mortars. A few miles further on, the evidence of soldiering became more apparent. I saw no enemy armour concentrations, no obvious signs of battle at all. But smoke and dust began to obscure the ground. It thickened rapidly until the confusion was alarming. Mac's reference points disappeared. In places even the horizon was obscured. The Yanks' problems of the previous day were understandable now. I wondered if Bomber Command had been hasty in stepping into the breach.

At least there were no enemy aircraft around. The pontoon buffs of the Luftwaffe clearly preferred wagering their pfennigs to their lives. We came in wholly focused on the job, on track and, at 15.50 hours, on time. Then, as if by prior arrangement with the Wehrmacht, the smoke parted over a few cratered fields where the TIs glowed yellow. I opened the bomb doors. Mac, prone over his bombsight, began his instructions. They were short-lived. A shower of reds appeared on the same ground. The Master Bomber rattled new orders over the R/T. 'Do not bomb the yellows. Do not bomb the yellows. Drop on the reds, please.'

'Hold it steady, skipper,' said Mac, concentrating more deeply than ever, his right hand poised to hit the release switches. Two, three, four times he said, 'Steady,' lengthening the second syllable with each repetition. Then came the first of the quick fire thumps from the bomb bay and the words that all aircrews longed to hear: 'Bombs gone and looking good. Bomb doors shut.' Twenty-four seconds later we could go home.[2]

Most of the bombing was accurate and effective but about halfway through the raids 77 Lancasters bombed short of the German positions; 65 Canadian troops in a quarry in which parts of the 12th Canadian Field Regiment were positioned were killed and many more wounded as some Lancaster bombs fell on their lines. Jack West on 115 Squadron recalled:

We arrived over our target just after midday and saw what we thought was a gathering of tanks and infantry. The pilot gave orders to dive down using the front turret guns first

P/O 'Selmo' Vernieuwe in the cockpit of Lancaster I NF913 PM-M at Elsham Wolds. The aircraft had been wrecked in a crash-landing at Elsham in 1943 and was later cut in half and joined to the rear of PH-H. Note the dinghy painted below the Swastikas following a sighting by Vernieuwe's crew of all seven of F/L R.M. Etchells' crew on a 156 Squadron Lancaster in the North Sea on 26/27 August following the raid on Kiel. Etchells' crew were brought ashore at Grimsby five days later after a rescue operation involving aircraft, a Danish fishing vessel and an ASR launch. NF913 completed almost 50 operational sorties before being lost with P/O Samuel Leo Saxe RCAF and crew on 7 March 1945 on a raid on Dessau.

and then using the rear and upper turrets. We levelled out at 200ft and I noticed that there was a lot of waving from the troops on the ground, whilst the wireless operator was taking a coded message which meant Abandon Mission and return to base. It seemed that most of the aircraft ignored this command and completed their mission before returning to base where we were told that we had not 'wiped out' a German Panzer division but instead had almost 'wiped out' an army of Canadian soldiers.

The error may have been caused by the yellow identification flares which were ignited by the Canadians. It was unfortunate that the target indicators being used by the Pathfinders were also yellow. Bomber crews claimed that the Canadians used the yellow flares before any bombs fell in the quarry; the Canadians said that the bombs fell first. The Master Bombers tried in vain to prevent further bombing in the wrong area but about 70 aircraft bombed the quarry and other Allied positions nearby over a 70-minute period. Thirteen Canadians were killed and 53 were injured, and a large number of vehicles and guns were hit. Two Lancasters were lost.

Another 155 Lancasters and four Mosquitoes of 5 Group made two separate attacks on the harbour at Brest. Since Cherbourg remained the only major port in American possession, the harbour facilities were needed so that supplies and reinforcements for the land battle could be unloaded from ships sailing directly from the US. Brest itself was defended by about 38,000 German troops. The bombers hit the *Clemenceau* and the cruiser *Gueydon* and left them slowly sinking. Two Lancasters were shot down by flak. When the Germans in Brest capitulated on 18 September the Americans found that everything had been totally destroyed by the pounding from land, sea and air and the Germans had wrecked anything remotely useful. Not until November, when Antwerp was finally opened to sea transport, would the Allies solve their problem of having enough ports of entry into the continent.

In August, mass raids on targets in the Normandy battle area and further afield continued. On Tuesday the 15th, 1,004 aircraft attacked night-fighter airfields in Holland and Belgium in preparation for a renewed night offensive against Germany. A force of 110 Lancasters and four Mosquitoes heavily cratered Volkel and put both runways out of action. Three Lancasters failed to return. On the night of Wednesday 16/17 August, Stettin was attacked by 461 Lancasters and Kiel by 348 aircraft – 195 of them Lancasters. The raid on Kiel was only partially successful, and two Lancasters and three Halifaxes were lost. At Stettin much damage was caused to the port and industrial areas. Five Lancasters were lost. On the night of 18/19 August, there was a devastating raid on Bremen, and the synthetic-oil plant at Sterkrade was also bombed. Heavies and Mosquitoes attacked a railway station and yards

at Connantre, east of Paris, and an oil depot and fuel-storage depot at Ertvelde-Rieme. The attack on Connantre marked the end of the long series of attacks on the French and Belgian railway networks. In 1,069 sorties that night four aircraft – two of them Lancasters – were lost. On 19 August, 52 Lancasters of 5 Group bombed a submarine oil-storage dump at La Pallice without loss, and when Halifaxes of 4 Group and Lancasters of 5 Group attacked ports and E-boat bases on the French coast on the 24th, all aircraft again returned safely. Next day, when 161 aircraft attacked five V-1 launching sites in the Pas-de-Calais, a Lancaster was shot down in the raid on Vincly.

Lancaster NF913 PM-M of 103 Squadron had been wrecked in a crash-landing at Elsham Wolds in 1943; it was later cut in half and joined to the rear of PM-H, going on to fly almost 50 operational sorties before being lost on 7 March 1945 on a raid on Dessau. It was the aircraft allocated to P/O 'Selmo' Vernieuwe, a Belgian, and his crew who flew their first op on 103 Squadron on the night of 25/26 August 1944, when 1 Group Lancasters hit the Opel factory at Rüsselsheim. Thirty-six aircraft left Elsham, 18 from each of the squadrons. The only one to turn back was 576 Squadron's *Mike Squared* which had a faulty rear turret. In many ways Rüsselsheim was a text-book attack, with the Pathfinders marking the target accurately and much of the factory being damaged in a raid which lasted just 10 minutes. It was also a night when the night-fighters were active, very active as far as the new crew at Elsham was concerned. P/O Vernieuwe later reported eight encounters with fighters, three of which turned into full-blooded combats. As he was approaching the target, a Me 410 was dealt with by the mid-upper gunner, Sgt Relf, who sent it down in flames into cloud and claimed it probably destroyed. On the homeward route a Fw 190 attacked and it was sent down in flames by Tommy Quinlan, the Welsh rear gunner, who saw it crash and explode on the ground and therefore claimed it destroyed. A later engagement with an unidentified aircraft proved inconclusive. During one of the combats, the Lancaster plunged from 18,000ft to 2,000ft during which 'a few rivets popped'. It was a good start for this crew's first operational trip. The attacks on Rüsselsheim and Darmstadt cost 22 Lancasters, including the two Deputy Master Bombers who were shot down.

It was during this same raid that a second 103 Lancaster – PM-T – flown by F/O Ryerse, shot down a Ju 88 and claimed a Me 110 as 'damaged'. The following night, 26/27 August, with rivets replaced, PM-M and crew were back on operations again

P/O Stanley Beeson's crew at No 1 LFS at Hemswell in 1944. They joined 550 Squadron and were lost on 28 August on the raid on the V-1 site at Wemaers-Cappel when PA991 *E-Easy* was hit by flak and crashed at Bollezeele. Back row, L–R: F/Sgt E. McQuarrie RCAF; P/O Stanley Beeson; Sgt Horace Sydney Picton (rear gunner) (killed in action); F/Sgt Derrick Neal (POW). Front: Sgt J.R. 'Red' Hewlett (flight engineer) (POW); Sgt J.K. 'Kenny' Norgate (POW); Sgt James Arthur Trayhorn (mid-upper gunner) (killed in action). S/L K. MacAleavey, who replaced McQuarrie that night and flew as the navigator, was also killed.

when 372 Lancasters and ten Mosquitoes of 1, 3 and 8 Groups were dispatched to Kiel and 174 Lancasters of 5 Group went to Königsberg, an important supply port for the German Eastern Front and the capital of East Prussia. At Kiel the Pathfinder marking was hampered by smokescreens but heavy bombing was reported in the centre of the town and the resulting fires were fanned by a strong wind. Seventeen Lancasters were lost over Kiel, all of which were probably shot down by Tame Boars. Returning home, the pilot of *N-Nan* on 550 Squadron, which was on its 94th operation, saw what he believed was the moon glinting off the runway at North Killingholme; F/Sgt R. Hofman RAAF lowered the undercarriage and gently put *N-Nan* down but it was the Humber estuary! The crew could not inflate their dinghy but they were close to the shore and all seven men swam to safety and raised the alarm. Transport was sent from North Killingholme for them and they arrived back at the airfield in time for their post-op meal. No one was very sorry to see the back of *N-Nan*, which was referred to on the squadron as 'that old cow'. Its main failings were its engines and its inability to climb to a respectable height, and it had a nasty habit of having engine failures, particularly on ops.

The eastern part of Königsberg was bombed and four Lancasters failed to return. There were no combats for PM-T this time but Ryerse's crew did spot a dinghy in the North Sea believed to contain the crew of a missing bomber. A message was later received from Group HQ at Bawtry Hall saying the sighting helped the ASR to locate the dinghy, which contained the seven members of a Canadian crew who had ditched. A few days later, PM-M was sporting 13 bomb symbols, two swastikas

Lancaster ED470
M-Mother, hit by
flak over Leipzig on
23 September 1944.

and a miniature dinghy painted on the nose. Vernieuwe and his crew later went on to fly Elsham's more famous *Mike Squared* and brought it back from the raid on the Wanne-Eickel refinery in November with serious flak damage. It was soon after that that Elsham's veteran retired. There were a few more trips for Vernieuwe's crew before they completed their tour in January 1945, including one where they just made it into the emergency airfield at Woodbridge in Suffolk with two engines out of action.

On 27 August, 243 aircraft were detailed to bomb the Rhein-Preussen synthetic-oil refinery at Meerbeck near Homberg. This historic raid was the first major operation by Bomber Command to Germany in daylight since 12 August 1941, when 54 Blenheims had attacked power stations at Knapsack near Cologne for the loss of ten aircraft. At Homberg the bombing was based on Oboe marking but 5–8/10ths cloud produced difficult conditions, though some accurate bombing was claimed through gaps in the cloud. There was intense flak over the target but no bombers were lost.

The summer of 1944 saw the full weight of Bomber Command, together with the fighter bombers of the

2nd Tactical Air Force, launched against the ski sites from where the V-1 bombs were launched. Most of these were spotted by reconnaissance aircraft, and within hours Lancasters and Halifaxes would begin raining high-explosive bombs on the sites. That summer Bomber Command carried out a total of 176 attacks on launch sites or V-1 storage dumps in northern France. The attacks ended only when Allied troops finally occupied the launch zones. They cost Bomber Command a total of 114 aircraft: 88 Lancasters, 22 Halifaxes and four Mosquitoes. The very last of those aircraft, on 28 August, was *E-Easy* on 550 Squadron at North Killingholme. It was one of a small number of 1 Group aircraft which took part in a concerted attack on 12 sites that day. The squadron's target was the Wemaers-Cappel site near Amiens. It was the 30th and final operation for the crew of *E-Easy* and should have been a milk run. The aircraft was on its way back to England when it was caught by a radar-predicted flak battery near Dunkirk. *E-Easy* received a direct hit and went down in flames. Four members of the crew – the pilot P/O Stanley Beeson, the navigator, bomb aimer and engineer

– all escaped by parachute and became prisoners of war, but the wireless operator and both gunners were killed.[3]

On the night of 29/30 August, 402 Lancasters and one Mosquito of 1, 3, 6 and 8 Groups attacked parts of Stettin that had escaped damage in previous attacks, and 189 Lancasters of 5 Group carried out one of their most successful attacks of the war when the target was Königsberg again. Only 480 tons of bombs could be carried by the Main Force because of the distance from England but severe damage was caused to the city and thousands of people were 'de-housed' in a single night. There was heavy fighter opposition in the target area and 15 Lancasters failed to return. Twenty-three Lancasters were lost over Stettin.

With the Allied armies advancing into France, the Chief of Air Staff once more gained control of Bomber Command, which resumed its area-bombing campaign and also mounted a new precision-bombing campaign against oil and transportation targets. On the afternoon of 5 September in good visibility, 313 Lancasters, 30 Mosquitoes and five Stirlings carried out the first of a series of heavy raids on the German positions around the 400-year-old town of Le Havre, where the enemy was still holding out after being bypassed by the Allied advance. One of the French residents who survived the raid said that the Lancasters 'turned the sky black'. Harry Yates on 75 Squadron recalled:

All the way in, the stream was tighter than on my previous daylight raids. Now, on our bombing run, everything looked good. Fecamp came into sight to port and the Seine estuary ahead. Below us, visible in every detail to the naked eye, lay Le Havre. The Master Bomber carried a heavier burden than usual that day. But he came through on the R/T as composed as ever: 'Come in, Main Force,' he said. 'Bomb the reds.'

We followed the first wave. I saw their bombs falling with remarkable accuracy as, indeed, we all knew they must. Mac began to line up for the drop. That familiar voice and those same old, spare phrases guided my hands and feet. Then the Master Bomber – the MC to us – cut in again. He wanted a fresh marking and ordered the Pathfinders to drop green TIs. Mac, however, already had the aiming point square in his sight. He was too single-minded and downright stubborn to heed the descending greens. Fifteen brisk thumps and *O-Oboe*'s simultaneous struggle upward told us that our work here was done, bar the camera run.

Air and ground crew of F/O Colin Henry's Lancaster on 12 Squadron at Wickenby. At 0710hrs on 14 October while outbound to Duisburg, Henry ditched *Q-Queenie* off the Lincolnshire coast between Mablethorpe and Grimsby after an engine fire. He and his crew were picked up by an ASR launch.

It was claimed that the raid was accurate and no aircraft were lost but when the raid was over at least 2,000 French civilians were dead. Another 60 Lancasters and six Mosquitoes of 5 Group bombed gun positions outside Brest, whose garrison was also still holding out. Again no aircraft were lost. On 6 September, Emden was bombed in daylight by 181 heavies escorted first by Spitfires and then American Mustangs, for the loss of one Lancaster on 7 Squadron, flown by the Deputy Master Bomber, which suffered a direct hit from a flak shell. The bombing was accurate and Emden was seen to be a mass of flames. This was the final Bomber Command raid of the war on the city. Le Havre also was bombed again, by 344 aircraft, over 300 of them Lancasters. All the bombers returned safely but two Lancasters on 101 Squadron crashed at Ludford Magna.

On 8 September, 304 Lancasters, 25 Mosquitoes and four Stirlings on 149 'East India' Squadron at Methwold in Norfolk, the last Stirlings in Bomber Command to carry out a bombing operation, were dispatched on an early-morning run to Le Havre to bomb the German positions. Three Lancasters, including one flown by the Master Bomber, W/C H.A. Morrison RCAF on 405 'Vancouver' Squadron, were lost (the whole crew evaded capture) and a fourth crash-landed at Tangmere on return.

The Penrose family with their adopted air crew in the back garden of their home in Hull before all the airmen went to the Hull City police annual dinner where they were guests of honour. Back row, L–R: Bill Marriott (bomb aimer); Geordie Young (rear gunner); Sgt Fred Hesketh (flight engineer); F/O Colin Henry (pilot); Sgt George Heywood (navigator). Front: Amy and Jack Penrose and their son F/Sgt J.K. 'Penny' Penrose, the crew's wireless operator.

Roy Abbott, who joined the RAF in 1942 and trained as a flight engineer, and was on his fifth trip on Stirlings, recalls:

Cloud over target was heavy so we were ordered by the master bomber (call sign 'Carfax') to descend below cloud base. At one stage we caught a brief glimpse of the target indicators, which the Pathfinder force had already dropped, but the cloud quickly closed in before we could use them as our aiming point. Three runs we made at that target from a height of 3,000ft, then the Master Bomber was shot down and crashed in the target area. '"Carfax II" to "Press-on": marmalade, marmalade.' That was the Deputy Master Bomber to Main Force. 'Marmalade' was the code to abandon bombing for fear that we might bomb French civilians. We had been in the target area too long and were running short of petrol, so I suggested to the skipper that we should land at the nearest airfield, which happened to be a fighter base of the American Air Force. The Americans were very hospitable and in reply to their 'How much?' I said, '1,000 gallons.' A few days later I was told off by the Squadron Leader because he had a bill from the Americans for those 1,000 gallons. I never did find out who paid for that petrol.

After this the Stirlings were pensioned off and the squadron was equipped with 27 brand-new Lancasters, hot from the factory. We viewed the change-over with some trepidation as it meant that we would be flying much

deeper into enemy territory. My next targets were generally in the industrial Ruhr Valley and included oil refineries at Dortmund, Homburg and Castrop-Rauxel; factories at Essen (Krupp), Leverkusen (Bayer), Homburg and Oberhausen: railway yards at Cologne, Koblenz, Neuss, Saarbrücken, Hamm, Siegen, Solingen, Vohwinkel and Krefeld; the airfield at Bonn; troops at Heinsberg; the towns of Duisburg and Nuremberg. Our squadron was fitted with special navigational equipment for bombing through cloud. It was known as 'G-H'. All Lancasters carrying 'G-H', and that included ours, were painted with yellow tails for identification. The idea was for each 'G-H' Lancaster to lead a formation of four other bombers not fitted with 'G-H' and instruct them when to release their bombs. On one occasion all four Lancasters formating on us were shot down.

Another daylight raid, on eight different strong points at Le Havre on 10 September, involved 992 heavies and all the bombers returned safely. A follow-up raid the next day also saw the return of all 218 aircraft involved when 105 Halifaxes, 103 Lancasters and ten Mosquitoes, escorted by 26 squadrons of fighters, pounded three synthetic-oil plants in the Ruhr area. Two Pathfinder Lancasters were lost on the Nordstern raid and two more were lost to flak or to 'friendly' bombs at Castrop-Rauxel and Bergkamen. At Le Havre, bombs were dropped by just over 170 aircraft before the Master Bomber ordered the final wave to cease bombing because of smoke and dust. All of the bombers returned safely.

'September 1944,' recalls Clayton Moore:

was to be a busy one, both for 83 Squadron and for our crew. We were allocated PB368 as a replacement for our ill-fated ND464 (OL-S) in which we had completed nine raids, including the opening of the invasion. Because we were by then one of the more senior crews operating with the squadron, we qualified for a Lancaster fresh from the manufacturers and the new OL-S was a delight to fly. We soon had her trimmed to our liking and we were delighted with her performance during a series of day and night training trips that we made with her. Our first op in OL-S took place on Sunday September 10th when we attacked Mönchengladbach. Our new Lancaster performed perfectly and the 3½-hour trip proved mostly uneventful.

On the following evening [11/12 September] we were again called upon to join the squadron in an attack, this time on Darmstadt, an industrial target about twelve miles south of Frankfurt. The first half of our flight presented no serious problems, although the increased activity and determination of the German defences in the area of the target was very much in evidence. The drift to increased German defences was one that I had noticed ever since the opening of the second front. The Fatherland was no longer the aggressor, but we were being made to face up to the tenacity and fighting skills of a cornered animal. Nevertheless, it wasn't until we had carried out our attack that the real troubles began, although these were not entirely due to the action of the enemy.

Some 226 Lancasters and 14 Mosquitoes of 5 Group carried out the area attack on Darmstadt. During a period of 51 minutes, 399 tons of HE and 580 tons of incendiaries were dropped by 218 bombers. Twelve Lancasters failed to return.

'The following evening [Tuesday 12 September],' Clayton Moore continues:

found us on the battle order for the third consecutive night. This time, our objective was to be Stuttgart, a town deep inside Germany and one which we had attacked on two previous occasions. The briefing was unusually early, because a new tactic was to be tried. We were to take off at 5.40 p.m. and hedge-hop in daylight almost to the Allied lines before climbing to our operational height, by which time darkness would have fallen. The tactic was designed to avoid radar detection so that the deployment of the German fighter squadrons would be delayed through lack of any advance warning or our approach. After take-off, we at once set course for Beachy Head, following as closely as possible the contours of the terrain and keeping it just a few feet beneath our heavily laden bomb bays. As we flew down England, the big black Lancasters were all around us, bobbing and weaving like porpoises as they skimmed over and around each obstacle that they encountered. Old Bill was thoroughly enjoying the task and I was glad that he had in the past indulged his fondness of low flying at each given opportunity. As a result, he was expert at handling a fully laden Lancaster at 'zero' feet and this caused me to regret having once accused him – in a light-hearted manner – of suffering from acrophobia.

The sun was just beginning to touch the horizon as we dropped still lower and roared out over the English Channel.

The Henry brothers from Armidale, New South Wales – Gavin, David and John ('Mk I, Mk II and Mk III') – at Elsham Wolds after a raid on Cologne. It needed special dispensation from the king to allow all three pilots on 103 Squadron to fly together on the same raid. It was Gavin's first op, brother David's 23rd, and John's 30th and last operation. All three brothers survived the war but John, the eldest, was killed later in a road accident.

A few minutes later, the final segment of the solar disc disappeared as we rose over the coastline and headed into France, still hugging the ground. Now the landscape that slid out from beneath my turret presented a marked contrast to the early autumn tints of the English countryside we had left. Not because this was a foreign country, but because it bore the scars that had been left by the fierce battle that had raged there in recent weeks. Apparently deserted and lifeless, the fast-receding panorama closely resembled my concept of a moonscape, with a profusion of ugly bomb and shell craters spoiling what must once have been a scene of rural tranquillity. Roofless and wrecked farm buildings dotted the landscape, and even the trees and hedgerows had suffered from the carnage as it passed. Tank tracks were everywhere,

The crew of *S-Sugar* who served on 103 and 166 Squadron at Elsham Wolds. Back row, L–R: C. Jackson (rear gunner); K. Cave (bomb aimer); P. Chesterton (pilot); A. Robson (mid-upper gunner). Front: H. Ackroyd (flight engineer); Jack Bullock (WOp); Ted Milling (navigator).

Lancasters caused severe destruction in the western districts of the city. Four Lancasters were lost on Stuttgart, and at Frankfurt 17 Lancasters failed to return.

Main Force daylight raids on Gelsenkirchen, Osnabrück and Wilhelmshaven followed, and night attacks resumed on 15/16 September when over 480 Lancaster and Halifax crews were detailed to bomb Kiel. Four Halifaxes and two Lancasters were lost. On the following two nights, Bomber Command supported the landings by British, American and Polish airborne troops at Arnhem and Nijmegen in Holland. Operation Market Garden, as it was called, ended in heroic failure and the British 1st Airborne Division was destroyed at Arnhem. At the end of September, 'Bomber' Harris wrote a letter to Winston Churchill pointing out that the Germans were still fighting hard and their defence would be even greater when the Allies reached the Rhine and crossed into Germany, so this was the time to use overwhelming Allied air superiority and 'knock Germany flat'. However, his wish for an all-out attack was rejected by Churchill, who did not rate the contribution of the air force as high as Harris did. The C-in-C maintained that bombing German cities to destruction could still win the war.

Night raids on Bremerhaven and the twin towns of Mönchengladbach and Rheydt by forces of Lancasters were carried out on the nights of Monday 18 and Tuesday 19 September respectively. Just over 200 Lancasters of 5 Group, carrying fewer than 900 tons of bombs, devastated the centre of Bremerhaven and the port area, rendering 30,000 people homeless. These attacks were followed by night raids on Neuss, the Dortmund-Ems Canal, Münster and Karlsruhe, as well as five daylight raids on the Calais area. When, on the night of Wednesday 27 September, 217 Lancasters and ten Mosquitoes were dispatched to attack Kaiserslautern, about 25 miles west of Mannheim, Bill Siddle's crew found themselves on the battle order for their final operation. Clayton Moore recalled:

some leading to an abandoned and burnt-out wreck. The only living creature that I saw was a solitary cow trotting aimlessly in panic as it tried to escape the roar of our engines. Siddle reported that a Lanc had crashed into a tree ahead and to starboard. I rotated my turret to watch the burning hulk come into view, but couldn't read the squadron letters. What appeared to be the lifeless body of a crew member lay a few yards away from the wreckage, with flames enveloping his flying suit. I could see nothing of the other six and thought it unlikely that there could possibly be any survivors.

Soon the daylight began to fade and we and the others started the long climb to operational height. During the hedge-hopping part of the flight, I had been keeping a close watch on the sky above us, trying to visualise the abuse to which a gaggle of diving enemy fighters could put us, since we were without a fighter escort. Fortunately, we were spared such maltreatment and were soon soaring into the friendly darkness of our otherwise hostile environment. The remainder of the trip was fairly straightforward, with no hazards to be met other than those we were used to. Because of the new tactic, we arrived back at base just after midnight and were de-briefed, fed and bedded down soon afterwards. There were no squadron losses.

That night the northern and western parts of the centre of Stuttgart were wiped out and at Frankfurt the bombing by 378

I had hoped to land an easy target with which to finish off my second tour, but the briefing served to show that Kaiserslautern was an industrial target and that we could expect to find it well defended. This we did and we had a few anxious minutes over the town before we managed to discharge our load. The weather presented problems during the outward flight, with violent electrical storms in abundance for much of the trip. As we turned for home, the darkness was frequently broken by brilliant flashes of lightning beneath us and these served to illuminate other aircraft in our vicinity. We were still over Germany when a storm cloud came into my view. Its interior was glowing continuously from the effect of the electrical charges within it and I could clearly see other Lancasters silhouetted against its steep sides as it towered several thousand feet above us. As we drew level with the cloud, I noticed that my guns were crawling with glowing electrical charges and that the Perspex cupola of my turret was similarly decorated with shifting rivulets of fire which closely resembled miniature streaks of lightning. Glancing to either beam, I could see that the trailing edges of the big elevators and rudders were also outlined in dancing white light. The phenomenon, known as St Elmo's fire, was not new to me, but I had never before seen it so densely displayed. I witnessed a blinding flash, accompanied by the most ear-splitting CRACK! I had ever heard, at once followed by total darkness and a deathly silence. In almost the same instant, I became aware of a tumbling sensation, as if my turret was falling. My analysis of the situation was that we had suffered a lightning strike. I was about to remove the flying helmet and bale out when I discerned a faint yet familiar voice in my earphones.

'Is everybody all right?' it asked.

It was the voice of Bill Siddle requesting a report from each crew position in turn. The static was gone from my earphones now and I could at last see and recognise my immediate surroundings. What was more, the sensation of falling out of control had left me. Had Bill's call come just a split second later than it had, my helmet would have been removed and I would have been descending into Germany. Ours was the only aircraft to have returned in a damaged condition and the squadron had again suffered no losses on the raid.

A crew conference was held at which the opinion seemed uppermost that we should continue together until the war ended. There was little doubt that the Allies were winning the fight and most believed that victory was little more than a few weeks away. I agreed – with reluctance – to comply with the decision but luck had played a significant part in our survival and I considered it unlikely that such good fortune could be expected to remain with us for many more trips, especially if we were expected to get involved in the daylight bombing campaign. On the morning of Thursday October 6th I made my way over to the Office Block after breakfast and told the wing commander that I wished to take a rest from operations. I was posted away from the

Lancaster *Winnie*, showing that the Americans were not the only air force to decorate their bombers with nose art and other paraphernalia.

squadron immediately. A DFC soon followed. I boarded the train for Lincoln without having the chance to say goodbye to my old crew. As my train pulled away from the platform I wondered how they would take the news that I had left them so abruptly.

During the final year of the war over a third of Bomber Command sorties were flown in daylight. Loss rates on these were lower than on the night-time attacks, averaging below 1 per cent. October 1944 witnessed one of the Command's major daylight bombing campaigns of the war. It was directed towards the Dutch peninsula of Walcheren with the aim to flood and destroy the German coastal gun batteries which dominated the approaches to the vitally important supply port of Antwerp. On 17 and 23 September, two major bombing attacks had already been carried out on gun emplacements at Flushing, Westkapelle and Biggekerke, but with little success. The Allies had only scant information on the exact locations of the guns, which led to many dummy positions being bombed on these two raids. During September, the Allies contemplated an airborne attack on Walcheren, but not enough troops were available and German opposition was considered still to be too strong. Consequently, at the beginning of October, it was decided to flood Walcheren through air attack, as this was believed to be the only solution to eliminate the threat of the German gun positions on the island. On 2 October, two B-17s escorted by 15 Mustangs dropped leaflets informing the population of the impending bombing attacks and advising the Zeelanders to seek a good place to

A 'Tallboy' spin-stabilised, deep penetration bomb which, when dropped from the optimum height of 18,000ft, took 37 seconds to reach the ground, where it impacted at 750mph and penetrated 25ft into the surface before exploding.

hide. Next day, the sea wall at Westkapelle was subjected to a heavy bombing raid by 247 Lancasters and seven Mosquitoes attacking in eight waves between 1156hrs and 1505hrs, causing a 40m-wide breach in the sea wall. The successful raid earned the bomber crews big newspaper headlines, the *Daily Mirror* using the headline 'RAF Sink Island' in letters 2in high across the front page.

On 7 October, a further force of 120 Lancasters and two Mosquitoes dropped 730 tons of high explosives on the sea walls near Flushing, because aerial reconnaissance had revealed that the water only slowly took possession of the peninsula as a result of the 3 October attack. Again, the attacking force suffered no losses and the dyke was successfully breached. During the second half of the afternoon on 11 October, 160 Lancasters and 20 Mosquitoes of 1 and 8 Groups attacked the

Fort Frederik Hendrik battery at Breskens, and another 115 Lancasters of 5 Group pounded gun positions near Flushing with 612 tons of high explosives. Both attacks went well in the face of heavy flak, although more than half of the Breskens force had to abandon the raid because their target was obscured by smoke and dust. A 101 Squadron Lancaster was shot down by flak and crashed at Heille with the loss of its eight crew; it was actually the first aircraft going in to attack the Breskens gun batteries. A further 60 Lancasters and two Mosquitoes of 5 Group successfully breached the sea walls at Veere without loss. Next day, 86 Lancasters and ten Mosquitoes of 1 and 8 Group dropped 531 tons of high explosive plus 88 marker bombs on four gun positions near Breskens, destroying two, without loss. On 17 October, 47 Lancasters and two Mosquitoes of 5 Group attacked the breach in the sea wall at Westkapelle after reconnaissance again had shown that the sea water had not advanced far enough. All aircraft returned safely, but the gap in the dyke was not widened.

There had been no operations for 48 hours when crews were roused around 1.00 a.m. on the morning of 14 October. Among those who were on the battle order that day was Colin Henry, a New Zealander, and his crew on 12 Squadron at Wickenby. They had come together at their OTU at Castle Donington and had arrived at Wickenby in May after surviving a hair-raising time at Sandtoft on 1667 HCU, where they failed to complete no fewer than six cross-country exercises because of the condition of the worn out Halifax I aircraft they were expected to fly. Bob Marriott, the bomb aimer, was also a New Zealander. The rest were British: mid-upper gunner 'Mart' Martin, rear gunner Geordie Young, flight engineer Fred Hesketh, navigator George Heywood and wireless operator Ken 'Penny' Penrose. Ken was from Hull and, as the rest of the crew hailed from distant parts, he took them home with him on occasions to see his parents, Jack and Amy. They became six extra sons for the Hull couple and whenever the crew had leave they would catch the train to Grimsby, then to New Holland and take the ferry to Ken's home city. At the briefing crews were told that they were to be part of Operation Hurricane, a demonstration of the might of Allied air power against one target. They were to be part of a 1,013-bomber raid – 519 of them Lancasters – by the RAF on Duisburg in the morning. The 8th Air Force would follow with an attack by 1,250 heavy bombers and Bomber Command would return at night with a third heavy attack.

At Kelstern *Q-Queenie* on 625 Squadron crashed only minutes after taking off, killing two of the eight men on board. *Q-Queenie* was the regular aircraft of F/O Lloyd Albert Hannah RCAF and his crew, who were halfway through their tour. For this raid they planned to take with them the bulk of a new crew which had just arrived on the squadron. One of the newcomers was flight engineer Sgt Robert Bennett, who takes up the story:

This was to be our first op and Sgt D.R. Paige our pilot, me, the bomb aimer, rear gunner and wireless operator were to fly with Hannah, his navigator and mid-upper gunner for experience. We took off at 06.28 and we had just got airborne when an engine caught fire. We had been in the air for less than two minutes when the order was given to bale out. Seven of us got out but Hannah did not and was killed when the Lancaster crashed in flames at Bradley's farm,

Little Grimsby. We jumped at about 600ft but our bomb aimer was killed when his parachute failed to open. He came down in the next field to me.

Two weeks after the crash, Bennett and F/O D.R. Paige RCAF were operating again and were to complete 24 ops before they once more had to parachute to safety on 23/24 February 1945. While bombing Pforzheim in *O-Oboe* they were hit by a large number of incendiaries dropped by another aircraft in the bomber stream. Sgt Jack Bettany, the wireless operator, tackled a number of the burning bombs and threw several overboard, sustaining serious burns in the process. However, the fire spread to the starboard wing tanks and, once over the Rhine, the crew baled out, all landing safely this time. Sgt Bettany was awarded the Conspicuous Gallantry Medal for his actions.

After their pre-flight bacon and eggs, Colin Henry and the crew of *Q-Queenie* had been driven out to their distant dispersal and, after a final cigarette, climbed on board for what was to be their 23rd operation. Everything seemed fine as the engines were started one by one, and finally *Q-Queenie*, which was carrying a 4,000lb 'cookie' and canisters of incendiary bombs, moved out to the end of the main runway. At 0640hrs *Q-Queenie* got the green light from the control caravan; Henry, with his flight engineer at his side, opened the throttles and the heavily laden Lancaster started to roll. That is when their problems began, as Ken Penrose recalls:

We were about three-quarters of the way down the runway when the port outer engine suddenly caught fire. We were too far into the take-off to abort and Colin managed to get her into the air. I sent out a distress call and told the station what had happened. We were refused permission to land but told to go out to sea, jettison our bombs and ditch. The fire was quickly spreading. By the time we reached the coast it had burnt the cowlings off the engine and was starting to spread along the wing towards the fuel tanks. We knew we hadn't got much time, either before the tanks went up or the wing fell off, so we got just off Mablethorpe and decided that was far enough and dropped our full bomb load in the sea. The cookie went off and I later heard it blew in most of the windows in Mablethorpe. I sent out our position and we all went to our ditching positions and the pilot put her down in the sea. She broke in half just near the H_2S dome and started to sink immediately but we all got out very smartly. Looking back, it was very impressive. We just went out bang, bang,

bang! We found that luckily the dinghy had come out of the starboard wing the right way up so most of the crew didn't even get their feet wet.

As we were getting out, the flight engineer had forgotten to take off his parachute harness and when he went through the hatch the bulldog clip caught on the retaining wire on the exit hole and, as you can imagine, even my shoulder under his backside could not get him out. This started a bit of a flap as the Lancaster was starting to sink quickly and we had to pull him back in, undoing the bulldog clip at the same time. We had all been in the dinghy for about 15 seconds when the aircraft disappeared. Once in the dinghy, I pulled the emergency box on board, opened it and found the small Very pistol. We couldn't see the coast so I fired off one of the red cartridges in the hope that an Air Sea Rescue launch would be on its way. We must have been in the dinghy for about 25 minutes when I saw a light flashing in the distance. It was in Morse and it was saying 'help coming'. So I fired off another flare. [Later, he was to discover the light had been flashed from the top of Grimsby's Dock Tower.] Eventually the ASR launch arrived with a scrambling net over the starboard side and we were all helped on board. I don't think I have ever been as glad to see anyone as I was when that launch arrived.

Once back in Grimsby, *Q-Queenie*'s crew gave most of their flying gear to their rescuers. Flying jackets and boots were prized possessions, even if some of them had been immersed in sea water. Back in Grimsby they were given a hot drink and within an hour an RAF bus arrived to take the crew back to Wickenby. They returned just before the first of 12 Squadron's aircraft started arriving back from Duisburg. Their own trip had included half an hour in the air, 90 minutes in a dinghy, half an hour in a launch and an hour in a bus. Soon after they got back an official inquiry was opened into the loss of *Q-Queenie*. It was completed in the afternoon and the crew immediately sent on a week's leave. They were to complete their tour, and 'Penny' Penrose finished his war instructing on Ansons in Cumberland.[4]

Q-Queenie was one of 15 Lancasters plus two Halifaxes that failed to return from Duisburg. Roy Abbott on 149 Squadron recalls:

It was the first 1,000-bomber raid I had taken part in. Imagine, 1,634-engined Lancasters and Halifaxes, each carrying about six tons of bombs all concentrating on one town. The first of the two raids on Duisburg was in daylight. We were well escorted by an umbrella cover of several hundreds of long range Spitfire and Mustang fighters. When I got back to base my name was on the battle order again. No time for sleep; just time for a meal while the plane was re-fuelled and bombed up; we were soon in the air again. This time we could see the fires burning in Duisburg long before we got there. Into that inferno a further 1,005 bombers dropped their loads.

Five Lancasters and three Halifaxes were lost and five more 'Hallys' were written off in crashes.

In an attack on Stuttgart on 19/20 October, *A-Able* on 12 Squadron was attacked by a fighter and dived out of control. The pilot, F/O W.J. Buchan, ordered his crew to bale out but managed to regain control, but not before two of his crew had taken to their parachutes. Six Lancasters were lost from the 565 that set out in two separate forces, 4½ hours apart. In another attack, 263 Lancasters and seven Mosquitoes of 5 Group attacked Nuremberg. Two Lancasters failed to return.

On 21 October, 75 Lancasters of 3 Group carried out accurate visual bombing of a coastal battery at Flushing. One Lancaster, on 75 Squadron RNZAF, was a victim to flak over the target, all of its crew perishing. Two days later, 92 Lancasters of 5 Group attacked gun battery positions at the harbour town of Flushing, but visibility was poor and consequently the bombing was scattered. Four aircraft were shot down by moderate but accurate flak in the target area, with another 20 returning with severe battle damage. That night Bomber Command dispatched 1,055 aircraft, including 561 Lancasters, to Essen to bomb the Krupp works. It was the heaviest raid on the already devastated German city so far in the war and the number of aircraft was also the greatest to any target since the war began. The force dropped 4,538 tons of bombs including 509 4,000-pounders on Essen. More than 90 per cent of the tonnage carried was high explosive because intelligence estimated that most of the city's housing and buildings had been destroyed in fire raids in 1943. Five Lancasters and three Halifaxes failed to return from the raid. On the night of the 24th/25th, 25 Lancasters and nine Halifaxes sowed mines in the Kattegat and off Oslo. Next day, daylight raids were carried out on Essen and Homberg, where sky-markers guided the heavies to their aiming points. Two Lancasters were lost on Essen.

On the 26th, crews went to briefings to learn what their target was. Harry Yates, on 75 Squadron at Mepal, recalls:

The chatter from ten crews hardly compared to the usual uproar before Briefing. But the sudden silence and rising tension as the CO and his team entered the room was as arresting as ever. He sprang onto the platform. The curtain swept aside, revealing the map of Germany massively daubed in red around the Ruhr Valley. But this was not to be a third consecutive run to Essen. The CO said:

'Gentleman, your target for today is a large chemical complex on the east bank of the Rhine at Leverkusen. You will see that it lies about eight miles north of Cologne. It is of vital importance to the German war machine and to the industries of the Ruhr. The weather in the target area has been deteriorating for twenty-four hours and is not expected to improve in time for the attack. In other circumstances the operation might be scrubbed. But a Group decision has been reached and you are to be part of a force of exactly one hundred aircraft. You are to bomb through cloud using "G-H" radar. This will be the squadron's second operational use of "G-H". It is vitally important that we can report a good concentration.'

Opposition in the locality was expected to be slight. 'G-H' was impervious to the weather but German fighters weren't. Our gravest danger would arise if we strayed south over the heavy AA batteries of Cologne, which remained as formidable as any.

We were first of the ten into the air. The murk at low level had thinned somewhat. One might have expected some weak sunshine and a suggestion of blue sky to develop. But we rose into greyness that still obtained at 8,000ft over Woodbridge where we had been due to rendezvous with our nominated 'G-H' leader, a Waterbeach kite. By the time the Dutch coastline appeared on Bill's H_2S set, conditions were deteriorating fast. We were being tossed around like the proverbial cork on the ocean and formating was less of a concern than the fear of collision. Everyone but Bill and Mac was on watch, straining to make out the first, vague outline hardening to black and giving us perhaps two or three seconds in which to react.

'Only five minutes to the German border,' announced Bill over the intercom. But as he spoke we emerged into dazzling sunshine and a vivid blue sky. Huge anvils of cumulo-nimbus rose before us like sky gods mindful only of their own

towering pride and importance and perfectly oblivious of our trespass. Then a Lancaster broke cloud ahead, her back, wings and tailplane shining with light, her props spinning faithfully. Another appeared, just as magnificent, and a third. These were not mere bombers, crude forms of steel and oil. They were guiding beacons of the spirit. With them flew our pride, our hope, our purpose.

Within two or three minutes the entire raiding force had appeared around us. I immediately began to search for the nearest available 'G-H' aircraft. Up to this time, formation flying had meant little more to us than a tight and disciplined stream. As such, it had become almost second nature and was the subject of some sharp, inter-Group rivalry. The pilots of 3 Group were proud of their achievement in mastering this art. They were well aware of the condescension from certain quarters towards Main Force crews and were not slow to point out when, for example, 5 Group were content with a gaggle. But 'G-H' was a harder taskmaster, requiring us to form up in a three- or five-aircraft vic. The moment for opening bomb doors and for release followed on from the vic leader's actions. It was a simple, visual mechanism that, effectively, placed all responsibility for the accuracy of the raid on the shoulders of the 'G-H' navigator ahead.

Flying from a temporary Russian base, 28 Lancasters of 9 and 617 Squadrons attacked the battleship *Tirpitz* in Kaa fjord, north Norway, on 15 September 1944. The smoke-screen failed to prevent some accurate bombing and the battleship was hit by one of 13 'Tallboys' dropped. This and other damage rendered the ship unserviceable for sea action and it never went to sea again, being used instead as a floating gun battery. Unfortunately, this was not realised by Allied Intelligence.

The battleship *Tirpitz* capsized in Tromsø fjord by a 14,000lb 'Tallboy' dropped during the raid by Lancasters on 29 October 1944.

I took up station, levelling out at 20,000ft for the final, outbound leg. By the standards of Hamel (7,000ft) or Le Havre (12,000ft) this altitude may seem high for a single target. But the weight of AA fire over Germany made safety the prime consideration, particularly since vic-flying brought us much closer to other kites and multiplied the dangers if one blew up.

Our leader's bomb doors opened and we followed suit. We began our run above a sea of white cloud that looked both glacial and turbulent. Everything was set. Now that *S-Sugar* was riding steady in her vic with no fighters and no real flak barrage to negotiate, this was turning out unexpectedly easy, a veritable milk-run.

From behind his window Mac watched the bombs spill from the lead aircraft and pressed the tit.

'Bombs gone,' he announced to the usual, thumping accompaniment and violent lift. It certainly didn't feel usual to drop on an unseen target in virtually clean air. Nor did it seem so to go through the motions with a photograph of solid cloud – but we did.

I pointed *S-Sugar* down into the claustrophobic gloom again and we turned onto our first homeward leg. Thus we remained, weary now but cocooned at least from fighters.

We broke through the cloud base only a couple of miles seaward of the English coast. It had been like coming out of a long, dark tunnel. The boys even cheered.

So we were back. Number thirteen was lucky for us. In debriefing, the Intelligence Officer did his best to draw some useful information from us. But a blind drop above solid cloud with little opposition was hardly promising material.[5]

At the early morning briefings on Saturday 28 October, 733 aircraft were detailed to bomb Cologne, and 277 aircraft – including 86 Lancasters – were tasked to bomb several German gun batteries at Dishoek, Domburg, Oostkapelle, Westkapelle and seven others to the east and west of Flushing. Enormous damage was caused at Cologne, where the clouds were thin enough for the Main Force to bomb visually in two separate waves. Three Lancasters and four Halifaxes failed to return. Crews returning from the other raids reported weather good, clear visibility, target identified and bombed visually at between 7,000ft and 8,000ft. Results were described as 'excellent'. *Y-Yoke* on 90 Squadron at Tuddenham piloted by F/O Robert J.C. Higgins, which was shot down into the sea, was the only aircraft that failed to return. Four parachutes were seen to emerge from the aircraft. Two landed in the sea; the others were seen to land on the ground and, tragically, they were straddled by the bombs. Only the Australian bomb aimer, F/Sgt Francis Frederick Austerbury, was found. His body was discovered on a beach near Ritthem very close to Flushing. He was laid to rest in the Bergen-op-Zoom war cemetery. That night, 237 Lancasters and seven Mosquitoes of 5 Group were detailed to attack the U-boat pens at Bergen but the area was found to be cloud covered. The Master Bomber tried to bring the force down below 5,000ft but cloud was still encountered so he ordered the raid to be abandoned after only 47 Lancasters had bombed. Three Lancasters failed to return.

Next day, 358 aircraft – 194 of them Lancasters – dropped 1,600 tons of bombs in good visibility on 11 gun positions at Walcheren. It was believed that all the targets were hit and only one aircraft, a Lancaster III on of 582 Squadron, was lost. On the 30th, 102 Lancasters and eight Mosquitoes successfully attacked the gun batteries again. A Mosquito on 627 Squadron was lost over the sea when a TI prematurely exploded in the bomb bay. The following day, the ground attack on Walcheren commenced, supported by hundreds of Typhoon fighter-bombers attacking various strong points near Flushing and Westkapelle. The island garrison surrendered after a week's heavy battle with Canadian and Scottish troops, including commandos who sailed their landing craft through the breaches in the sea walls created earlier by Bomber Command.

An oil refinery at Wesseling was also the target for 102 Lancasters of 3 Group on 30 October, when bombing was carried out in cloud using 'G-H'. That night, Cologne was the target for 905 aircraft, including 435 Lancasters, in another Oboe-marked area bombing attack, as Harry Yates recalls:

Our course took us eastward, straight towards the Ruhr. Düsseldorf, Krefeld, Essen, Hamm, Duisburg, Gelsenkirchen, Dortmund … any one of them could have been the target. Already, the Mossies were pitching in. Up to a hundred miles away they were sowing confusion and uncertainty. The German command centres would be swamped with reports of raids that did not exist. How they reacted and where they dispatched their fighters could make the difference between life and death for some of us. When the stream turned late and decisively to the south-east the die was cast. Cologne lay ahead of us, hidden by a thin but extensive layer of cloud and maybe also by mist up from the river. As yet we had not seen a single orange flare. But of searchlights there were perhaps a hundred. They lanced the cloud, transforming it into a lustrous and mobile white sea. The heavy flak, which had been sporadic, developed into a sustained barrage. Perhaps hundreds of tons of AA shells were being thrown up. The barrels of those guns must have been glowing like hot coals. All this attention was directed at the Pathfinders, but it failed. The TIs went down and splashed red in the centre of the drop zone, well concentrated and distinguishable despite the cloud cover. Bombs from the first wave began to follow. But apart from some bright, quick fire explosions, nothing else could be seen. It was certainly not as clear to the eye or as apocalyptic as Bremen or Kiel.

However, I did witness the end of two aircraft ahead of us. One of them carved a fiery trail down into the cloud. There seemed to be a reasonable chance that someone could get out. The other offered no lifeline. It took a direct hit in the bomb bay and in the blink of an eye was a ball of shocking white and orange, expanding violently outwards across a large area of sky and then petering into a sickeningly slow drift to earth. I averted my eyes, not out of professionalism but in humility. Sometimes the cynical view was right: it was all a matter of luck. Every kite had its bomb doors open. One

Two members of 625 Squadron with Browning machine guns removed from their Lancaster.

small splinter of flak hitting *S-Sugar*'s cookie or incendiaries would bring the same end to us.

Mac had squeezed his long frame under my seat and down into the nose some time ago. Now he informed me that we were right on track. For the moment, I only had to keep her steady. For his part, Tubby [F/Sgt Denys Westell, the flight engineer] still had a stock of 'Window' to feed down the chute. No sooner had he risen from the tip-up seat beside me than we entered a furious hailstorm of red-hot metal. An instant later there was an explosion beneath us, unseen but no distance away. The aircraft convulsed on the shockwave. Something very solid and fast-moving smashed upward, not into the bomb bay but the cockpit. A Perspex panel above me blew out, spreading icy turbulence everywhere. A flood tide of fear surged over me. 'It' was happening to me again and I was utterly helpless. But the seconds that followed were reassuringly normal. No crystal arrows filled my eyes. *S-Sugar* flew on in perfect equilibrium and I struggled to catch up.

Tubby reappeared all agog, asking if I was all right. Then he turned his attention to the floor less than an arm's reach to my right. It had been ripped open. Just above was his tip-up seat, upright now though he had not left it so. He pulled it down and ran his right hand around a raised and jagged hole, plumb centre and about five inches in diameter. He stared at it for a while, fascinated and horrified in equal measure. 'Oh, Shh … ugar,' he said finally. It was as near to

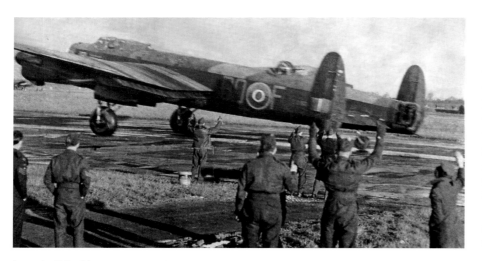

Lancaster *F-Fox Ad Extremum 'Press On Regardless'* on 550 Squadron taking off from North Killingholme on its 100th operation on Düsseldorf on 2 November 1944.

The next evening the skies over Cologne were filled with the sound of 475 bombers – 331 of them Lancasters – and 18 Mosquitoes in another Oboe-marked area bombing attack. Flak was slight and this time just two Lancasters failed to return. On the first and second days of November two daylight raids on the Rhein-Preussen oil plant in the Meerbeck district of Homberg were attempted and in between a night raid was carried out on Oberhausen by 288 aircraft, 74 of them Lancasters. The first raid on Homberg by over 220 Lancasters and two Mosquitoes of 5 Group was relatively unsuccessful. The marking was scattered and only 159 of the Lancaster crews attempted to bomb. One of the Lancasters failed to return. On 2 November, 184 Lancasters of 3 Group carried out a successful 'G-H' attack on the oil plant, large fires and thick column of smoke being seen. For Roy Abbott it was his 18th operation:

an obscenity as I ever heard fall from his lips. No one was hurt, no vital equipment damaged. Mac got the bombs away. We executed the 28-second camera run though we knew the photograph would be a wash-out (without it the raid wouldn't count towards our tour). Before we parted from the fury of Cologne we saw other aircraft burning. We knew that many among the five and a half thousand young airmen who came to this turbulent place would never leave and we knew how lucky we were not to be among them.

On the cold and draughty run to the Dutch border our only fighter of the night put in an appearance. Its silhouette was unfamiliar. It passed directly in front of us, climbing out of nowhere at an impossible speed and angle.

'Jet, skipper … jet!' yelled Geoff, who had a clear look at it but no chance to react on the trigger.

However, this novelty provided only a secondary talking point. The boys came one by one to the cockpit to stand in the gale, which was quite moderate this time and gape … first at the roof, then the floor and, finally, Tubby's seat. It didn't take long for curiosity to spiral down into scatology. 'F*** me,' said Bill, 'they're firing suppositories at us now.'

Someone thought that if ever our bombs hung up Tubby could drop a cookie of his own on the Hun and another offered to remove the Elsan to the cockpit so all *S-Sugar*'s future crews could sample her plumbing while over the Fatherland. *S-Sugar*'s was the only damage sustained by any of 75's aircraft on the raid. But thirty-eight bitter losses were logged elsewhere, a high toll at this stage of the war.[6]

Our crew had originally been allocated *T-Tommy*, but this plane was lost over Germany when being flown by another crew, so we were re-allocated *Z-Zebra* for the raid on Homberg. When the bomb doors opened a terrific explosion rocked the aircraft as a shell burst beneath the port mainplane, but she righted herself and continued steady. Petrol poured from No.2 tank in a long stream behind the aircraft, but eventually the tank sealed itself and the petrol leak ceased. In all we lost a mere 50 gallons, equivalent to 12½ minutes flying time. Petrol consumption on the Lancaster was about one gallon per minute per engine. *Z-Zebra* was badly shot up. Of the 20 aircraft sent out by 149 squadron, all returned, but 18 of them with flak damage. There was a lot of damage to the tail on *Z-Zebra* and the rear gunner was lucky to escape injury, so for the next raid, (a daylight on Cologne) we changed to *Q-Queenie*. We had an engine put out of action by flak on our bombing run, but we got back to base alright on the other three. The aircraft was full of holes, including a particularly large jagged hole within twelve inches of my feet. We got hit by flak again on raid No.23 (Siegen) and No.25 (Cologne again). On the Siegen raid flying *G-George* the flak caused the plane to lurch violently, but damage was slight. Damage was more severe on the Cologne raid in repaired *Q-Queenie*. Our compass was put out of action which made navigation back to base somewhat difficult. In all we had been hit by flak on four raids, but the only injury suffered by our crew was a minor one to the mid-upper gunner as a flak splinter damaged his cheek.

On the night of 2/3 November, 561 Lancasters, 400 Halifaxes and 31 Mosquitoes supported by 37 RCM (radio counter measures) sorties and 51 Mosquito fighter patrols took part in a heavy raid on Düsseldorf. More than 5,000 houses were destroyed or badly damaged in the northern half of the city, plus seven industrial premises destroyed and 18 seriously damaged. This was Bomber Command's last major raid on the city. Eleven Halifaxes and eight Lancasters failed to return, four of these crashing behind Allied lines in France and Belgium. The next major night raid was on 4/5 November, when Bochum and the Dortmund-Ems Canal were the objectives for the Main Force. Some 749 aircraft of 1, 4, 6 and 8 Groups attacked the centre of Bochum and more than 4,000 buildings were left in ruins or seriously damaged. Three Lancasters from the 174 dispatched by 5 Group failed to return from the raid on the Dortmund-Ems Canal and 23 Halifaxes and five Lancasters were missing from the raid on Bochum. Most of the night's bomber losses were due to *Nachtjagd* fighters.

On 6/7 November, 235 Lancasters and seven Mosquitoes of 5 Group again attempted to cut the Mittelland Canal at Gravenhorst, and Lancasters of 3 Group took off at dusk to attack the marshalling yards at Koblenz. On the Mittelland Canal raid crews were confronted with a cold front of exceptional violence and ice quickly froze on windscreens. Only 31 Lancasters bombed before the Master Bomber abandoned the raid due to low cloud. Ten Lancasters failed to return.

Harry Yates on 75 Squadron flew his 18th operation and recalls the raid on Koblenz:

Though an important crossing point of road, river and rail, Koblenz had, like Solingen, rarely if ever attracted the attentions of a main bomber force. Now, though, the Germans were pouring troops, tanks and supplies through the town, mostly by rail. Their objective was to halt the advance of Patton's tanks. Ours was to shut the door first. To accomplish this, 3 Group was employing 170 Lancasters carrying a variety of bomb and incendiary types, including 8,000- and 4,000-pounders. Our bombing height would be 18,500ft, time on target 19.30 hours. What followed four hours later was a text-book night attack. The run-in was complicated only by one or two fighters engaging in the stream. But their efforts bore little resemblance to the running battles of even the recent past. The skies were dark but clear and the marshalling yards were easy to identify visually. PFF Mosquitoes laid their flares on Oboe and the main force was called in. The

Lancaster *F-Fox Ad Extremum 'Press On Regardless'* on 550 Squadron at North Killingholme, taken after the Lancaster's 100th operation. In the cockpit is the skipper, F/L David Shaw from Thornton in Fife.

area in and around the yards quickly became a contagion of orange-centred, concussive rings as the blockbusters hit, and bright pins of light where the incendiaries landed. The town's AA batteries replied with all possible ferocity but we saw no one in trouble. As we turned away, the tormented ground was throwing up smoke and dust 10,000ft into the air. It was impossible to believe that the yards remained open for the Wehrmacht's business.[7]

One of the transportation targets selected for destruction in the final year of the war was the Dortmund-Ems Canal system. Several bombing raids were carried out on this target during the second half of 1944, which had the result of preventing smelting coke from the Ruhr mines reaching three important steelworks at Brunswick and Osnabrück, which in turn proved a devastating setback to the German war industry. One of these raids took place on the night of 6/7 November, when 235 Lancasters and seven Mosquitoes of 5 Group were detailed to cut the Mittelland Canal at its junction with the Dortmund-Ems

Canal at Gravenhorst on the Dutch-German border. Only 31 aircraft bombed the canal junction before the Master Bomber ordered to abandon the raid due to unsuccessful marking of the target. Ten Lancasters were lost and another crash-landed in England due to poor visibility.

On the afternoon of 16 November, 1,188 aircraft of Bomber Command bombed Düren, Heinsberg and Jülich, where a high concentration of troops, including the 'Hermann Goering' Division, had been reported. Over 1,200 American bombers made attacks on targets in the same area, all with the aim of cutting communications behind the German lines so that the American First and Ninth Armies could advance in the area between Aachen and the Rhine in Operation Queen.

Harry Yates, flying *S-Sugar* again, one of 25 on 75 Squadron to take off from Mepal, recalled:

Number 3 Group was sent against Heinsberg itself. At 15.30 hours the 9th Army troops, dug in a few miles to the west, would have watched the first of 125 Lancasters sail into the flak barrage above the town, bombs spilling into the void. We were operating under the umbrella of a hundred Mustangs and Thunderbolts, so enemy fighters were of minimal concern. But our bombing height of 12,000ft was ideal for the ground defences who had a real go at us. We came in towards the rear of the stream. Huge, swirling clouds of smoke began to obscure the town well before Mac was ready to drop our 4,000lb cookie and 9,000lb of GPs. In the sky ahead hundreds of black smears marked the spent efforts of the AA gunners. But I saw not a single aircraft in trouble. And so it turned out. We were the last to touch down at base. Two aircraft came home ahead of us with some damage, but nothing too serious. Everything had gone smoothly. Even the East Anglian weather afforded us a four-hour, afternoon break in an otherwise foggy day.[8]

The bombing offensive against German cities and oil targets continued unabated with raids in November on Homberg, Wanne-Eickel, Harburg and Dortmund among others. It was mainly a night of good visibility on 21/22 November, and 274 Lancasters and nine Mosquitoes of 1 and 8 Groups attacked the local railway yards and lines at Aschaffenburg with some success. Two Lancasters were lost. Another 273 aircraft, 79 of them Lancasters, attacked an oil refinery at Castrop-Rauxel for the loss of five Halifaxes. A raid on another oil refinery at Sterkrade was not successful. Two other forces attacked the Mittelland and Dortmund-Ems canals again. Two Lancasters on 49 Squadron at Fulbeck, one flown by Group Captain C.T. Weir DFC, the station commander, were lost.

The aqueduct was bombed by several successive waves of aircraft, each wave arriving over the target at a different height. Weir was first in the bottom wave of 40 Lancasters that were briefed to attack from 14,000ft. However, when they neared the target shortly before nine o'clock, the area was under cloud and the Master Bomber, who was directing the operation from a lone Lancaster low down near the aqueduct, called the bottom wave down through the cloud to 4,000ft. Weir found the aqueduct – it had been marked by the Pathfinders – and released his bombs. He heard the bomb aimer report, 'Bombs gone'. A second later the aircraft exploded. Weir, sitting at the controls, was aware only of a great white flash and a roaring noise in his ears. He remembered no more than that. Three hours later, at about midnight, he regained consciousness, lying in the mud of the Ems-Weser Canal a few yards from what remained of the aqueduct. His only injuries were a gash in each leg and a broken jaw. Stretched out in the mud beside him was his parachute – fully opened and, but for two torn panels, undamaged. He climbed out of the mud and sat on the canal bank trying to sort out in his mind what had happened. Shivering, he saw that he was soaked to the skin. And then he saw why. He had landed in the canal before the

S-Sugar R5868 the century-maker on 467 Squadron RAAF at Waddington is suitably acknowledged after its sortie on 11/12 May 1944 to Bourg-Leopold, Belgium. *(Imperial War Museum)*

following aircraft had breached the aqueduct. Unconscious, he had floated there until, as the bombs poured down and the aqueduct collapsed, the water had gushed away, the canal had all but dried up and he had been left like a stranded fish in the mud. The canal bottom, the banks and the surrounding fields were pockmarked with hundreds of deep ugly holes. For several hundred yards around, the area had been saturated with the thousand-pounders of more than 100 Lancasters. It gave him a curious eerie feeling to realise that he had lain throughout the bomb deluge in the midst of it all, unconscious and unaware of the man-made earthquakes, the blasts and the flying steel that had not touched him.

Not far away the wreckage of two Lancasters was still burning in a field. Weir's Lancaster had been hit by a 1,000lb bomb dropped from an aircraft in one of the bomber streams attacking from a higher level. The other Lancaster, which must have been following close behind, had been caught in the explosion and had been destroyed in the same second. Of the 14 men in the two aircraft, Weir alone had survived. He had been blasted bodily out of his seat, up through the Perspex cockpit roof, and, as he hurtled away amid a shower of wreckage, the blast had ripped his neatly folded parachute canopy from the seat-type pack hanging behind his thighs and blown it open. He had collected the gashes on the insides of his legs from the control column as he left the seat and his jaw was broken as he burst head-first through the roof. Weir was at liberty for just 24 hours before being captured by a German policeman.[9]

On the night of 26/27 November, 270 Lancasters and eight Mosquitoes of 5 Group went to Munich. Bomber Command claimed that this was 'an accurate raid' in good visibility 'with much fresh damage'. Two Lancasters crashed in England just after take-off, and one Lancaster crashed in France on the way home. The following night, over 350 Lancasters and ten Mosquitoes were detailed to bomb Freiburg, a minor railway centre, and 290 aircraft were tasked to bomb Neuss on the western edge of the Ruhr.

At Freiburg, 1,900 tons of bombs were dropped in the space of 25 minutes and over 2,000 houses were destroyed, more than 2,000 people killed and over 4,000 injured. Freiburg was not an industrial town and had not been bombed before, but it was believed that many German troops were billeted in the town and that these could threaten American and French units advancing in the Vosges, only 35 miles to the west. Of those who died, it is believed that only 75 were German troops. One Lancaster failed to return. AVM Sir Edward Rice, AOC

1 Group, later sent a congratulatory message to his squadrons on their efforts in dispatching 312 aircraft against this target, the most ever sent in one raid by any of Bomber Command's eight Groups. 'It was,' he said, 'a splendid achievement.' At Neuss the central and eastern districts were heavily bombed and many fires started.

Bomber Command went back to Neuss the following night, when Essen also was targeted by another force of 316 aircraft. This time 150 Lancasters of 1 and 3 Groups would carry out a mainly 'G-H' attack at Neuss. Harry Yates, on 75 Squadron, recalls:

The quantity of petrol being pumped into each kite: 2,000 gallons. Back to the Valley, said the station wiseacres. And they weren't wrong.

'Gentlemen, your target for tonight,' said the CO emphatically, 'is the industrial town of Neuss in the south-west of Düsseldorf.'

It was not difficult to read the underlying message. No mention of anything more specific than industry – this was to be a blitz. That meant, in effect, targeting the working-class population in their homes. Just another job, we would say to ourselves, and think no more about it, if possible.

It was our first night operation for a fortnight and the latest I ever took off for Germany. Loaded with Benzedrine we lifted off into fine skies at 02.52 hours. That meant arriving over Neuss at the indecently early hour of 05.00 when, with any luck, most sensible crews would have concluded that their night's watch had passed uneventfully and a warm bed awaited. It was also the first full *Wanganui* attack we had ever flown. The method employed single flares dropped twice en route by Oboe to guide the stream into the bombing run. Then green flares were sent down to mark the aiming point. Our Pathfinders managed to deposit two clusters. Mac and every bomb aimer early in the stream were left with nothing visible through the ghostly green cloud and no clue as to which markers to bomb. Bill took the fastest *Gee* fix of his life. One cluster, he thought, might be too far south. That was enough for Mac who duly unleashed his blockbuster and the 1,000 and 500lb GPs on the other one. By the time we left the target area behind, something of Neuss was already apparent. The cloud began to be suffused by a widening, dull red light in and around which were stabs of intense whiteness as the cookies and blockbusters exploded. Down there the town was ablaze, but little of it was visible to us. Down there was terror, for were we not the *terrorfliegern*?[10]

Next day, 29 November, over 300 Main Force aircraft took off for Dortmund but bad weather interfered with the operation and the bombing was scattered. Six Lancasters, including one that was involved in a mid-air collision with another Lancaster, were lost. Just two men survived from the six aircraft lost. On the 30th, Bomber Command dispatched two forces, as Harry Yates recalls:

Shortly after breakfast we took our seats in the briefing room. The attack was to be on a coking plant at Osterfeld, not far from Essen. Only sixty aircraft, all 3 Group, were scheduled for the trip. Number 75 Squadron was to send eighteen of these. At about the same time sixty more Lancasters from other 3 Group squadrons would attack the nearby plant at Bottrop.

We lifted off at 10.50 hours for a rendezvous over Woodbridge forty minutes later. Cloud conditions were unremitting for the entire journey out, as forecast. We crossed into enemy skies above 10/10ths cloud with the tops at 10,000ft. Under all this somewhere ahead of us was the vast industrial complex that we must now destroy. We came in well compacted, probably too well, and passed very close to a great, black island of smoke spread out over the cloud tops. That was Bottrop on fire. The batteries began to spit up their 88mm shells. The sky became dirty. We rocked and bored through it with just Mac uncaring and at his window watching for the 'G-H' leader to open bomb doors.

'Two minutes to target,' announced Bill, although his function was purely advisory at this stage.

The flak showed no sign of slackening.

'Open bomb doors, skip,' Mac said, seeing our leader do so.

'Bomb doors open,' I replied, pulling the lever down to my left. And now we would wait again.

Suddenly, ahead of us in the stream, a vic of three kites was consumed in a prodigious burst of flame which immediately erupted outwards under the force of a secondary explosion. The leader had been hit in the bomb bay; the others were too close. No one could have survived, I knew. There was no point in looking for parachutes. I flew on straight and level, Tubby standing beside me, both of us dumbstruck by the appallingly unfair swiftness and violence of it all. But there was still that deeply drawn breath of relief that somebody

else, and not oneself, had run out of luck. And hard on the heels of that was a pang of guilt. One grieved for whoever was in the kites and wondered if friends might not be coming home. If so, there was nothing to be done for them now. We stared more fixedly than ever at our own leader's bomb bay. His bombs fell away and Mac was quick on the button.

The attack was probably successful, though it was impossible to tell with any certainty. Some black smoke curled into the cloud but there was nothing else to indicate the effect on the coking plant. I flew *M-Mother* home with my thoughts still of the poor devils in that doomed vic.[11]

An attack on Duisburg on the last night of the month was hampered by cloud cover and three Halifaxes, including two that were involved in a mid-air collision, were lost.

When, on 2/3 December, 504 bombers were dispatched to Hagen a Halifax and a Lancaster were lost when they crashed in France. Two nights later there was a 'spoof' on Dortmund as 892 heavies set out for Karlsruhe and Heilbronn in the Ruhr. A handful of *Nachtjäger* engaged the bomber stream on their run in to the target at Heilbronn and shot down 12 Lancasters. On 5/6 December, Bomber Command sent 497 aircraft on a raid to Soest, which was successful, with most of the bombing occurring in the northern part of the town where the railway installations were situated. About 1,000 houses and over 50 other buildings were completely flattened, killing over 280 Germans and foreign workers. On the night following, Bomber Command were given three targets. Over 470 Lancaster and a dozen Mosquito crews were detailed to bomb Leuna near Merseburg for Bomber Command's first major raid on an oil plant in eastern Germany. Another 453 aircraft were briefed for the first major raid on Osnabrück since August 1942. Their target was the railway yards. Over 250 Lancasters and ten Mosquitoes of 5 Group were given the town centre and railway yards at Giessen as their objectives. At Osnabrück the railway yards were only slightly damaged but four factories, including the Teuto-Metallwerke munitions factory, were hit and over 200 houses were destroyed. Seven Halifaxes and one Lancaster failed to return.

Among the 550 Squadron crews at North Killingholme that night was Sgt Frank Woodley, a mid-upper gunner. It was Woodley's crew that survived a hair-raising ride back from Merseburg, a trip which was to last 8 hours 40 minutes. One of a bomber crew's great fears was being hit by 'friendly' bombs, something which happened all too often in bomber streams and accounted for an unknown number of aircraft lost during the last three years of the war. Over Merseburg their 550 Squadron Lancaster, *N-Nan*, began to rock as it was hit repeatedly by 3lb incendiary bombs scattered from a Lancaster just above them. From his turret Woodley noticed that one of the bombs had stuck near the wing root and had ignited. There was a very real danger it would lead to a major fire and so the flight engineer decided the only way to tackle the bomb was to chop a hole in the fuselage and extinguish it by hand. With Woodley giving directions, this is just what the flight engineer did, now aided by the wireless operator and, in Woodley's words, 'snuffed out' the bomb. No sooner had they done this than there was a crash and an almighty gale blew through the Lancaster. The bomb aimer's dome in the nose had been simply sheared off, the crew later coming to the conclusion it had been carried away by a 1,000lb bomb dropped by another of the 1 Group Lancasters out that night. The bomb aimer had just left the position, otherwise he could have been decapitated by the bomb. At the same time the pilot found that *N-Nan* was becoming more difficult to control. The damage happened as their aircraft was turning out of the stream, across the path of other aircraft coming in to bomb. The loss of the nose cone meant the Lancaster was acting like a giant scoop. It was also having difficulty in maintaining height and Woodley remembers that fearfully cold ride back across the breadth of Germany with *N-Nan* gradually filling with snow sucked in through the damaged nose. They did make it back to North Killingholme where, much to their amazement, they found part of the tail had been damaged by another 1,000lb bomb – and another 50 incendiary bombs, none of which had been in the air long enough to ignite, were stuck in the aircraft. *N-Nan* was declared a write-off.

On Monday 11 December, Lancasters of 8 Group and 3 Group carried out daylight raids on the Urft Dam and the railway yards and the benzol plant at Osterfeld. These raids appeared successful and just two aircraft were lost. *Y-Yorker*, skippered by Harry Yates, was one of 16 Lancasters on 75 Squadron that took part in the raid on Osterfeld:

> Intervention by the enemy was minimal. No Luftwaffe fighters came up to challenge our escort. And for some extraordinary reason, the guns that brought destruction on 30 November did not even achieve half the rate of fire. As usual with these daylight 'G-H' ops over 10/10ths cloud, the lead-in was anything but uniform. Three of the Mepal crews

N.L. Howlett with air and ground crew on *Howlett's 'Hooligans'* on 626 Squadron in late 1944.

reported an undershoot by between one and seven miles. One crew reckoned they were led twenty miles off track. Another bombed Essen. There were no losses among the 150 Lancasters that flew to the Ruhr that day.[12]

Next day, 140 Lancasters of 3 Group carried out a 'G-H' raid on the Ruhrstahl steelworks at Witten. Fw 190s intercepted the force in the target area and they struck at the first wave of vics, shooting down six Lancasters. Then they raced for the cloud cover and safety as Spitfires came in. Two other Lancasters collided right in front of F/O Alex Simpson on 75 Squadron. In the resultant explosion one of the Merlins was detached. Still fully cowled and with the propeller milling, it came straight for Simpson's cockpit. He plunged his Lancaster downward and somehow the engine passed clean over. In another collision both Lancasters managed to stay airborne.[13] That night 540 aircraft attacked Essen. Six Lancasters were missing in action.

When 13 December broke it was under a very heavy frost. Towards mid-morning thick fog enveloped stations in Norfolk and visibility deteriorated from 2,000yd to less than 80, so operations were scrubbed. The fog would hamper operations across East Anglia for two weeks, but that night 52 Lancasters and seven Mosquitoes of 5 Group took off for Norway to attack the German cruiser *Köln*. However, by the time they reached

Repairs being carried out to a Lancaster on 300 'Masovian' Squadron.

Oslo fjord, the ship had sailed so instead other ships were bombed. On 15/16 December, 327 Lancasters and Mosquitoes of 1, 6 and 8 Groups raided the northern part of Ludwigshafen and the small town of Oppau where two important chemical factories, the vast BASF (Badische Anilin- und Soda-Fabrik) and I.G. Farben Industrie, were situated, at the point where the Neckar and Rhine rivers meet. The raid was very successful, with 450 HE bombs and many incendiaries falling on the premises of I.G. Farben Industrie. The plant ceased production of synthetic oil completely 'until further notice'. One Lancaster was lost.

Two nights later, Ulm received 1,449 tons of bombs during a 25-minute raid by 317 Lancasters and 13 Mosquitoes of 1 and 8 Groups, and Munich and Duisburg were also hit. Two Tame Boars each destroyed a Lancaster over Ulm. One square kilometre of the city was completely engulfed by fire and 29 industrial premises were badly damaged. This was Bomber Command's first and only raid on the old city. A third Lancaster was lost after it collided and crashed at Laon in the Aisne. Bomber Command claimed 'severe and widespread damage' in the old centre of Munich and at railway targets, which were attacked by 280 Lancasters and eight Mosquitoes of 5 Group. Eight Lancasters, including one which crashed and

exploded at Worthing in Sussex killing all the crew, were lost on the Munich attack.

On 18/19 December, 236 Lancasters of 5 Group attacked Danzig (Gdynia) on the Baltic coast and caused damage to shipping, installations and housing in the port area. In the Gotenhafen and Danzig areas three crews of I./NJG5 claimed all four Lancasters that failed to return from the Main Force raid. Another 14 Lancasters of 5 Group dropped mines in Danzig Bay.

On 21 December, 113 Lancasters of 3 Group again attempted to bomb the railway yards at Trier in two waves. Cloud covered the target and crews were unable to observe results but a large column of smoke eventually appeared and the second wave caused heavy casualties. All aircraft returned safely. That night, a total of 475 heavies attacked the Cologne/Nippes marshalling yards, which were being used to serve the German offensive in the Ardennes, railway areas at Bonn and the hydrogenation plant at Pölitz near Stettin, while four Lancasters of 5 Group flew a diversionary raid to Schneidemühl. Sixteen Lancasters of 617 Squadron, each carrying a 'Tallboy', formed part of an all-5 Group force of 207 Lancasters that attacked Pölitz. The Dam Busters were not happy about the choice of target or that it was at maximum range and the weather was bad. At some stations the fog was so thick that aircraft could not be seen taking off. Post-raid reconnaissance showed that the power-station chimneys had collapsed and that other parts of the plant were damaged, but overall the bombing results were not impressive. Of the 11 'Tallboys' dropped, it was thought that at least three had fallen in the target area, probably to the north of the plant. It was considered afterwards that W/C John Woodroffe, the Master Bomber, had assessed the markers as being closer to the target than they were, so leading much of the attack astray.[14] Four Lancasters were lost, two of them crashing in Norway, killing everyone on board. Five more crashed in fog from Wick to Lincolnshire attempting to land back at their bases.

Thick cloud at Cologne and Bonn prevented accurate raids: photo reconnaissance revealed that only a few bombs fell on the yards at Nippes and none on the railway area at Bonn. No aircraft were lost on these raids, but when 23 Lancasters on 525 Squadron that took part in an attack on Bonn returned, Kelstern was fog-bound and they had to use Ludford, Fiskerton and Sturgate, all equipped with FIDO.

There was a near disaster at Wickenby on 22 December when a 626 Squadron Lancaster developed an engine problem on take-off for a raid by 166 Lancasters and two Mosquitoes on Koblenz. With more Lancasters taking off, the pilot, F/O Reginald Roderic Preece, was ordered to head for Leeming in North Yorkshire, but his aircraft suddenly went out of control and crashed in the bomb dump bays. There was a huge explosion as the Lancaster with its full load went up. The station records speak of an 'enormous' crater and the good fortune the whole camp had that the Lancaster did not hit the dump itself. No trace of any of the seven men on board was found.

On 23 December, a high pressure front pushed through enough moisture from the skies over the Ardennes after four days of fog, snow and freezing rain and 153 Lancasters of 3 Group attempted to bomb the railway yards at Trier through cloud. Another 27 Lancasters and three Mosquitoes carried out an attack on the Cologne/Gremberg railway marshalling yards to disrupt enemy reinforcements for the Battle of the Bulge. The force attacking Cologne was split into three formations, each led by an Oboe-equipped Lancaster with an Oboe Mosquito as reserve leader. S/L Robert A.M. Palmer DFC* on 109 Squadron and his crew at Little Staughton led the first formation in *V-Victor*, an Oboe-equipped Lancaster borrowed from 582 Squadron on the station. The raid went very badly. During the outward flight two Lancasters on 35 Squadron collided over the French coast killing everyone on board. At the target Palmer and *V-Victor* came under intense AA fire. Smoke billowed from *V-Victor* and Fw 190s knocked out two of the engines and destroyed a tail fin, but Palmer carried on and completed the bombing run. The Lancaster then went over on the port side and went down, splitting in two. There was only one survivor. Palmer's Lancaster was followed down by four more Lancasters on 582 Squadron. In April 1945, the award of a posthumous VC was made to S/L Robert Anthony Maurice Palmer.

On Christmas Eve 1944, 97 Lancasters and five Mosquitoes of 1 and 8 Groups attacked the marshalling yards at Cologne/Nippes and 104 Lancasters of 3 Group bombed Hangelar airfield near Bonn, losing one aircraft. Roy Abbott on 149 Squadron recalled: 'Our particular target was the officers' mess.' The nights following were much quieter, with no Main Force operations on Christmas night because of bad weather. On Boxing Day the weather at last improved and allowed Bomber Command to dispatch 294 aircraft, including 146 Lancasters, against German troop positions near St-Vith. Two Halifaxes failed to return. Next day, 200 Lancasters and 11 Mosquitoes attacked the railway yards at Rheydt for the loss of

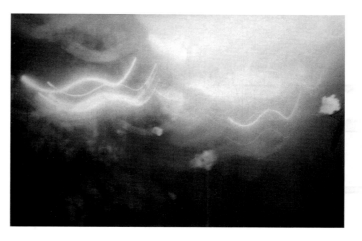

Ludwigshafen on 15/16 December 1944 from 19,000ft. A total of 327 Lancasters and 14 Mosquitoes of 1, 6 and 8 Groups attacked the northern part of Ludwigshafen and the small town of Oppau in which two important I.G. Farben chemical factories were situated. Severe damaged was caused to the factories, with 450 HE bombs and many incendiaries falling on the target. One Lancaster was lost. (*Via 'Pat' Patfield*)

one Lancaster and a Mosquito. That night, 328 aircraft – 66 of them Lancasters – attacked the marshalling yards at Opladen 3½ hours after an earlier raid by Mosquitoes. Two Lancasters failed to return. Next day, 167 Lancasters of 3 Group attacked the Cologne/Gremberg marshalling yards. Bombing was accurate and no aircraft were lost. S/L 'Jock' Calder led 16 Lancasters of 617 Squadron in the afternoon of 29 December to bomb the E-boat pens at Waalhaven, near Rotterdam. The clear winter skies made marking unnecessary and, with no enemy fighters to concern them, 617 Squadron flew without any escort. There was only light flak over the pens so the Lancasters flew over at heights of between 16,000ft and 18,000ft to drop their 'Tallboys'. One bomb undershot the target by 200yd and another was to the west, but the rest hit the target. Two separate forces that bombed railway yards in Koblenz returned without loss.

On the night of 29/30 December, 324 Lancasters and 22 Mosquitoes of 1, 6 and 8 Groups attacked Scholven-Buer, with the loss of four Lancasters. Another 197 aircraft attacked the railway yards at Troisdorf but most of the attack missed the target. The day of 30 December dawned grey and frosty but fog-free. Snow was falling gently. Thirteen Lancasters on 617 Squadron set out to bomb the U-boat pens at IJmuiden, but the raid was abandoned because of the weather. The night following, 470 aircraft – including 93 Lancasters – attacked the Cologne/Kalk railway yards, and 154 Lancasters and 12 Mosquitoes of 5 Group bombed a German bottleneck in a narrow valley at Houffalize in Belgium, one Lancaster crashing in France. One Halifax and a Lancaster were lost on the raid on Kalk.

Lancaster on 218 'Gold Coast' Squadron.

On the 31st, 3 Group were detailed to carry out a 'G-H' raid on the railway yards at Vohwinkel. At Mepal, Harry Yates awoke to the last day of the old year to find that he was on the battle order. Just this op and he would complete his tour:

'The target today,' the CO announced, a certain, routine note in his voice, 'is marshalling yards at Vohwinkel in, er …' He straightened the curtain pull while seventeen crews shifted uneasily in their seats. Nobody was too sure of the exact location of Vohwinkel. It was not a name we had heard before. When the curtain swept back a ghastly groan went up. '… the Ruhr. This is a 3 Group effort. One hundred and eighty aircraft will attack at 14.00 hours from a bombing height of 19,000ft.'

For the thirtieth and last time I watched for the hour it took Briefing to unfold. The manifold aspects of bomb load, take-off time, intelligence, weather, call signs … every blessed detail was provided and dutifully digested. The choice of target – marshalling yards deep in the German supply system – was almost certainly linked to events in the Ardennes. The rationale was the same as that for the recent attacks on Trier, Rheydt, Gremberg and Koblenz.

All Briefings for a Ruhr op required a close study of the 88mm gun concentrations around the valley. The map was scarred with red, so much of it that I wondered why we continued with the delusion that quiet skies were waiting for us somewhere out there. They weren't. It was impossible to reach any Ruhr target without running a gauntlet of shellfire. One could never say so publicly, of course. That would have brought an instantaneous reprimand from the Station Commander for lack of patriotism or for damaging morale.

'N-Nan,' shouted the WAAF driver. 'That's us,' I replied.

A chorus of 'Good luck' rang out from the other crews as we jumped down from the tailgate.

'And to you,' replied the boys.

Tim Blewett was in the truck with his crew on his way out to The Paper Doll.

'You lucky devils,' he said. 'We've still got two dozen to go.' Then recognising that a second dickey was starting his account with us today as he himself had done three weeks earlier, he acknowledged Rob Aitchison, a keen young New Zealander,[15] with a wave and said, 'See you in the smoke.'

Checks were accomplished well within the fifteen minutes I generally allowed. Everyone was set. I waved to the boys on the chocks and swung Nan round to the perimeter track. Good practice recommended the cautious skipper to tarry at dispersal until the rest had rolled by on their way to the caravan. But there were only seventeen in the queue and someone had to get the show on the road. It was 11.30 hours exactly when, first in the order and with Nan's Merlins singing and every rivet and panel raring to go, the Aldis flashed green beside us.

'Let's go, boys,' I said, as I had from the very beginning. Whatever was to happen in the coming hours, those words would never fall from my lips again.

Halfway down the long, main runway Tubby had the throttles locked. Nan bore on, tail up and going light. At 110mph on the airspeed indicator I lifted her off. The undercart was all but fully retracted as we cleared the airfield boundary and, in a wide, climbing circuit, cut our way into dense and turbulent cloud at 1,200ft.

Two Lancasters failed to return.

N-Nan and Harry Yates' crew re-crossed the English coast at 5,000ft and broke cloud at 1,500ft over Chatteris to await their turn to land. It was raining steadily and visibility was poor. The wet streets glistened below. N-Nan received the all-clear to enter the circuit. As they crossed the Old Bedford River ahead of them the flarepath shone.

'N-Nan you are clear to land,' came a WAAF voice from Control. They had made it.

That night – New Year's Eve – as Harry Yates' crew celebrated until two in the morning, 149 Lancasters and 17 Mosquitoes of 1 and 8 Groups attacked the railway yards at Osterfeld where three Lancasters failed to return. One of the missing Lancasters was *Q Squared* on 626 Squadron at Wickenby, which lost a second Lancaster when *X Squared* crash-landed on return at Manston. *Q Squared* was piloted by F/O Richard O. Beattie, one of five Canadians on the crew who were on their 15th operation. They had just dropped their bombs and were turning for home when they were hit by a blast of cannon fire. Sgt Herbert Harrison, the flight engineer, believed that it came from below and astern because it almost certainly killed F/O William Henry Pogson, the Canadian rear gunner, and put two engines out of action. 'There was no warning and it is likely that we were the victim of a twin-engined night-fighter equipped with *Schräge Musik*. The fire in the engines quickly spread to the wing tanks and Dick Beatty gave the order to bale out.'

Harrison, a Yorkshireman, remembers sitting on the edge of the forward hatch debating whether to go. He had never used a parachute before; he knew the theory, but the practice, 18,000ft above Germany, was a different matter. His next recollection was of spinning through the air wondering what he had to do next. He took hold of the D-ring, there was a terrific crack and he found himself floating in the freezing air above an area he later discovered to be between Aachen and the Ardennes. He caught a brief glimpse of his blazing Lancaster and then plunged through a thin layer of cloud and never saw it again. He landed in a wooded area in deep snow in what appeared to be a country estate. It just gave him that impression. Everything was so orderly. 'I wandered around like a lost sheep for a bit, trying to decide what to do,' he said. 'Then, quite by chance, I found a bicycle, just leaning against a wall. It was a weird and wonderful thing. There were no tyres, just cable wrapped round the wheel rims and bolted on. But it worked.'

Herbert Harrison spent four days on the run in the snow-covered Ardennes and at one stage he chewed leaves to try to satisfy his hunger pangs. Hunger, cold and sheer fatigue had dulled his senses when, finally, he was taken in by the American 1st Division and, after a brief interrogation, thrown into a cellar. Next day he was taken to Liege and his

Loading incendiaries.

story was finally verified. He was later flown home in a Dakota from Brussels with a party of American Red Cross nurses for company. His wife was staying with her parents and had received no notification that he was missing. When he arrived he was still in the clothes he had escaped in and was covered in mud. His eyes were bloodshot. His mother-in-law refused to believe what had happened and was convinced that he had been on a big drinking binge! Harrison later discovered that only three others of his crew had survived – Beattie, who had broken his back when he landed, and the navigator and bomb aimer. A memorable reunion in Lincoln ended in the city's police cells! Herbert Harrison flew 12 more ops on 101 Squadron at Ludford before the war ended.

December had proved costly for 166 Squadron. Four aircraft were lost, one over Karlsruhe on 4/5 December, and two over Cologne on Christmas Eve. The fourth was *X-X-Ray* flown by F/O Stanley Roy Hanna RCAF. Coming back from the Karlsruhe raid, it hit trees in Brocklesby Park and burnt out. There were no survivors. In another incident, a Lancaster was damaged in an attack on the Leuna refinery near Leipzig, and five of the crew baled out over France. The pilot, F/O Lee, and his navigator remained with the aircraft and brought it back to Manston in Kent. Both were later decorated and all seven of the crew were reunited within a short period at Kirmington.

LANCASTER LEGACY

I always say there were three great aircraft in the Second World War on the British side and our three were the Lancaster, the Mosquito and the Spitfire. There was only one, in my mind; as soon as you got into it as a pilot you knew it was a great aircraft. That was the Lancaster. You got in and you said 'This is a pilot's aeroplane'. It looks right, it feels right. You looked around it, the view is great, it just sits right on the ground. It's one of these aircraft. Much the same can be said of the Mosquito and the Spitfire, but they look a bit more lethal to the pilot. They remain great aeroplanes. And, of course, all three were blessed with having the Merlin engines which were incredibly reliable.

Captain Eric 'Winkle' Brown RN

On New Year's Day 1945, 152 Lancasters and five Mosquitoes of 5 Group took off just before dawn for the Dortmund-Ems Canal at Ladbergen, which the Germans had once more repaired. The weather caused the first casualties. At Bardney two Lancasters on 9 Squadron crashed on take-off. One Lancaster was lost from the 146 aircraft of 3 Group that successfully attacked the railway yards at Vohwinkel. No aircraft were lost from the force of 105 Halifaxes of 4 Group and 18 Lancasters and 16 Mosquitoes of 8 Group that attempted to bomb a benzol plant at Dortmund. Bombing was scattered and none of the bombs hit the plant. At the Dortmund-Ems Canal bombing was carried out from just 9,000ft. A third Lancaster of 9 Squadron was lost when *D-Dog* was shot down. Immediately after the bombing run *U-Uncle* was hit by a salvo of two 88mm shells and the Lancaster was set on fire. F/Sgt George Thompson, the 24-year-old wireless

The crew of AR-O on 460 Squadron RAAF at Manston in Kent after an emergency landing following a mid-air collision with another Lancaster while en route to Munich on 7/8 January 1945. Back row, L–R: Dave Fellowes (rear gunner); D. Collett,(navigator); F/O Arthur 'Whitty' Whitmarsh DFC RAAF (pilot); Jack Wilson RAAF (WOp). Front: A. Shepherd (engineer); Ken de la Mare (mid-upper gunner); Jock Turnbull (bomb aimer).

operator, saw that the mid-upper gunner was unconscious in the blazing turret. He battled through the flames and managed to drag the gunner clear and extinguish his burning clothing with his bare hands. Severely burnt himself, Thompson then noticed that the rear gunner was also trapped and unconscious in the flames. Again he went back and extracted the gunner and extinguished his clothing with his hands. Thompson then crawled forward through the badly holed fuselage and reported the situation to the pilot. Such were his injuries, including by now frostbite, that the pilot couldn't recognise his valiant wireless operator. F/O Denton RNZAF crash-landed *U-Uncle* 40 minutes later and Thompson was rushed to hospital but he died three weeks later, on 23 January, of pneumonia. For his brave and selfless actions, Thompson was awarded a posthumous VC. Denton was awarded the DFC.

On 2/3 January, 514 Lancasters and seven Mosquitoes of 1, 3, 6 and 8 Groups bombed Nuremberg and 389 aircraft raided Ludwigshafen. At Nuremberg, with the help of a rising full moon and in clear visibility, the Pathfinders carried out good ground-marking, and the centre of the city, in particular the eastern half, was destroyed. Over 1,800 people were killed and over 400 separate industrial buildings were destroyed. For Roy Abbott on 149 Squadron it was the last but one raid of his tour of operations: 'The night-fighters came up in force to meet us and had no difficulty in picking off bomber after bomber by the light of searchlights and the full new moon. I felt a bit lucky to be able to tell the tale, but survival was not all luck; much depended on the skill and training of the crew.' Four Lancasters were lost at Nuremberg, and Lancaster X KB700, better known as *Ruhr Express* on 419 Squadron, piloted by

F/L A.G.R. Warner RCAF, overshot the runway at Middleton St George on return. An anonymous RCAF airman recalled:

Ruhr Express was one of the first Lancasters built in Canada. It was just finishing up its last trip – I think it had already completed a tour or two – before going on display back in Canada. I remember the pilot of it once saying that if it wasn't going to the Ruhr, then it would always have engine trouble or something like that and would end up having to turn back. It couldn't go any place but the Ruhr. At about this time our base was under construction. All the machinery was pulled off the runway and parked. Anyway, when the planes were coming back after a raid – it was still dark – the *Ruhr Express* landed and overshot the runway a bit. The pilot stopped and turned the Lanc around to taxi back down the tarmac, but in doing so he ran up one of the construction diggers, straddled the wheels, ran under the body and got jammed. Well, the plane caught fire! Everybody got out OK, but the *Ruhr Express* went up in flames. The unused flares and ammunition caught fire too and we witnessed the greatest fireworks display on the base! Flares going off in every direction; its very last trip![1]

At Ludwigshafen the Main Force used the area of the two I.G. Farben chemical factories as their aiming point. The bombing was accurate, with about 500 HE bombs and 10,000 incendiaries falling inside the limits of the two factories, and 1,800 people were bombed out.

On 3 January, the Main Force went out in daylight again when almost 100 Lancasters of 3 Group made 'G-H' accurate attacks through cloud on the Hansa benzol plant at Dortmund and another benzol plant at Castrop-Rauxel.[2]

No night-bombing operations were carried out by the heavies on 3/4 January, but on the following night 347 Lancasters and seven Mosquitoes of 1, 5 and 8 Groups controversially attacked Royan at the mouth of the River Gironde. Upwards of 800 French civilians were killed. Four Lancasters were lost and two more collided and crashed behind the Allied lines in France. The number of aircraft lost in mid-air collisions is not known but it is generally accepted that the numbers were few. On 5/6 January, 23 Halifaxes and eight Lancasters failed to return from the raid on Hanover, the majority shot down by an effective Tame Boar operation, while two Lancasters were lost on a raid on a bottleneck in the German supply system in the Ardennes in a valley at Houffalize in Belgium.

Left: F/O Arthur 'Whitty' Whitmarsh DFC RAAF in the cockpit of *O-Oboe*.

Below: Lancaster engulfed by fire.

The next night, over 600 aircraft set out to bomb an important German rail junction at Hanau and the marshalling yards at Neuss. Many of the bombs dropped by 468 Halifaxes and Lancasters of 1, 4 and 6 Groups at Hanau and 147 Lancasters of 1 and 3 Groups at Neuss missed the targets and fell in surrounding districts. Hanau was reported to be '40 per cent destroyed', while in Neuss over 1,700 houses, 19 industrial premises and 20 public buildings were destroyed or seriously damaged. Four Halifaxes and two Lancasters failed to return from the raid on Hanau.

On 7/8 January, Bomber Command returned to area bombing with the final major raid on Munich, which was memorable for the sheer beauty flying along Lake Constance

Top: 101 Squadron in spring 1945.

Above: Air and ground crew pose for the camera behind Lancaster III PA995 BQ-V *The Vulture Strikes!* on 550 Squadron at North Killingholme.

The raid on Munich was followed by one on Krefeld and three visits to Saarbrücken. A series of oil locations were then targeted by Bomber Command and on 13/14 January over 200 Lancasters attacked the Pölitz plant near Stettin and reduced it to a 'shambles' for the loss of just two Lancasters. The following night, over 570 Lancasters took off for a raid on the synthetic-oil plant at Leuna and caused severe damage throughout the facility. On the 15th, a benzol plant at Recklinghausen was the target for over 80 Lancasters of 3 Group; again the bombing appeared to be 'excellent' and no aircraft were lost. Another 60-plus Lancasters of 3 Group attacked the Robert Müser benzol plant at Bochum and again all aircraft returned. Just over 300 aircraft destroyed 44 per cent of the built-up area of Magdeburg on the 16th/17th, when four separate targets were hit on this cold winter's night. Over 230 Lancasters and Mosquitoes raided the synthetic-oil plant at Brüx in Czechoslovakia and oil production was set back severely. Only one Lancaster failed to return from the raid on Brüx. The Krupp benzol plant at Wanne-Eickel was also attacked, by 138 Lancasters of 3 Group. No results were known and one Lancaster was lost. Another 328 Lancasters caused much damage to the northern half of the Braunkohle-Benzin synthetic-oil plant at Zeitz near Leipzig. Ten Lancasters were lost on the raid.

The next Main Force op was on 22/23 January, when the benzol plant in the Bruckhausen district of Duisburg was attacked by 286 Lancasters and 16 Mosquitoes for the loss of two Lancasters. Another 152 aircraft carried out an area-bombing raid on Gelsenkirchen. All aircraft that were dispatched returned safely from Gelsenkirchen. On 28 January, 153 Lancasters of 3 Group attacked the railway yards at Cologne/Gremberg in conditions of good visibility. Three Lancasters were shot down and a fourth crashed in France. Night operations resumed on 28/29 January, when the target for 602 aircraft – 258 of them Lancasters – was the Stuttgart area. F/O Peter Birt's crew on 460 Squadron RAAF at Binbrook were on standby after returning from leave. The day before, as F/Sgt Doug 'Ben' Benbow, the WOp, who was from Adelaide, recalls:

Our kite *A-Able* had been taken up by a new crew on a training flight and who, on their return to base, had landed in the bomb dump, not doing much good to the bomb dump nor *A-Able* so we took off for Stuttgart in *S-Sugar*. No sooner were we airborne than Sgt Alf Fields (the flight engineer), who was from Halifax in Yorkshire, said to Pete: 'This isn't *S-Sugar*, it's *S-Shit*; bloody Packard-Merlins

with the snow-covered Alps crystal clear in the moonlight. The raid was carried out by 645 Lancasters and nine Mosquitoes of five Groups.

Bomber Command claimed a successful area raid, with the centre and some industrial areas of Munich being severely damaged. A Lancaster on 630 Squadron crashed after returning to East Kirkby soon after take-off. Two crew members were killed and five were injured. Ten Lancasters were shot down, three more went down after collisions and a 106 Squadron Lancaster flew into trees near Void-Vacon. A 467 Squadron Lancaster returning to Waddington was lost when it crashed near Eye in Cambridgeshire. All the crew perished in the crash.

instead of Merlins – we'll never keep up!' How true, we were twenty minutes late over target and a Junkers 88 came down on us from seven o'clock high at about 23.43 hours. It was either at or near full moon and it was possible to clearly see every detail of the Ju 88. F/Sgt Gordon Wilson (tail gunner) of Nowra said to Pete: 'Dive port Pete, dive port!' Down we went but as we sheered off the 88 hit us with cannon shells in the port inner motor and wing, both of which burst into flames. The thing that stands out in my memory is the flashes of the cannons – who could forget that?

Pete to Alf: 'Feather port inner Alf.'

Alf to Pete: 'She won't Pete.' The fire extinguishers had no effect. The port wing was extensively damaged and cannon shells were lacing through the fuselage.

A second 88 joined in the attack. Gordon to Pete: 'I got one Pete.' His last words. I was at the astrodome and saw it going down. Then a long burst of cannon shells took off the rear turret and Gordon with it. I stepped down from the astrodome and sent an SOS to Group with our call sign. Not, of course, for help but just to let them know the time of the attack. By now we were in a very steep dive from 18,000ft. Pete: 'Bale out, bale out.' Cannon shells were still hitting the aircraft. Alf, Ern Truman (bomb aimer) and Spen Symes (navigator) – both from Melbourne – went out the front hatch. I went to the rear and found the rear door shot to pieces and the broken hydraulic lines to the rear turret spraying burning oil across the open rear end.

F/Sgt Doug 'Scarlett' O'Hara (mid-upper) from Glen Innes was wandering about in shock, perhaps wounded, so I grabbed his arm and signalled forward. I scrambled over the main beam, gave Pete, who was still at the controls, the thumbs-up and then found the forward hatch cover had somehow blown diagonally across the hatch and jammed, so there was much kicking and pulling before forcing an exit. I was partially concussed by a cannon shell exploding nearby just as I exited. I was the last man out of the last bomber shot down in the last raid on Stuttgart in WWII.

The aircraft with full bomb load, 10,000lb of mixed high explosives and incendiaries, hit the deck and exploded at about 23.49 hours. I later learned that the remains of Pete and Scarlett lay nearby; the remains of the tail turret and some bits and pieces of Gordon were found some distance away. Spen, Ern and Alf landed in and about Stuttgart, whilst I landed in the top of a very tall pine tree many miles from Stuttgart. My chute had only just opened. I thought I had

suffered a bit of damage; as it turned out later, it was quite a bit. It will never be known if Scarlett had been wounded; he may have got out of the aircraft, but too late. There hadn't been any panic but I still remember the fear and the way

Left: F/O George Blacker (in the cockpit) brought *The Vulture Strikes!* home safely from its 100th operation on 5 March 1945 and successfully completed his own operational tour. Two nights later, on 7/8 March, on its very next trip, it was one of three Lancasters on 550 Squadron lost when F/O Cyril John Jones RCAF (killed in action) and crew failed to return from Dessau.

Below: Pilots on 424 'Tiger' Squadron RCAF with *Piccadilly Princess* at Skipton-on-Swale, Yorkshire, in 1945. *(Ralph Green)*

Pete stuck to the controls and that Gordon had shot down an 88. Our crew which, because of overall ability, had been considered the one most likely to survive had been taken out by two 88s because of those bloody Packard-Merlins.[3]

In February 1945, Bomber Command mounted no fewer than 40 raids. On 1 February, Lancaster I NG243 BQ-M2 on 550 Squadron – better known as *Mike Squared* – came close to disaster when the port inner engine caught fire as F/O Ken Sidwell was taking the bomber off the main runway at North Killingholme for a raid on Ludwigshafen. Flight engineer Jack Allen explained:

Right: Bomb aimers on 424 'Tiger' Squadron RCAF with *Victorious Virgin*.

Below: A 106 Squadron crew at Metheringham relax in the spring sunshine in 1945 near the station firing range. L–R: F/O J. Bell (pilot); Sgt D. Sykes (engineer); F/Sgt D. Peck (bomb aimer); F/Sgt S. Bottriell (navigator); F/Sgt E. Payne (WOp); Sgt G. Gorman (mid-upper gunner); Sgt P. Thomas (rear gunner).

A conrod had come through the side of the engine, there was a bit of a fire and fuel was blowing back onto the exhaust stubs. I was standing at the side of the skipper and I just kept my eyes on the dials. The boost was going up, down, up again and finally it started to go altogether so I pushed the feather button. Our skipper did really well. We were down to about 400ft over Habrough Station and the port wing dropped like a dose of salts. We were full of fuel and bombs and we almost went in near the station. If we had, I think we would have ended up in Grimsby in bits but the skipper kept us in the area. We were ordered to take our bombs out to sea and dump them in a designated drop area before returning to North Killingholme. We were an experienced crew, otherwise we might have been sent to the diversion airfield at Woodbridge with its extra-long runway. We only dropped the cookie and decided to take the rest back with us. The skipper put the Lancaster down like a duck landing on water. It was very impressive.

The following night, a new engine fitted, *Mike Squared* made a 6-hour 40-minute round trip to Wiesbaden on the Rhine. At the end of March, Sidwell, his tour over, went to a Heavy Conversion Unit. Within a few weeks he took a new crew out over the North Sea and the Lancaster was not seen again.

A new aircraft made its appearance at Kelstern in that cold January, a prototype of the new Lincoln bomber. It was flown to the Wolds base from Boscombe Down to find out how it could best be fitted into the T2 hangars then widely used on bomber airfields in Lincolnshire. The Lincoln, designed as a successor to the Lancaster, was a big aircraft and it was realised that trolleys would be needed to swing the aircraft into the hangars sideways. The trials were conducted in the presence of AVM A.P.M. Sanders, who had arrived from Bomber Command HQ at High Wycombe especially for the demonstration. It seemed both he and the officials from Avro, the Lincoln's builders, were happy, and the following day the aircraft returned to Boscombe Down. Lincolns were to be in squadron service in Lincolnshire later that year.

F/Sgt Harry 'Paddy' Kelso, an Irish air gunner with 101 Squadron, gives a graphic account of his experiences over Ludwigshafen, manning the Frazer Nash Type 50 mid-upper turret on Lancaster ME419 on his crew's first operation on 1/2 February:

At around 20.40, the navigator advised the Skipper, 'Target ahead fifteen minutes' and gave the bomb aimer a course on to the bombing run. Nerves were on edge and everyone was alert with no need for 'Wakey Wakey' tablets. Our Special Operator advised us of enemy fighters ahead as he had heard Ground Control issue the order to scramble. I must have checked my guns a dozen times during this period. We were now experiencing heavy flak as the Ludwigshafen defences knew where we were heading. Shells seemed to be exploding everywhere and searchlights were combing the sky all around the target. From my mid-upper turret I could get an all-round view and looking ahead I saw this massive concentration of bursting shells and weaving searchlights and thought, 'God Almighty, have we got to go through that?' There was also the frightening sight of exploding aircraft from direct hits.

We managed somehow to get on to our bombing run and I heard Jack say, 'Bomb doors open, steady, left, left, steady – right a bit, steady' and then after what appeared to be a long pause, 'Bombs gone.' The aircraft seemed to heave a sigh of relief as 12,000lb of bombs dropped away to fall on hapless Ludwigshafen. We still had to fly straight and level for another ten seconds (which seemed an hour) in order to photograph the results of our bombing. During this time shells were exploding all around us and we could do nothing but fly a level course. Searchlights seemed everywhere and it seemed a miracle they didn't find us. At last the pictures were taken and we headed for home thinking the worst was over. We had only just cleared the target area and were feeling quite exhilarated at having completed our first op when suddenly a stream of tracer came up from port quarter down. Bill Green in the rear turret saw the gun flashes and immediately gave the order, 'Fighter port quarter down – corkscrew port – go.' We both opened fire in the general direction of the fighter but I doubt if we hit anything as it was very difficult to aim accurately from a diving aircraft and we only had a few seconds to do so. On getting the command to corkscrew, Ted, our pilot, threw the Lancaster into the most violent downward turn I had experienced. We had practised corkscrews in training, but nothing like this. I think Ted must have thought he was flying the Tiger Moth on which he did his initial training. We jinked and weaved all the way down from 20,000 to 12,000ft where Ted saw a large bank of cumulus cloud into which we dived and managed to escape the fighter's attention.

On 2/3 February, Bomber Command mounted raids on Mannheim, Wanne-Eickel, Wiesbaden and Karlsruhe by 1,200

aircraft, about 250 being ordered to raid Karlsruhe. Owing to adverse weather conditions and extensive Luftwaffe night-fighter activity near and over the targets, operations were only partially successful. Twenty-one aircraft failed to return, including 11 from the Karlsruhe force, and another 13 crashed in liberated French territory. It was a bad night for 189 Squadron, with four aircraft failing to return to Fulbeck. Further raids followed, on the Prosper and Hansa benzol plants at Bottrop and Dortmund respectively, and on the plant at Osterfeld and

Left: Lancaster over Holland on 4 March when 128 Lancasters of 3 Group carried out a 'G-H' attack through cloud on Wanne-Eickel. No aircraft were lost. ('Spud' Taylor)

Below: F/O Vallance and crew on 626 Squadron at Wickenby in 1945.

187

On 14 March 1945, Lancasters of 5 Group attacked the Bielefed and Arnsberg railway viaducts in Germany. Twenty-eight Lancasters dropped 12,000lb 'Tallboy' bombs, and the 617 Squadron Lancaster of S/L C.C. Calder dropped the first spin-stabilised 22,000lb 'Grand Slam' bomb (pictured), at Bielefed. The Arnsberg viaduct, 9 Squadron's target, was later found to be undamaged but near misses at Bielefed created an earthquake effect which caused 100yd of the viaduct to collapse. The 'Grand Slam' bomb, 41 of which were delivered before the end of the war in Europe, contained approximately 11,000lb of Torpex D.

the Nordstern synthetic-oil plant at Gelsenkirchen. Bonn too was subject to an attack by 238 aircraft.

On 7 February, 100 Lancasters of 3 Group attacked the Krupp benzol plant at Wanne-Eickel again. Only 75 aircraft were able to bomb in wintry conditions, which scattered the force, and the results were unknown. One Lancaster failed to return. That night 464 aircraft[4] bombed Goch, and 295 Lancasters and ten Mosquitoes of 1 and 8 Groups attacked the small town of Kleve, about 5 miles west of the Rhine.[5] Both raids were mounted to prepare the way for the attack of the British XXX Corps across the German frontier near the Reichswald. Considerable damage was caused in Goch but most of the inhabitants had probably left the town. At Kleve, red and yellow 20mm and 37mm tracer shells were criss-crossing from the flak batteries outside the town. Flashes from the exploding blockbusters on the ground were blinding. John Gee on 153 Squadron at Scampton was nominated to take Richard Dimbleby and his engineer and recording equipment. In his commentary Dimbleby could not conceal from listeners the reality that Kleve had been 'utterly destroyed already' and was 'nothing more than a heap of rubble'.

'It is the most extraordinary sight I have ever seen in the air,' he began as the Lancaster started its bombing run:

Down go more target indicators, a medley of bright, indeed lovely colours, their reflections glowing on the thick white and dark grey clouds that are rolling up … Our bombs are going. The flak is bursting just under us. We are going over the top now. There's more fire. I don't know how we can

stand this. We are shaking with the flak. But how steady the crew and skipper are as they hold to their course. Our bombs are bursting there now, flash, flash, flash. I am sorry, I tried to be steady and contained on this commentary but it is more than I can do. It is a staggering sight to see in the sky.

One Lancaster was lost when it crashed on its run-in, blowing up with its full bomb load.

On 8/9 February, Bomber Command returned to attacks on synthetic-oil plants when Pölitz was bombed by 475 Lancasters and seven Mosquitoes. The attack took place in two waves, the first being marked and carried out entirely by the 5 Group method, the second being marked by the Pathfinders of 8 Group. The weather conditions were clear and the bombing of both waves was extremely accurate. Severe damage was caused to this important synthetic-oil plant. It produced no further oil during the war. Ten Lancasters were lost and one crashed near Hjortshög in Sweden. The pilot survived and was interned, but all of his crew died in the crash. F/O D.E.J. Chalkley's Lancaster was on its way back from the raid on Pölitz when it collided with a 192 Squadron Halifax over the Danish coast. The Lancaster survived and landed back at Kelstern with a chunk of the Halifax embedded in its wing. The Halifax crashed into the sea and all eight crew perished.

Two hundred Halifaxes carried out a smaller-scale raid on the Krupp benzol plant at Wanne-Eickel, and 151 Lancasters attacked the Hohenbudberg railway yards at Krefeld, but both of these raids were largely unsuccessful. Forty-seven RCM sorties were flown for the loss of a single Halifax on 192 Squadron, which collided with a Lancaster on 625 Squadron on the way home. The Lancaster made it back but the Halifax crashed into the sea with the loss of all the crew. Two other Lancasters were lost. Two Halifaxes on the Wanne-Eickel raid failed to return when one crashed near Dunkirk and the other was abandoned over Belgium. A Halifax on 426 'Thunderbird' Squadron crash-landed near Wetherby in Yorkshire, injuring the pilot and one of his crew. The other five men on board were killed.

There then followed a series of minor operations involving Mosquito bombers mainly while the Main Force was grounded, 9–12/13 February. Bomber Command, though, was merely building up for an operation that has since gone down in history as one of the most controversial bombing raids of the war.

By February 1945, the ancient city of Dresden was one of four cities close to the Eastern Front (the others were

Berlin, Chemnitz and Leipzig) that presented a formidable obstacle to the advancing Soviet Army. Dresden was a key communications and logistics centre which had long awaited the inevitable Bomber Command attack. Chemnitz and Dresden had largely been untouched by the bombing of the previous two years but Operation Thunderclap had been under consideration for several months and was planned to cause as much destruction, confusion and mayhem in Berlin, Leipzig, Chemnitz and Dresden as possible. The orders had been issued to Bomber Command on 27 January. Stalin was informed of the plan a week later, at the Yalta Conference, and he gave enthusiastic encouragement. Thunderclap was to have started with an American raid on Dresden on 13 February but bad weather over Europe prevented any US involvement until the 14th. Dresden was targeted now because it had become a vital communications and supply centre for the Eastern Front.

At the bomber stations in England the plan for Operation Thunderclap was made known to crews at briefings. Most looked upon Dresden as just one more operation towards the completion of their tour. 'I remember very clearly the briefing we had before the raid,' said Geoff Robinson, a flight engineer on 12 and 626 Squadrons at Wickenby:

We were told that the Russians had requested the raid because German troops were massing in the city in readiness for an attack on the Eastern Front. It was made quite clear to us that the raid was intended to help the Russians. At the time I didn't know Dresden was one of the most historic cities in Germany. My geography did not extend to that.

F/L Freddy Hulance, a pilot on 227 Squadron at Balderton in Lincolnshire, recalled:

As far as we were concerned it was another routine operation. There was no briefing about the industries in Dresden, about which all I knew was that that they made fine porcelain. It was a long trip, about ten hours, but everything worked that night. The weather was very good, there was a real lack of fighter interference, the marking was magnificent and frankly we were winning. I have to say that I never felt any remorse at all about it afterwards, because it was always them or us. The notoriety came along afterwards. Years later the BBC brought me in to pursue this agenda, but there I met a displaced person from Dresden who had survived the raid. He may have been Jewish but he

Lancaster RF188 PH-*U*-*Uncle* on 12 Squadron, which was lost on Nuremberg on the night of 16/17 March 1945, with F/O Keith Wearning Mabee RCAF's crew. Mabee and four of his crew were killed.

was destined anyway the day after the raid to be shipped to a Nazi death camp. As the result of the bombing he was able to get away and he wanted to shake me by the hand for giving him the rest of his life.

Dresden was bombed in two RAF assaults three hours apart, the first by 244 Lancasters of 5 Group and the second by 529 Lancasters of 1, 3, 6 and 8 Groups. One hour and 45 minutes later, 1 Group attacked with 500 aircraft. The delay ensured that all emergency services would probably have been called in from outside Dresden, so the attack would knock those out as well. It seemed to the *Jägerleitoffiziers* following the path of the bombers on radar that Leipzig was the likely target, but 50 miles from it the Lancasters turned towards Dresden. The Marker Force of Mosquitoes found that the cloud base was not too thick and the flares illuminated Dresden for the markers who placed their red target indicators very accurately on the aiming point. At 2213hrs, 244 Lancasters commenced the attack and it was completed by 2231hrs. As tons of explosives plummeted from the sky, an 8,000°C firestorm, similar to that created in Hamburg on 27/28 July 1943, tore through the heart of the Saxon capital, burning an estimated 25,000 to 40,000 Dresdeners alive. The second attack went in at 0130hrs on 14 February by another 500 aircraft of Bomber Command. So great were the conflagrations caused by the firestorms created in the great heat generated in the first attack that crews in the

Hamburg on 31 March 1945, when 469 aircraft – 361 of them Lancasters – attempted to attack the Blohm und Voss shipyards where the new type of U-boats were being assembled. Eleven aircraft, including eight Lancasters, were shot down.

profit. [Dresden] was a mass of munitions works, an intact government centre and a key transportation point to the East. It is now none of those things.

After four attacks on Wesel preparatory to the crossing of the Rhine by the Allies in the west, 254 Lancasters and six Mosquitoes of 5 Group were dispatched to Böhlen on 19/20 February, but the raid was unsuccessful, probably because the aircraft of the Master Bomber, W/C Eric Arthur Benjamin DFC*, was shot down by flak over the target and he was killed. On the night of 20/21 February, Bomber Command mounted three Main Force raids: 514 Lancasters and 14 Mosquitoes of 1, 3, 6 and 8 Groups set out to bomb the southern half of Dortmund, another 173 bombers raided the Rhenania-Ossag refinery about 20 miles south-east of the centre of Düsseldorf, while 128 aircraft attacked another Rhenania-Ossag refinery at Monheim, and 154 Lancasters and 11 Mosquitoes attacked the Mittelland Canal. Twenty-two aircraft failed to return. Worst hit was the Dortmund force, which lost 14 Lancasters. Only one Lancaster was lost on the raid on the refinery near Düsseldorf.

The following night, 1,110 sorties were dispatched to Duisburg and Gravenhorst on the Mittelland Canal and the first and only large raid on Worms. Eleven Mosquitoes and 362 Lancasters of 1, 6 and 8 Groups took part in the last area-bombing raid on Duisburg. Seven Lancasters were lost and three crashed behind the Allied lines in Europe. The area raid on Worms resulted in 1,116 tons of bombs being accurately dropped on the town's built-up area, of which 39 per cent was destroyed and 35,000 bombed out from a population of about 58,000 people. Some 64 per cent of the town's buildings were destroyed. Ten Halifaxes and a Lancaster failed to return. Bomber Command claimed that the raid on the Mittelland Canal by 165 Lancasters and 12 Mosquitoes of 5 Group rendered it '100 per cent unserviceable'. Nine Lancasters failed to return from the raid, four crashing in France and Holland.

second attack reported that the glow was visible 200 miles from the target.[6] For most of the participating air crew the Dresden raid of 13/14 February was another well-executed and very efficient area-bombing attack. Bomber Command lost six Lancasters over Dresden. An unknown compiler wrote, 'the raid was "first class", with many good fires raised' and added, 'Many crews felt that this was one of the most spectacular attacks in which they had participated'.

The raid on Chemnitz – a round trip of 8 hours 20 minutes – and on an oil refinery at Rositz, near Leipzig, involved 499 Lancasters and 218 Halifaxes attacking in two phases, three hours apart. Eight Lancasters and five Halifaxes were lost.

Later, Winston Churchill said, 'It seems to me that the moment has come when the question of bombing of German cities simply for the sake of increasing the terror, though under other pretexts, should be reviewed … The destruction of Dresden remains a serious query against the conduct of Allied bombing.' Harris responded angrily, claiming:

We have never gone in for terror bombing and the attacks which we have made in accordance with my Directive have in fact produced the strategic consequences for which they were designed and from which the armies now

F/L Roy Day was captain of *H-How* on 50 Squadron, part of the force that headed to the Mittelland Canal at Gravenhorst. Their Lancaster was to contribute to the breaching of the canal with 14 1,000lb bombs, with half-hour-delay fuses. The raid was a complete success, the canal being rendered completely unusable, but on the way back, German night-fighters mingled with the bomber stream, as Roy Day recalls:

The initial sighting was by the mid-upper gunner, Peter Macdonald, who reported a Mosquito passing overhead!

As soon as the bomb aimer/front gunner, Roy Skinner, identified it correctly as a Ju 88 I initiated a corkscrew down and to port. When we realised the Ju 88 was not giving chase I turned the tables and went in pursuit of him, saying over the intercom something like 'Shoot the bastard.' Roy Skinner opened up as soon as he got his sights to bear, but to no effect. The Ju 88 passed under us, which brought the mid-upper turret into action and Peter opened up. I could now see the aircraft and shortly after Peter opened fire I saw a flash amidships, and it must have been at this stage that he realised he was under attack and dived to port and we lost him. Fighter flares were seen along our track and as we continued on our flight home we saw many aircraft shot down by fighters. Thirteen Lancasters were lost; two were from our squadron. The percentage loss of just fewer than 8 per cent was almost the highest I experienced out of 23 sorties flown. We reported the incident as best as we could recall on return to base, Skellingthorpe, and some time later we were credited with 'Ju 88 damaged confirmed' – I proudly entered this remark in my log book. The day after the incident the aircraft, SW261, was taken to the butts and on test firing the front turret guns they were found to be grossly misaligned; which explained Roy's lack of success.

On Thursday 22 February, the Alma Pluto benzol plant at Gelsenkirchen and another oil target at Osterfeld were the targets for two forces in 3 Group of 85 and 82 Lancasters respectively. Over 360 Lancasters and 13 Mosquitoes of 1, 6 and 8 Groups carried out the first and only area-bombing raid on Pforzheim, a city of 80,000 people, on the night of 23/24 February. In just over 20 minutes 1,825 tons of bombs were dropped from only 8,000ft. More than 17,000 people were killed and 83 per cent of the town's built-up area was destroyed in 'a hurricane of fire and explosions'. Bomber Command's last VC was gained on the night of 23/24 February, the award going to Captain Edwin Swales DFC SAAF on 582 Squadron, operating as Master Bomber for the attack on Pforzheim. Swales' Lancaster was hit and crippled but he continued issuing aiming instructions to the Main Force. Finally, he ordered his crew to bale out and he kept the aircraft steady while each of his crew parachuted to safety. Hardly had the last crew member jumped when the aircraft plunged to earth. Captain Swales was found dead at the controls. Ten Lancasters were shot down by enemy fighters and jets were also active.

V-Victor on 625 Squadron with 80 bomb symbols (including at least 32 daylight ops) with air and ground crew early in 1945.

Lancaster I PB155 *K-Kripes* on 460 Squadron RAAF, with F/O Sam Cox at the controls, fought off a Me 262 while flying at a height of 7,500ft in the target area. F/O Ben Curren, the rear gunner, sighted two glows, light orange in colour and evenly spaced, approaching the bomber at a high speed from the port quarter (nearly port beam) at an approximate range of 1,200yd. The glows banked round to the port quarter and at 800yd (approximately) the rear gunner ordered a corkscrew port and both air gunners opened fire. The fighter followed the bomber in the corkscrew and closed to 400yd. At this stage the fighter broke away fine port quarter down and both gunners saw a small fire in the starboard wing. The fire was seen to grow larger as the fighter went down and the silhouette of the aircraft could be seen. P/O Don Crosby, the mid-upper gunner, recognised the wing as that of a Me 262. Both air gunners followed the burning aircraft to the ground, where it exploded on impact. The explosion was also seen by the pilot and P/O Alex Tod, the WOp. Throughout the attack, both gunners fired continuously, the rear gunner using the GGS (Gyro Gun Sight) and the mid-upper the reflector sight, and claimed the fighter as destroyed.

On 24 February, 340 aircraft – 26 of them Lancasters – set out to attack the Fischer-Tropsch synthetic-oil plant at Bergkamen just north of Kamen, and 166 Lancasters and four Mosquitoes of 5 Group took off to breach the Dortmund-Ems Canal near Ladbergen in northern Germany. However, both

F/O Bob Purves (left) usual pilot on P4-V *Vicious Virgin/Baby*, 'B' Flight, 153 Squadron, S/L McLaughlin and W/C Francis Sidney Powley DFC AFC, the CO. Powley was a Canadian from Kelowna in British Columbia who had joined the pre-war regular air force on a short service commission in 1936. On the night of 4/5 April 1945, he and S/L Gee's usual crew on RA544 P4-U perished on a mine-laying sortie in the Kattegat. Powley and F/L Arthur Joseph Winder's crew were each victims of Major Werner Husemann, a 32 *Abschüsse* ace of I./ NJG 3.

target areas were covered by cloud and the raid on Kamen, which relied on Oboe and H$_2$S markers, resulted in ineffective bombing. At Dortmund-Ems, the Master Bomber instructed crews to return to base with some of the bombs. Having escaped destruction, a 'G-H' attack was ordered on the Fischer-Tropsch plant on 25 February by 153 Lancasters of 3 Group. One Lancaster was lost and the plant still functioned. So, after night raids on Dortmund, Mainz and Gelsenkirchen, and on Mannheim and Cologne, on the night of 3/4 March, Kamen and the Dortmund-Ems Canal once again appeared on the battle order. Some 234 aircraft – 201 Halifaxes of 4 Group, and 21 Lancasters and 12 Mosquitoes of 8 Group – were detailed to bomb the synthetic-oil plant, while 212 Lancaster and ten Mosquito crews of 5 Group were to attack the aqueduct, safety gates and canal boats on the Dortmund-Ems Canal at Ladbergen. The Dortmund-Ems Canal was breached in two places and put completely out of action, but seven Lancasters were shot down.

Near the target area the gunners on the 619 Squadron Lancaster flown by W/C S.G. Birch claimed to have shot down a V-1 flying bomb which was probably aimed at the port of Antwerp.

On 5/6 March, Operation Thunderclap was resumed when a 760-strong force was detailed for a fire raid to destroy the built-up area, industries and railway facilities of Chemnitz. At North Killingholme, F/L Peter Sarll, commanding 'C' Flight on 550 Squadron, taxied *Mike Squared* out. Sarll had flown Blenheims early in the war, surviving the attacks on the Maastricht bridges. He sported a huge handlebar moustache and was noted for the hunting horn he always carried on ops, which he would sound as the aircraft was going into the target and again on leaving. At first this proved to be somewhat disconcerting for his crew, five of whom were Australians and had not come across this particular form of British eccentricity before. No bomb symbols were painted on the sides of *Mike Squared* and the same was true for several other Lancasters on 550 Squadron. The feeling among many of the squadron's crews was that it was not prudent to advertise the number of raids on their Lancasters in case they were shot down. 'The last thing you wanted was the wreckage of a Lancaster with 90 trips painted on the side when there was a lynch mob around pointing to the number of times you had dropped bombs on them,' said F/Sgt Frank Pritchard, the crew's Australian mid-upper gunner. 'It was much better to let them think it was your first time out.'

Bomber Command lost 13 Lancasters and eight Halifaxes shot down en route to and from Chemnitz. In a separate attack, 248 Lancasters and ten Mosquitoes of 5 Group targeted the synthetic-oil refinery at Böhlen, which was covered by cloud. Four Lancasters were shot down, one force-landed at Nuneaton and one other came down in the River Witham near Boston, killing four of the crew. Attacks on Salzbergen, Wesel and Sassnitz on Rügen Island in the Baltic followed. Then it was the turn of oil refineries at Hemmingstedt near Heide and Harburg, while Dessau, Hamburg and Kiel too were also hit before the attacks switched to day and night assaults on other German towns and cities and oil targets again.

On 7/8 March, Bomber Command mounted three major raids: on Dessau in eastern Germany, Hemmingstedt and Harburg. Small forces of Ju 88G-6s and Bf 110G-4s were directed to the Dessau force of 526 Lancasters and five

Mosquitoes, but the majority of the *Nachtjagd* crews were unable to get at the bomber stream due to conflicting ground control instructions and heavy jamming by 100 Group, rendering the SN-2 radar all but useless in the target area. From the 531 aircraft attacking Dessau, 18 Lancasters were lost, but the raid was devastatingly successful, with the town centre, residential, industrial and railway areas all being hit.

Left: Peter Sarll (centre, back row) and his crew on 550 Squadron who flew *Mike Squared* on 5/6 March 1945. Back row, L–R: Russ Longmire (navigator); Jock Murray (engineer); Peter Sarll (pilot); Alf Coombes (WOp); John Seppelt (bomb aimer). Front: Peter Robertson (rear gunner); Frank Pritchard (mid-upper gunner).

Below: Enemy action caused this damage to Lancaster PA226 on 429 Squadron.

193

Lancaster B.III PB532 HW-S2 *Santa Azucar* on 100 Squadron at Elsham Wolds in April 1945 with Eric Richmond's crew. The Lancaster's name resulted from the South American origins of its original pilot, F/L O. Lloyd-Davies. Note the 'crochet' bomb symbols. Back row, L–R: Frank Ockerby (WOp); Bob Mellows (bomb aimer); Arthur Jackson (engineer); Eric Richmond (pilot); Arthur Rose (navigator); Colin Gowans (mid-upper gunner); Jimmy Wallace (rear gunner). The ground crew includes the WAAF who drove *Santa Azucar*'s crew bus.

Some 234 Lancasters and seven Mosquitoes of 5 Group carried out an accurate attack on the oil refinery at Harburg. Fourteen Lancasters were shot down; four of 189 Squadron's 16 Lancasters were lost on the raid. Twenty-two-year-old and 6ft 3in tall F/O Roussel 'Russ' Stark RAAF was piloting *M-Mother* on 49 Squadron and took off from Fulbeck for his crew's 17th operation. Sgt Joe 'Dixie' Dixon from Liverpool manned the four Brownings in *M-Mother*'s rear turret:

We watched our bombs drop right in the middle of the fire. Everybody was laughing and joking about the state of the fire when Russ spotted a Fw 190; there were a few planes going down in flames. We ran into a lane of flares which were lighting the sky up and Russ and Paddy Gilbert, the mid-upper gunner, saw a Ju 88 pass underneath us. When I spotted it I opened fire and got it. We did some manoeuvres then flew level; Russ and Paddy watched the Ju 88 going down in flames. I was temporarily blinded by my own gun fire when something hit my turret and set it on fire. I told Russ and started getting ready in case we had to get out. A few seconds later Russ gave orders to jump out immediately. I tried to get out and had a bit of a fight to do it as my turret was now well alight, Russ kept control of the plane to give us a chance to get out. I believe Paddy and Gus Lovett, the wireless operator, were killed outright. When I did finally get out and I was floating down, seconds later I saw our

plane hit the ground in flames. Russ and Ralph Bairnsfather, the bomb aimer, must have died at once.

After I got down I found somewhere to stay the night. The following day I set out walking at about 6 a.m. After going 25 miles Fred Brennan, the engineer, caught up with me; he had slept in the same wood as I did but neither of us had known it. Fred told me that the last he saw was Ralph fastening his straps ready to jump. He must have lost his life by only a second or two. Later on in a Stalag (I was taken PoW on 8 March) I met Johnnie Yeoman, the navigator. He was badly scratched because he fell in an awkward place near some barbed wire. He had been unable to see what was going on from his place in the plane. I honestly believe Russ could have saved his own life at the risk of the others, but he thought of his crew first. He gave his life to try and save the rest of the crew. Three other mothers gave their sons in the cause of freedom so we can only hope that Russ, Ralph, Gus and Paddy lie peacefully together at rest.[7]

On 8/9 March, Bomber Command forces attacked Hamburg and Kassel, and on the 9th the north and south plants of the Emscher Lippe benzol plant at Datteln were bombed by 159 Lancasters of 3 Group. No other aircraft were lost.

Next day, it was again the turn of the Scholven-Buer synthetic-oil refinery, which was attacked by similar numbers of Lancasters in 3 Group. No aircraft were lost. Another daylight raid, on Essen by 1,079 aircraft – 750 Lancasters, 293 Halifaxes and 36 Mosquitoes – of all bomber groups, took place during the bright afternoon of Sunday 11 March. It was the largest attack of the war. Two days later, 354 aircraft attacked Wuppertal and Barmen without loss. Bomber Command had now dispatched 2,541 sorties by daylight to Ruhr targets in a three-day period. Benzol plants at Herne and Gelsenkirchen were bombed on the night of 13/14 March and the next day, 169 Lancasters of 3 Group carried out 'G-H' attacks through cloud on plants at Datteln and Hattingen near Bochum. On the following night, 230 aircraft of 6 and 8 Groups attacked Zweibrücken to block the passage through the town area of German troops and stores to the front. The same task was carried out at Homberg by 161 aircraft of 4 and 8 Groups. The raids were successful. Two Halifaxes were lost on the Homberg raid. Another 244 Lancasters and 11 Mosquitoes of 5 Group bombed the Wintershall synthetic-oil refinery at Lützkendorf near Leipzig. Eight Lancasters were lost on Lützkendorf. In a separate operation, 16 Lancasters on 9 and 617 Squadrons attacked the viaduct at Arnsberg without loss.

On 15/16 March, 267 bombers made an area attack on Hagen and another 257 Lancasters and eight Mosquitoes raided the Deurag oil refinery at Misburg on the outskirts of Hanover. Six Lancasters and four Halifaxes failed to return from Hagen, which suffered severe damage in the centre and eastern districts. Four Lancasters were lost on the Misburg raid, where the main weight of the raid fell south of the oil refinery. The night following, 231 Lancasters of 1 Group and 46 Lancasters and 16 Mosquitoes of 8 Group were detailed to attack the southern and south-western districts of Nuremberg in the last heavy Bomber Command raid of the war on the city. Würzburg, an old cathedral city famous for its historic buildings and which contained little industry, was the destination for another 225 Lancasters and 11 Mosquitoes of 5 Group. Altogether, 1,127 tons of bombs were dropped with great accuracy in just 17 minutes, destroying 89 per cent of Würzburg's built-up area. Estimates were that between 4,000 and 5,000 people were killed. In all, 24 Lancasters were lost over Nuremberg and six were lost over Würzburg on the night of 16/17 March, mainly to German night-fighters, which found the two bomber streams on the way to the targets.

On 17/18 March, when there was no Main Force activity, a sweep by 66 Lancasters and 29 Halifaxes was made over northern France to draw German fighters into the air, and formations of Mosquito bombers carrying 'cookies' visited targets in Germany. On 18 March, 100 Lancasters of 3 Group carried out 'G-H' attacks on benzol plants at Hattingen and Langendreer. Both raids appeared to be accurate and no aircraft were lost. That night, 324 aircraft of 4, 6 and 8 Groups carried out an area-bombing raid on Witten in good visibility. Some 1,081 tons of bombs were dropped, which destroyed 129 acres, 62 per cent of the built-up area, and severely damaged the Ruhrstahl steelworks and the Mannesmann tube factory. Eight aircraft, including six Halifaxes, were lost. Another 277 Lancasters and eight Mosquitoes of 12 and 8 Groups carried out an area raid on Hanau. Fifty industrial buildings and 2,240 houses were destroyed, and an estimated 2,000 people were killed. Just one Lancaster was lost.

Benzol plants in Germany were attacked on successive days and nights, 17–19 March. On the last of these, 79 Lancasters of 3 Group attacked the Consolidation benzol works at Gelsenkirchen. All except two aircraft returned. *P-for-Peter* on 90 Squadron at Tuddenham, piloted by F/O Paine,

force-landed near Mönchengladbach with three engines on fire after being hit by flak approaching the rendezvous point and again within seconds of the bomb run. On 19 March also, 19 Lancasters of 617 Squadron attacked the viaduct at Arnsberg a few miles north of the Möhne Dam, and 18 Lancasters on 9 Squadron, the bridge at Vlotho (Bad Oeynhausen) near Minden. At Arnsberg, five Lancasters carried 'Grand Slams' and the other 14 had 'Tallboys'. The first bomb was a direct hit on the viaduct and the rest went into the centre of the smoke that gushed up. When the smoke cleared, crews could see that a 40ft gap had been blown in the centre of the viaduct and there was a pile of rubble on the riverbed. However, the attack on Vlotho was not successful.

Next day, when 153 aircraft bombed Recklinghausen and 99 Lancasters of 3 Group attacked the railway yards at Hamm, both without loss, 14 Lancasters of 9 Squadron targeted the railway bridge at Arnsberg. On 21 March, it was the turn of the Arbergen railway bridge near Bremen to suffer the same fate. Twenty Lancasters on 617 Squadron were detailed; two scored direct hits with their 'Tallboys' and two piers collapsed. Another one was thrown 15ft out of alignment and earthquake shock threw a span off another pier. Flak got a direct hit on F/L Barney Gumbley DFM's Lancaster on the run up and the New Zealander and his crew were killed. Next day,

A Lancaster bracketed by flak bursts on the operation to Bremen on 22 April 1945, when 767 aircraft – 651 of them Lancasters – were sent to bomb the city in preparation for the attack by the British XXX Corps. The raid was hampered by cloud and by smoke and dust from bombing as the raid progressed, and the Master Bomber ordered the raid to cease after 195 Lancasters had bombed. The whole of 1 and 4 Groups returned without bombing. Two Lancasters were lost. The city soon capitulated after three days of ground attack. (*'Spud' Taylor*)

Above: Lancaster LM577/HA-Q *Edith* got its name on 622 Squadron, which operated the Lancaster for five months. 218 Squadron received the aircraft after depot maintenance and by 28 April 1945 had raised the sortie total to 84. *Edith* also flew 14 food-dropping and POW-collection sorties and had more hours' flying time on it than any other Lancaster in the squadron. (*Imperial War Museum*)

Right: Lancasters en route over the Alps to Berchtesgaden on 25 April 1945. Hitler's home, referred to as the 'Chalet' by the RAF, was the target for 359 Lancasters – including 33 each carrying a 12,000lb 'Tallboy' bomb – and 14 *Oboe* Mosquito and 24 Lancaster marker aircraft.

82 Lancasters of 5 Group attacked a bridge near Bremen, and 20 aircraft on 617 Squadron were detailed to bomb a bridge at Nienburg. W/C 'Johnny' Fauquier DSO** DFC sent four aircraft in to bomb while the rest circled and the first two earthquake bombs hit each end of the bridge. The span lifted bodily, still intact, into the air and a third bomb seemed to hang there for a second and at that very moment another bomb hit it right in the middle. Early the next morning, 128 Lancasters of 1 and 5 Groups were detailed to attack two more railway bridges in the Bremen area. The first three bombs dropped by 617 from 16,000ft at Bad Oeynhausen scored direct hits. The next two

were near misses followed by what looked like another direct hit. On the Bremen operation two Lancasters were lost.

On 20/21 March, no fewer than three feint attacks took place in support of the Main Force attack on the synthetic-oil refinery at Böhlen, just south of Leipzig, by 235 Lancasters and Mosquitoes. The 'Window' Force left the Main Stream soon after crossing the front line and made for Kassel, which was bombed. Further on, when closer to the true target, another 'Window' Force broke off and bombed Halle. The third feint force was provided by the Mandrel screen which, after the passage of the heavies, re-formed into a 'Window' Force and attacked Frankfurt with flares. The Main Force's zero hour was set at 0340hrs on the 21st. At about the same time, 166 Lancasters headed for the oilfield at Hemmingstedt in Schleswig-Holstein far to the north of the first target. This attack was to commence at 0430hrs and, together with the other attack, involved Bomber Command's main effort. The 'spoofs', diversionary attacks and counter measures helped keep losses down to nine Lancasters. One was lost on the attack on Hemmingstedt. Three aircraft of the support and minor operations forces were also lost. *Nachtjäger* claimed 11 Lancasters shot down.

On 21 March, 178 aircraft carried out an accurate attack on the rail yards at Rheine and the surrounding urban area for the loss of one Lancaster. One hundred and sixty Lancasters of 3 Group raided the railway yards at Münster and a railway viaduct nearby for the loss of three bombers. Another 133 Lancasters of 1 Group and six Mosquitoes of 1 and 8 Groups headed for Bremen and accurately bombed the Deutsche Vacuum oil refinery without loss. That same evening 151 Lancasters and eight Mosquitoes of 5 Group raided the Deutsche Erdölwerke refinery at Hamburg, and 131 Lancasters and 12 Mosquitoes of 1 and 8 Groups, a benzol plant at Bochum. Five Lancasters were lost; four of them over Hamburg. The following day, raids on railway targets continued, 227 Lancasters and eight Mosquitoes of 1 and 8 Groups raiding Hildesheim railway yards. Some 263 acres – 70 per cent of the town – was destroyed and 1,645 people were killed. Four Lancasters were lost. One hundred Lancasters of 3 Group carried out a 'G-H' attack on Bocholt, probably with the intention of cutting communication. All returned safely. Another 102 Lancasters of 5 Group in two forces attacked bridges at Bremen and Nienburg without loss. The bridge at Nienburg was destroyed, though no results were observed at Bremen.

On Saturday 23 March, 128 Lancasters of 1 and 5 Groups attacked and hit bridges at Bremen and Bad Oeynhausen, losing two of the Lancasters. At 1530hrs 80 Lancasters bombed the little town of Wesel, which was an important troop centre behind the Rhine front in the area about to be attacked by the 21st Army Group massing for the Rhine crossings at dawn. More than 400 tons of bombs were dropped on the German troops and many strong points were destroyed. Five hours later, only a short time before Field Marshal Sir Bernard Montgomery's zero hour, 195 Lancasters and 23 Mosquitoes of 5 and 8 Groups followed it up with another attack to complete the work of the afternoon. In exactly nine minutes, well over 1,000 tons of bombs went down from 9,000ft on those troops who had crept back into the ruins to await the British commandoes' attack. In all, more than 1,500 tons of bombs were dropped in the two attacks – a weight of bombs which had already almost completely wiped out cities eight times the size of Wesel. The effect on the defenders was devastating and the British Army was crossing the river in assault craft, aided by searchlights, before the bombers had left the area.

On 24 March, attacks were made in good weather on the railway yards at Sterkrade, the town of Gladbeck and the Harpenerweg and Mathias Stinnes benzol plants at Dortmund and Bottrop, because the Ruhr industries were still supplying fuel and munitions for the fighting front, which was now just 15 miles away. Three Lancasters were lost on the Dortmund raid. Next day, Bomber Command flew 606 sorties against the main reinforcement routes into the Rhine battle area, and Hanover, Münster and Osnabrück were heavily hit. A 166 Squadron Lancaster was brought down after being struck by a bomb over the target at Hanover. On 27 March, 268 Lancasters and 8 Mosquitoes attacked Paderborn without loss, and two other forces of 150 Lancasters and 115 Lancasters attacked benzol plants in the Hamm area and an oil-storage depot at Farge respectively. All aircraft returned safely from these attacks. Two days later, 130 Lancasters of 3 Group carried out a 'G-H' attack on the Hermann Goering benzol plant at Salzgitter.

On the 31st, 469 aircraft – including 361 Lancasters – attempted to bomb the Blohm und Voss shipyards at Hamburg, where new types of U-boats were being assembled. The target area was completely cloud covered and most of the bombs that were dropped fell over a wide area of southern Hamburg and Harburg. The leading Pathfinder and other Lancaster formations became heavily engaged by at least 30 Me 262 jets of JG 7 over the target area, but had plentiful fighter escort protection and lost few bombers to the flashing jet assaults. At the rear of the stream, however,

a gaggle of Canadian Lancasters, ten minutes late over target and thus without escort, suffered heavily. For five minutes of non-stop assaults five Lancasters were shot down (three other 6 Group aircraft had already fallen victim to fighter attacks) in a total of 78 individual attacks, with 28 crews later reporting actual engagements. Canadian gunners claimed four Me 262s destroyed, three probables and four others damaged. F/L J.L. Storms on 427 Squadron piloted his aircraft back from Hamburg with one shattered aileron and a 5ft chunk of one wing missing, but finally achieved a safe return, thereby earning himself a DFC. The Canadians' 'Day of the Jets' had been the Group's greatest daylight battle of the war.

During March, Bomber Command had flown a record 53 day and night operations. April began with two attacks by Lancasters and Mosquitoes on what were believed to be military barracks near Nordhausen, east of Göttingen, which were in fact being used to accommodate a number of

Berchtesgaden from 18,000ft on 25 April 1945. Those who bombed the 'Chalet' mostly missed. A mountain peak between the *Oboe* ground station and the aircraft had blocked out the bomb-release signal. Since *Oboe* signals went line of sight and did not follow the curvature of the earth, the further the target, the higher one needed to be, and the *Oboe* Mosquitoes flew at 36,000ft because of the Alps. Crews heard the first two dots of the release signal and then nothing more. They were unable to drop and brought the markers back.

Right: Berchtesgaden under attack.

Far right: A Lancaster on 15 Squadron at Mildenhall dropping food to the Dutch during Operation Manna in May 1945. From 29 April to 7 May, Lancasters flew 2,835 food sorties and delivered 6,672 tons of food to the starving Dutch people during Operation Manna. RAF Mildenhall's Lancasters alone dropped enough supplies for around 50,000 people. (*Via Harry Holmes*)

Opposite: 6 Group RCAF Lancaster with stores for the Dutch. During Operation Manna (29 April–7 May 1945) Lancasters flew 2,835 food sorties, PFF Mosquitoes made 124 sorties to 'mark' the dropping zones and Bomber Command delivered 6,672 tons of food to the starving Dutch people. (*Imperial War Museum*)

concentration camp prisoners and forced workers employed in a complex of underground tunnels where various secret weapons were constructed. Three Lancasters were lost on the raids. On 4/5 April, synthetic-oil plants at Leuna, Harburg and Lützkendorf were bombed. Severe damage was caused to the Rhenania oil plant at Harburg and Bomber Command claimed 'moderate' damage at Lützkendorf. A total of 11 Lancasters and Halifaxes were lost. A benzol plant at Mölbis near Leipzig was attacked by 175 Lancasters and 11 Mosquitoes of 5 Group on 7/8 April, and the following night 440 aircraft of 4, 6 and 8 Groups carried out the last raid on Hamburg by Bomber Command aircraft when they targeted oil-storage tanks in the shipyard areas. Seventeen Lancasters on 617 Squadron blasted the U-boat shelters in the already devastated city with 'Grand Slam' and 'Tallboy' bombs. Three Halifaxes and three Lancasters failed to return from the raid. The Lützkendorf oil refinery, which had escaped serious damage the previous night, was rendered 'inactive' by 231 Lancasters and 11 Mosquitoes of 5 Group. Six Lancasters failed to return.

On the afternoon of 9 April, 57 Lancasters of 5 Group attacked oil-storage tanks (40 aircraft) and U-boat shelters (17 aircraft of 617 Squadron with 'Grand Slams' and 'Tallboys') at Hamburg. Both attacks were successful. Two Lancasters were lost from the raid on the oil tanks. Sgt Ted Beswick was the mid-upper gunner on Lancaster QR-Y on 61 Squadron at Skellingthorpe and describes what happened:

My skipper was W/O Ivor Soar and after checking everything was OK with QR-Y, took off from Skellingthorpe at 14.28

hours. After navigator W/O Walter McLean gave him a westerly course to fly, we all settled down in our crew positions while the heavily loaded aircraft gradually climbed up to our operational height of 16,800ft. One hour after take-off while flying over the Irish Sea, east of Belfast, we formed up with 50, 61 and 617 Squadron aircraft and then set course for Hamburg. QR-Y was in the leading gaggle of aircraft as we ran up to the target area and I could see from my mid-upper turret the black puffs of smoke from the predicted flak box barrages. The flak gradually reached the bomber stream and soon we were all flying through a sky that seemed to be full of exploding shells. The squadron started the attack at 17.30 hours and while on our bombing run the aircraft was hit by shrapnel in the nose and our bomb aimer, Sgt Harold Heppenstall, was slightly wounded in his left arm. He was able to carry out his duties and dropped our bomb load at the aiming point.

We had just left the target area when the Lancaster behind us, QR-J piloted by F/L Paul Greenfield DFC, suddenly reared up violently and exploded in a ball of fire.[8] Seconds later a German twin-engine Me 262 jet fighter appeared flying past the pall of smoke and flaming wreckage of QR-J. Our rear gunner, F/Sgt Jimmy Huck, opened fire immediately but I couldn't get my guns down low enough so I yelled out to the skipper, 'Corkscrew port go' but due to the close proximity of other aircraft he responded with a shallow dive. On levelling out, the Me 262 was about 75 yards off our port beam and turning to line up on another Lanc some 200 yards ahead. I opened fire with my twin 0.303 Brownings and saw a number of strikes along the fuselage followed by black smoke from the cockpit area. The fighter then rolled over and dived out of sight. 'Taffy' Rees, our flight engineer, yelled out over the intercom, 'We've got him' and this was confirmed by the skipper and rear gunner. It's funny how time slows down in combat situations. From start to finish the whole incident happened in less than 30 seconds. The Me 262 fighter pilot was JG 7 *Staffelkapitän* Hauptmann Franz Schall. He was the second highest scoring Me 262 ace of the war with 16 kills. German records show that Hauptmann Schall died the following day in a flying accident.[9]

Incidentally, this was not the end of our troubles, for we then discovered that we had a 1,000lb bomb hung-up in the bomb bay. Over the sea we tried to get rid of it but could not dislodge it until we landed when the shock of landing did the trick. Unfortunately by this time we were on the ground and could clearly hear it rattling around, live, inside the bomb bay. We turned off the runway and held there while the armourers came running to us. They rigged a winch inside the aircraft and dropped cables through apertures in the fuselage floor. The bomb doors were then wedged open just enough for an armourer to get an arm through, make the bomb 'safe' and attach winch cables. Then they lowered it down. We may have had a 'shaky-do' that day but afterwards our hats went off to those armourers!

On the night of 9/10 April, 591 Lancasters and eight Mosquitoes bombed the Deutsche Werke U-boat yards at Kiel. The pocket battleship *Admiral Scheer* was hit and capsized, and the *Admiral Hipper* and the *Emden* were badly damaged. Three Lancasters were lost. Another Lancaster was lost on 10 April when 230 bombers attacked the Engelsdorf and Mockau rail yards at Leipzig. *Nachtjagd* pilots claimed six Lancasters, a Halifax and a Mosquito on 10/11 April, when 307 Lancasters and eight Mosquitoes of 1 and 8 Groups attacked the rail yards in the northern part of Plauen and 76 Lancasters and 19 Mosquitoes bombed the Wahren railway yards at Leipzig. All the bombers returned safely from the raid on Plauen but seven Lancasters were lost on Leipzig where the eastern half of the yards was destroyed. More attacks on rail yards took place on 11 April when Nuremberg and Bayreuth were the targets for the Halifaxes of 4 Group, who in part were supported by Pathfinder Lancasters and Mosquitoes of 8 Group. On 14/15 April, 500 Lancasters of 1 and 3 Groups and 12 Mosquitoes of 8 Group took part in an operation on Potsdam

Canadian Lancaster *Princess Patricia* is loaded up with supplies for a Manna operation. (*Imperial War Museum*)

moorings. The following night, over 370 Lancasters and more than 100 Halifaxes carried out a heavy raid on the port area and the U-boat yards at Kiel but the bombing was 'poor' and two Lancasters failed to return.

Bomber Command's last bombing operations were the obliteration of Wangerooge and a failed attempt at destroying Hitler's 'Eagle's Nest' at Berchtesgaden in daylight on 25 April by 359 Lancasters and 16 Mosquitoes, and U-boat fuel-storage tanks at Yallo (Tønsberg) in Oslo fjord and part of Oslo fjord itself (off Horten) mined on the following night. At Fiskerton, Dickie Parfitt, a bomb aimer on 576 Squadron, recalled:

In the briefing room the target on the blackboard was Berchtesgaden, Hitler's hideaway home, or more precisely, the barracks. The Bombing Leader warned us that the height of the mountains needed careful attention and, with the barracks in a valley, we would need to take care not to bomb each other. The Met man gave us clear weather and the Intelligence Officer gave us some useful info. The CO told us we would be flying in a group gaggle like geese. The Flight Commander would be flying and leading with his wing tips painted red. We would fly close but not in formation. If any of us saw a German fighter, we had to fire a Very pistol then we would all close in for cover and greater firepower. Mustangs from Fighter Command would escort us. This was intriguing. We were to cross the Channel over Dover and return the same way. This had never happened before. We were well up front at 20,000ft with hundreds of Lancasters all around us. The Mustangs joined us, shepherding the loose aircraft, getting them into line. I had been up in my front gun turret since well into France. Soon we would be crossing over the German fighter zones where they were most active. Somebody fired a Very pistol. The pilots closed in tight and then I noticed out to port a Lancaster on its own. I imagined the navigator saying that he was right and on course or possibly they had engine trouble. As I watched, two German jet fighters approached the aircraft and took it out. It spiralled down and its bomb load exploded. Needless to say, that made everybody nervous. The gaggle leader ordered everybody to tighten up. As we approached the Alps, I dropped down into my bombing hatch. The flak had started and was getting fiercer. We were now over the Alps with not a lot of clearance. This was going to be tricky. I would be on the target quickly with no room for error and we wouldn't want to go round again. The Master Bomber

just outside Berlin. Although Mosquito bombers of the LNSF (Light Night Striking Force) had attacked the 'Big City' almost continually, this was the first time the Reich capital had been attacked by heavies since March 1944. One Lancaster was lost to an unidentified night-fighter over the target.

On Friday 13 April, 34 Lancasters on 9 and 617 Squadrons had set out to attack the *Prinz Eugen* and the pocket battleship *Lützow* in harbour at Swinemünde, but the raid was abandoned because of cloud at the target. A return raid two days later was equally unsuccessful when cloud again covered the target. Finally, on 16 April, a raid on the *Lützow* by 18 Lancasters on 617 Squadron was more successful. One Lancaster was lost and all except two aircraft were damaged, but 15 Lancasters managed to get their 'Tallboys' and 1,000lb bombs away. A near miss tore a large hole in the bottom of the pocket battleship and the *Lützow* sank in shallow water at her was on the air giving instructions. I called, 'Bomb doors open.' Suddenly we were there. We weren't far off track, so a few corrections and then, 'Bombs gone.' The flak was coming from everywhere – the valley and the mountain sides. Without the bomb load we soon climbed away and headed back home.[10]

F/L Peter Sarll on 550 Squadron, who took *Mike Squared* on its last bombing operation, made sure it was a fitting finale with three runs over the target area before being satisfied that the right target had been selected. F/O H.G. Payne on 460 Squadron RAAF ran into moderate heavy flak and crashed in the vicinity of Salzburg. All seven crew survived but the bomb aimer was wounded. F/O Wilfred Tarquinas De Marco and crew on 619 Squadron had just released their bomb load and were holding steady for the bombing photograph when the Lancaster was hit by anti-aircraft fire and brought down at about 0530hrs. Three of the crew were able to escape by parachute, although Sgt F.J. Cole discovered that when he reached for his parachute it was 'pulled'. F/Sgt J.W. Speers, himself wounded, gathered Cole's chute in his arms and then let it go as he kicked Cole out! F/Sgt Arthur Shannon broke a leg when he landed into a hostile reception but was mercifully spared by the intervention of an SS officer. De Marco and the rest of the crew were killed. Lancaster I RA542 JO-Z and F/O A. Cox's crew on 463 Squadron RAAF failed to return from the 5 Group raid by 119 Lancasters on the oil-storage plant at Tønsberg, south of Oslo. Cox put the Lancaster down at Såtenäs airfield in Sweden and the crew were interned.

By the end of the war in Europe there were 58 squadrons wholly equipped with Lancasters, including all 14 squadrons of 1 Group, the 11 squadrons of 3 Group, 17 squadrons of 5 Group, eight squadrons of 6 Group RCAF and eight squadrons of 8 PFF Group (including two detached from 5 Group). The Lancasters then turned to repatriating British ex-prisoners of war (Operation Exodus), each Lancaster carrying 24 passengers and dropping food to the starving Dutch (Operation Manna). Ted Stones, a member of a ground crew on 550 Squadron at North Killingholme, flew on one of those trips:

The crew just asked me if I wanted to go with them and I agreed without thinking. We went skimming across the North Sea heading for the clock at The Hague. When we arrived there, I noticed it was just one o'clock. Then we

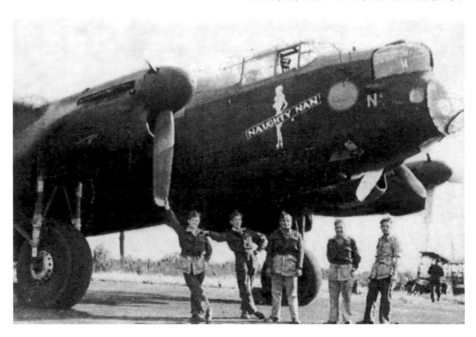

Naughty Nan! on 101 Squadron at Elsham Wolds after VE Day trips to pick up former POWs.

turned for Rotterdam and dropped our supplies, which included chocolates, salt, wheat and bags of peas into the stadium. As we were leaving we could see the Jerry ack-ack crews standing by their guns with their arms folded, just looking up at us. I also remember there were a lot of houses with flat roofs and written on them was 'Welcome RAF'.

Seventeen-year-old Arie de Jong wrote:

There are no words to describe the emotions experienced on that Sunday afternoon. More than 300 four-engined Lancasters, flying exceptionally low, suddenly filled the western horizon. One could see the gunners waving in their turrets. A marvellous sight. One Lancaster roared over the town at 70ft. I saw the aircraft tacking between church steeples and drop its bags in the south. Everywhere we looked, bombers could be seen. No one remained inside and everybody dared to wave cloths and flags. What a feast! Everyone is excited with joy. The war must be over soon now.[11]

After one food drop over Holland, F/O Handley spotted a ditched American Flying Fortress in the North Sea. There was a dinghy alongside and the Lancaster managed to attract the

Above: Lancaster B.I R5868 *S-Sugar* on 467 Squadron RAAF, which completed its 137th and final bombing sortie on 23 April 1945 to Flensburg, at Kitzingen airfield, Germany, on 7 May 1945 for the repatriation of British ex-POWs during Operation Exodus. *S-Sugar* is now on permanent display at the RAF Museum, Hendon, London.

Right: 467 Squadron crew during Exodus on 7 May 1945.

attention of a nearby ship, guiding the rescuers to the scene and helping save the lives of the ten-man crew of the Fortress. Many Lancasters also flew ground crews and VIPs on 'Cooks Tours' of the devastated cities of Germany. Repatriation of POWs and British troops in Italy was not completed until October–November 1945.

Plans were made for 12 squadrons of Lancaster VIIs with tropicalised power plants and one Mosquito squadron to operate in the Far East against Japan as 'Tiger Force'. Operation

Shield, the code name for the bombing of Japan, never came to fruition as the Japanese surrendered unconditionally on 15 August after the dropping of atomic bombs on Hiroshima and Nagasaki on 6 and 9 August respectively.

Could the Lancaster have dropped the bombs that ended the war? In September 1943, Dr Norman Ramsey, a gifted physicist and engineer organiser on the Manhattan Project, was instructed to find an aircraft with a bomb bay that could carry a weapon weighing as much as 9,500lb. At the time, there were no aircraft in the US inventory with a bomb bay that could contain this 17ft bomb. Only the Boeing B-29, by combining its two 12ft bomb bays into one, and the Avro Lancaster, could carry the weapon. In October 1943, Ramsey travelled to Canada to meet with Roy Chadwick, who, as luck would have it, had crossed the Atlantic to view Lancasters being built at the Avro Canada works in Toronto. Ramsey seized the chance to show Chadwick some preliminary sketches of both the gun and the implosion weapon casings. Chadwick assured Ramsey that the Lancaster could accommodate either bomb and promised whatever support might be needed. (Three of the B-29s that were used in the Manhattan Project were fitted with a single-point bomb release modelled after the British F-Type heavy-bomb mechanism used to drop the 'Tallboy' and 'Grand Slam' bombs.) When Ramsey returned to the US, he recommended that the Lancaster should be seriously considered as the delivery carrier. Apparently Major General Leslie Groves had not yet asked for US Army Air Forces support and the Chief of the USAAF, General Henry H. 'Hap' Arnold, who had been personally briefed on the programme's importance by the Army Chief of Staff, made it clear to Groves that if any atomic bombs were to be dropped in combat, a USAAF-crewed B-29 would deliver them. With that proviso firmly established, Arnold willingly endorsed the Manhattan Project's request for USAAF assistance.

On Saturday 23 August 1947, Roy Chadwick and Avro's chief pilot, Sidney Thorn, died in the crash of the Avro Tudor II. On investigation it was discovered that the aileron controls had been disconnected the night before and then wrongly reattached on the morning of the test flight. Any movement by the pilot would bank the aircraft in the opposite direction to the one intended. Sir Arthur Harris wrote to Roy Chadwick's daughter, Margaret, following her father's death:

Your father never received a tithe of the recognition and honours due from the nation for his services. The Lancaster

took the major part in winning the war with its attacks on Germany. On land, it forced the Germans to retrieve from their armies half their sorely needed anti-tank guns for use as anti-aircraft guns by over a million soldiers who would otherwise have been serving in the field. The Lancaster won the naval war by destroying over one third of the German submarines in their ports, together with hundreds of small naval craft and six of their largest warships. Above all, the Lancaster won the air war by taking the major part in forcing Germany to concentrate on building and using fighters to defend the Fatherland, thereby depriving their armies of essential air and particularly bomber support. But the Lancaster was Roy Chadwick and it was he who did all that for his country.

Above: Ex-British POWs file past Lancaster III PB935 F2-Z on 635 Squadron at Lübeck on 11 May 1945 during Operation Exodus. In the distance is a Gloster Meteor jet fighter. *(Imperial War Museum)*

Left: Four Lancasters on 300 Polish Squadron in formation over Lincolnshire in late spring/early summer 1945.

Appendix 1

BOMBER COMMAND LANCASTER CREW VICTORIA CROSS RECIPIENTS

	Squadron	Action	Award
Nettleton, Acting S/L John Dering, pilot	44	17.4.42	28.4.42
Gibson, Acting W/C Guy Penrose DSO* DFC* pilot	617	16/17.5.43	28.5.43
Reid, Acting F/L William RAFVR pilot	61	3/4.11.43	14.12.43
Cheshire, W/C Geoffrey Leonard DSO** DFC RAFVR pilot	617		8.9.44
Thompson, F/Sgt George RAFVR WOp	9	1.1.45	20.2.45§
Palmer, Acting S/L Robert Anthony Maurice DFC RAFVR pilot	109	23.12.44	23.4.45§
Swales, Captain Edwin DFC SAAF, 'Master Bomber'	582	23/24.2.45	24.4.45§
Bazalgette, Acting S/L Ian Willoughby DFC RAFVR 'Master Bomber'	635	4.8.44	17.8.45§
Jackson, Sgt (later W/O) Norman Cyril RAFVR flight engineer	106	26/27.4.44	26.10.45
Mynarski, P/O Andrew Charles RCAF mid-upper gunner	419	12/13.6.44	11.10.46§
			§ Posthumous award

Appendix 2

AIRCRAFT SORTIES AND CASUALTIES 3 SEPTEMBER 1939–7/8 MAY 1945

Aircraft Type	Sorties	Lost (% of sorties)	Operational Crashes (% of sorties)
Lancaster	156,192	3,431 (2.20)	246 (0.16)
Halifax	82,773	1,884 (2.28)	199 (0.24)
Wellington	47,409	1,386 (2.92)	341 (0.72)
Mosquito	39,795	260 (0.65)	50 (0.13)
Stirling	18,440	625 (3.39)	59 (0.32)
Hampden	16,541	424 (2.56)	209 (1.26)
Whitley	9,858	317 (3.22)	141 (1.43)
Manchester	1,269	64 (5.04)	12 (0.95)

Appendix 3

SUMMARY OF PRODUCTION

	Prototype	Mk.I	Mk.II	Mk.III	Mk.VII	Mk.X	Totals
A.V. Roe & Co. Ltd (Manchester area)	3	835		2,140			2,978
A.V. Roe & Co. Ltd (Yeadon area)		53		642			695
Armstrong Whitworth Aircraft		911	300	118			1,329
Austin Motors		150			180		330
Metropolitan-Vickers		941		139			1,080
Vickers-Armstrong (Castle Bromwich)		300					300
Vickers-Armstrong (Chester)		235					235
Victory Aircraft (Canada)						430	430
Total by Marks	3	3,425	300	3,039	180	430	7,377

(Mk.VI converted from Mk.III: 8)

Appendix 4

LANCASTER I SPECIFICATIONS

Dimensions: Span 102ft; tail-up length 69ft 6in; tail-up height 20ft 6in. Wing area 1,297sq.ft.

	Weight:	B.I Specials
Empty	37,000lb	35,500lb
Normal loaded	65,000lb	72,000lb
Max bomb load (w/o special mods)	18,000lb	22,000lb
Fuel (2,154gal)	15,509lb	1,675gal
Oil (150gal)	1,350lb	150 gal

Provision for 2 x 400gal overload tanks in bomb bay or fuselage.

Performance: Normal outbound, fully loaded, climb and cruise, 170mph up to 15,000ft then 160mph up to 22,000ft (normal ceiling). Max level 275mph, fully loaded at 15,000ft; 287mph empty; 245mph fully loaded at sea level. Max diving 360mph. After bombs gone: Normal cruising 200mph; normal stalling (50,000lb) 92mph. B.I specials 1,550 miles at 15,000ft at 200mph.

Range:
2,530 miles with 7,000lb bomb load
1,730 miles with 12,000lb bomb load
1,550 miles with 22,000lb bomb load

Engines:
Prototype Rolls-Royce Merlin X, 1,145hp
Mk.I Rolls-Royce Merlin XX, 22 or 24, 1,280hp/1,620hp
Mk.I (Special) Rolls-Royce Merlin 22 or 24, 1,280hp/1,620hp
Mk.II Bristol Hercules VI or XVI, 1,650hp
Mk.III Packard-Merlin 28, 1,300hp. Packard-Merlin 38, 1,390hp. Packard-Merlin 224, 1,640hp
Mk.VI Rolls-Royce Merlin 85, 1,750hp
Mk.VII Rolls-Royce Merlin 24, 1,620hp
Mk.X Packard-Merlin 28, 1,300hp. Packard-Merlin 38, 1,390hp. Packard-Merlin 224, 1,640hp

Armament (all turrets by Frazer Nash except for a few special installations such as the Rose-Rice rear turrets): Front turret – FN 5 2 x .303 Browning machine guns; mid-upper – FN 50 2 x .303 Browning machine guns; rear – FN 20 4 x .303 Browning machine guns; rear (later production) FN 121 4 x .303 Browning Mk.II machine guns; ventral (few a/c only) – FN 64 2 x .303 Browning Mk.II machine guns. Also some single .5 Browning guns fitted ventrally. B.II (FEs) fitted with FN 82 rear turret with 2 x .5 Browning machine guns.

Bomb load: Max 18,000lb (22,000lb on B.I (specials)). All Lancaster Is could carry 4,000lb bomb and from mid-1943 most could carry 8,000lb bomb. Later aircraft could carry a 12,000lb bomb (not to be confused with 12,000lb 'Tallboy').

Appendix 5

HALIFAX/LANCASTER COMPARISON AT THE END OF 1943

	Halifax	Lancaster
Number of sorties dispatched	12,382	19,338
Number of sorties attacking	11,080	17,923
Tons of bombs dropped	27,844	72,751
Number of aircraft missing	657	681
% of a/c attacking to sorties dispatched	89.5%	92.7%
% of a/c missing to sorties dispatched	5.3%	3.5%
Weight of bombs per attacking aircraft	5,635lb	9,070lb
Tons of bombs dropped for every a/c missing	42.58	102.05
Average sorties/month per 20 aircraft	96	112

Appendix 6

COMPARATIVE LANCASTER AND HALIFAX SQUADRONS

Operational Squadrons	Halifax	Lancaster
Sept 1942	10	10
Feb 1943	11	17
Aug 1943	15	23
Feb 1944	21	35½
June 1944	26	41
Dec 1944	27	58
May 1945	17	66

Appendix 7

LANCASTER SQUADRONS AT PEAK STRENGTH 1 AUGUST 1944

Squadron	Group	Location	First Lancaster Operation
7	8 PFF	Oakington, Cambridgeshire	8/9 July 1943 Cologne
9	5	Waddington/Bardney	10 September 1942
12	1	Wickenby, Lincolnshire	3/4 January 1943 'Gardening'
15	3	Mildenhall, Suffolk	14/15 January 1944 Brunswick
35 'Madras Presidency'	8	Graveley, Huntingdonshire	April 1944
44 'Rhodesia'	5	Waddington/Dunholme Lodge/Spilsby	3/4 March 1942 Heligoland Bight
49	5	Scampton/Fiskerton/Fulbeck/Syerston	30/31 May 1942
50	5	Swinderby/Skellingthorpe	May 1942
57	5	Scampton/East Kirkby	May 1942
61	5	Syerston/Skellingthorpe/Coningsby/Skellingthorpe	5/6 May 'Nickelling'
75 RNZAF	3	Mepal, Cambridgeshire	9/10 April 1944 Villeneuve-St-Georges
83	5	Scampton, Wyton, Coningsby	29/30 April 1942 Paris/Gennevilliers
90	3	Tuddenham, Suffolk	June 1944
97 'Straits Settlements'	5	Coningsby/Woodhall Spa/Bourn/Coningsby	20 March 1942 Ameland
100	1	Waltham/Elsham Wolds	4/5 March 1943 'Gardening'
101	1	Holme-on-Spalding Moor/Ludford Magna	20/21 November 1942 Turin
103	1	Elsham Wolds, Barnetby, Lincolnshire	21/22 November 1942 'Gardening'
106	5	Coningsby/Syerston/Metheringham	30/31 May 1942 Cologne
115	3	East Wretham/Little Snoring/Witchford	20/21 March 1943 'Gardening'

156	8 PFF	Warboys/Upwood	26/27 January 1943 Lorient
166	1	Kirmington, Lindsey, Lincolnshire	22/23 September 1943 Hanover
207	5	Bottesford/Spilsby	24/25 April 1942 Rostock
218 'Gold Coast'	3	Methwold/Chedburgh	September 1944
300 'Masovian'	1	Faldingworth, Lincolnshire	18/19 April 1944 Rouen
405 'Vancouver'	6 RCAF	Gransden Lodge/ Linton-on-Ouse	17/18 August 1943 Peenemünde
408 'Goose'	6 RCAF	Linton-on-Ouse, Yorkshire	7/8 October 1943 Stuttgart*
419 'Moose'	6 RCAF	Middleton St George, Co. Durham	27/28 April 1944 Friedrichshafen
428 'Ghost'	6 RCAF	Middleton St George, Co. Durham	July 1944
460 RAAF	1	Breighton/Binbrook, Lincolnshire	22/23 November 1942 Stuttgart
463 RAAF	5	Waddington, Lincolnshire	26/27 November 1943 Berlin
467 RAAF	5	Scampton/Bottesford/ Waddington	2/3 January 1943 'Gardening'
514	3	Foulsham/Waterbeach	3/4 September 1943 Düsseldorf
550	1	Waltham/North Killingholme	26/27 November 1943 Berlin
576	1	Elsham Wolds/Fiskerton	2/3 December 1943 Berlin
582	8 PFF	Little Staughton, St Neots, Huntingdonshire	9/10 April 1944 Lille
617	5	Scampton/Coningsby/ Woodhall Spa	16/17 May 1943 Dams raid
619	5	Woodhall Spa/Coningsby/ Dunholme Lodge/Strubby	11/12 June 1943 Düsseldorf
622	3	Mildenhall, Suffolk	14/15 January 1944 Brunswick
625	1	Kelstern/Scampton	18/19 October 1943 Hanover
626	1	Wickenby, Lincolnshire	10/11 November 1943 Modane
630	5	East Kirkby, Spilsby, Lincolnshire	18/19 December 1943 Berlin
635	8 PFF	Downham Market, Norfolk	22/23 March 1944 Frankfurt

* Operated the Lancaster from October 1943 to September 1944 when it converted back to Halifaxes.

(426 'Thunderbird' Squadron operated Lancasters in 6 Group RCAF at Linton-on-Ouse, June 1943 to May 1944)

(432 'Leaside' Squadron operated Lancasters in 6 Group RCAF at East Moor, October 1943 to February 1944)

Appendix 8

LANCASTER SQUADRONS FORMED LATE 1944–45

Squadron	Group	Location	First Lancaster Operation
149 'East India'	3	Methwold	17 September 1944 Boulogne
150	1	Fiskerton/Hemswell	November 1944
153	1	Kirmington/Scampton	7 October 1944 Emmerich
227	5	Bardney/Balderton/Strubby	11 October 1944 Walcheren
186	3	Tuddenham/Stradishall	18 October 1944 Bonn
170	1	Kelstern/Dunholme Lodge/ Hemswell	19/20 October 1944 Stuttgart
195	3	Witchford/Wratting Common	26 October 1944 Leverkusen
189	5	Bardney/Fulbeck/Bardney	1 November 1944 Homburg
150	1	Fiskerton	2 November 1944
431 'Iroquois'	6 RCAF	Croft	December 1944
424 'Tiger'	6 RCAF	Skipton-on-Swale	January 1945
434 'Bluenose'	6 RCAF	Croft	2/3 January 1945 Nuremberg
433 'Porcupine'	6 RCAF	Skipton-on-Swale	1/2 February 1945 Ludwigshafen
427 'Lion'	6 RCAF	Leeming	11 March 1945 Essen
138	3	Tuddenham	29 March 1945 Hallendorf
429 'Bison'	6 RCAF	Leeming	1 April 1945

NOTES

Chapter 1

1 *A WAAF In Bomber Command* by Pip Beck (Goodall Publications Ltd 1989).

2 *Lancaster: The Biography* by S/L Tony Iveson DFC and Brian Milton (André Deutsch 2009). Patrick Dorehill was awarded the DFC.

3 *Operation Millennium: 'Bomber' Harris's Raid on Cologne, May 1942* by Eric Taylor (Robert Hale 1997).

4 308 Wellingtons, 113 Lancasters, 70 Halifaxes, 61 Stirlings, 54 Hampdens and 24 Whitleys.

5 *A WAAF In Bomber Command* by Pip Beck (Goodall Publications Ltd 1989). The other Lancaster lost was on 106 Squadron at Coningsby. All the crew were killed.

6 By the end of the war the PFF had flown over 50,000 sorties with the loss of 3,700 air crew and 675 aircraft.

7 Fifteen Wellingtons, two Lancasters, a Halifax, a Hampden, a Whitley and a Stirling.

8 By the end of 1942 the U-boat menace was threatening Britain's Atlantic sea lanes and the 'Gardening' or aerial magnetic mine-laying campaign off the coastlines of the Third Reich was intensified and would last until the start of the Battle of the Ruhr in March 1943.

9 *Maximum Effort: The Big Bombing Raids* by Bernie Wyatt (The Boston Mills Press 1986).

10 Airborne Interception (AI).

11 The Stirlings were withdrawn from an original plan so only the higher-flying heavies would participate. Most of the force was from 5 Group. So far this was the largest number of Lancasters on one raid.

12 On the night of 16/17 May, W/C Guy Gibson, the CO of 617 Squadron, led 19 Lancasters to bomb the Ruhr dams. The Möhne and Eder dams were breached and the Sorpe was damaged. Eight Lancasters were lost. Of the 133 men who flew on the raid, 53 men were killed and three were captured. Barnes Wallis, the inventor of the Upkeep mine used on the raids, said it was 'the most amazing feat the RAF ever had or ever could perform'. Two days after the operation, Winston Churchill was given a standing ovation at the Trident Conference with Roosevelt in Washington. Gibson, who already had two DSOs and two DFCs, was awarded the VC for leading the 'Dam Busters' and many of the officers got DFCs and DSOs. Twenty-two veterans of the Dams raid were killed on ops later. Guy Gibson was killed on the operation on Rheydt on the night of 19/20 September 1944 when flying a Mosquito. He was controller for the raid. On 15 October, 18 Lancasters on 9 Squadron set out to attack the Sorpe dam and 16 aircraft dropped 'Tallboys' or other bombs from 15,000ft; hits were seen on the face of the earth dam but no breach was made. All aircraft returned safely.

Chapter 2

1 *A WAAF In Bomber Command* by Pip Beck (Goodall Publications Ltd 1989). On 29/30 July, P/O Cliff Shnier RCAF and his crew were killed.

2 *The Air War: 1939–1945* by Janusz Piekalkiewicz (Sudwest Verlag GmbH 1978).

3 *Q-Queenie* returned safely to Wyton with General Anderson, who also went to Hamburg two nights later with Garvey's crew in the same aircraft. *Claims To Fame: The Lancaster* by Norman Franks (Arms & Armour 1994).

4 On the night of 30/31 July, 273 aircraft were dispatched to bomb Remscheid on the southern edge of the Ruhr, which had not previously been bombed. The Oboe ground-marking and the bombing were exceptionally accurate and 83 per cent of the town was devastated, although only 871 tons of bombs were dropped. German defences were quickly overcoming the effects of 'Window' and 15 aircraft were shot down. A Lancaster crashed at Downham Market.

5 A total of 457 aircraft raided Mannheim. Bomber Command lost six Halifaxes and three Lancasters. DV198 was lost on tours on 10/11 April 1944.

6 F/Sgt Kirton was shot down over Mannheim on 5/6 September 1943 and was taken into captivity.

7 *Boots, Bikes & Bombers* by Eric Jones (unpublished manuscript).

8 *Barnes Wallis' Bombs: Tallboy, Dambuster & Grand Slam* by Stephen Flower (Tempus 2002).

9 Nuremberg was found to be free of cloud but it was very dark. The marking was to be mainly by 47 PFF H_2S aircraft that were to check their equipment beforehand by the dropping of a 1,000lb bomb on Heilbronn. At the target the initial PFF markers were accurate but a 'creep-back' quickly developed, which could not be stopped as only 28 aircraft were able to mark because so many Pathfinder aircraft were having difficulties with their H_2S sets. The Master Bomber could do little to persuade the Main Force to move their bombing forward as only a quarter of the crews could hear his broadcasts and it was estimated that most of the bombing fell in open country south-south-west of the city, but bombs were scattered across the south-eastern and eastern suburbs. Thirty-six aircraft failed to return, 11 of them Lancasters.

10 He had been granted his commission on the day of their second flight to Mannheim.

Chapter 3

1 *A Lancaster Pilot's Impression on Germany* by Richard 'Dick' Starkey (Compaid Graphics, Preston 1999 & 2004).

2 The mid-upper gunner who died was Sgt Gilbert George Provis, 27, of Ystrad Rhondda. On 22/23 March 1944, Manning and five of his crew were killed on the operation to Frankfurt. G/C Norman Charles 'Shorty' Pleasance, the Bardney station commander who accompanied the crew on the raid, was also killed.

3 Mullock was destined to go to Berlin on two further occasions during the battle, and eventually made five visits to the city out of his 22 ops. He was recommended for the DFC in August 1944.

4 Ted Ansfield was the sole survivor. He remained at large for four days before being captured and was later sent to *Stalag Luft I*.

5 *To Fly Lancasters* by Clive Roantree. Richardson and four of his crew were killed. Three others were injured.

6 'When the aircraft was examined next morning, the full extent of the damage was revealed. There was severe damage to the port fin and rudder, more than 50% was missing, and the port side of the fuselage had been riddled with bullets which stopped just before the wireless operator's position but had gone through the mid-upper gunner's legs. Material covering the port aileron had been ripped off and, most frightening of all, a cannon shell had exploded on the underside of the port mainplane creating a jagged hole approximately one foot in diameter. If the shell had exploded further forward it would have hit the fuel tanks and the aircraft would have "gone up".' *A Lancaster Pilot's Impression on Germany* by Richard 'Dick' Starkey (Compaid Graphics, Preston 1999 & 2004).

7 Twenty-year-old P/O Edward James Argent and four crew were killed in action on the night of 14/15 January 1944 over Brunswick.

8 Ray Meredith writing in *Thundering Through The Clear Air: No.61 (Lincoln Imp) Squadron at War* by Derek Brammer (Tucann Books 1997).

Chapter 4

1 The Main Force lost 16 heavies, 14 of them Lancasters. At Wickenby two crews on 12 Squadron failed to return. On 626 Squadron F/L W.N. Belford RAAF's aircraft was damaged over the target and ran out of fuel over the North Sea. The wireless operator managed to get off an accurate 'fix' for the rescue services and Belford managed to ditch. Coastal Command and air-sea rescue launches were sent out immediately and later that morning a dinghy containing Belford and his crew was spotted. The Coastal Command aircraft dropped a bigger dinghy for them and eventually they were picked up by a warship, HMS *Midge*. The crew were sent on leave when they returned to Wickenby. They were back in time to take part in a raid on Berlin

on 27/28 January and all except one of the crew was lost when their aircraft was shot down by a night-fighter.

2 Developed by Oberst Viktor von Lossberg of the Luftwaffe's Staff College in Berlin, Tame Boar (*Zahme Sau*) was a method used whereby the (Himmelbett) ground network, by giving a running commentary, directed its night-fighters to where the 'Window' concentration was at its most dense. Night-fighters were fed into the bomber stream (which was identified by H_2S transmissions) as early as possible, preferably on a reciprocal course. Crews then hunted on their own using SN-2 AI radar, which unlike early Lichtenstein AI could not be jammed by 'Window'; Naxos 7 (FuG 350) which homed into the H_2S navigation radar; and Flensburg (FuG 227/1) homing equipment. 'Long Window' made its appearance in July for jamming SN-2 radar (which previously was unaffected by 'Window').

Chapter 5

1 *A Lancaster Pilot's Impression on Germany* by Richard 'Dick' Starkey (Compaid Graphics, Preston 1999 & 2004).

2 Douetil landed heavily in a frozen field on the outskirts of Hanover and he subsequently became a POW in *Stalag Luft III* until it was overrun by the Russians in 1945. F/Sgt Fred Thomas Price was the only other crew member to survive. On 19 April 1945 he was gravely wounded when the POW column in which he was marching was attacked by RAF Typhoons. He died 11 days later.

3 F/L Crawford and his crew were all killed in action on the night of 15/16 March 1944 on the operation on Stuttgart.

4 'At some stage during our tour both Pat and Peto were taken ill and missed out on seven and five of our ops respectively. This meant that when we had finished they both had a few trips to complete their tour. On 22/23 March 1944, when the target was Frankfurt, they were making up the crew of F/Sgt R. Greig when his aircraft was shot down and all the crew were killed. It was Peto's 26th trip of his tour and it was Pat's 22nd. It was terrible for their luck to desert them at the last fence. Initially, they were both listed as missing and we were sure that they would eventually turn up. Some time later I received the news that they were dead and it was the end of all hope. JB421 *K-King* went down on her 37th operation with P/O A.L. Anderson's crew on the operation on Salbris on 7/8 May 1944. It was their first op and they were all killed.'

5 *A Lancaster Pilot's Impression on Germany* by Richard 'Dick' Starkey (Compaid Graphics, Preston 1999 & 2004).

6 *A Lancaster Pilot's Impression on Germany* by Richard 'Dick' Starkey (Compaid Graphics, Preston 1999 & 2004). The only Lancaster lost this night was *R-Robert* on 207 Squadron, which crashed in the target area at Clermont-Ferrand. S/L Dudley George Hart Pike MiD and his crew were all killed.

7 At another explosives factory at Angoulême on the night of 20/21 March, 20 Lancasters of 5 Group, including 14 on 617 Squadron,

successfully bombed the Poudrerie Nationale after Leonard Cheshire put his spot fires in the centre.

8   Between 8 February and 10 April 1944, the dozen French targets allocated to Bomber Command were destroyed or very seriously damaged. Nine of these targets were wiped out on independent 617 Squadron operations.

9   *A Lancaster Pilot's Impression on Germany* by Richard 'Dick' Starkey (Compaid Graphics, Preston 1999 & 2004).

10 F/L Tom Blackham DFC was shot down on Mailly-le-Camp on 4/5 May. He evaded for a time, was picked up and later incarcerated in the notorious Buchenwald concentration camp. P/O Stewart Godfrey was assisted by Mme Deguilly of Romilly-sur-Seine before being passed to a Resistance Group. He died when the Wehrmacht attacked their camp on 24 June. He has no known grave. Walton, Jones, Wilkins, Ridd and Sgt William Dennis Dixon were killed. So too was P/O Cyril Edward Stephensen RAAF, the 'second dickey'. *RAF Bomber Command Losses of the Second World War* Vol.5 1944 by W.R. Chorley (Midland 1997).

11 Dick Starkey's crew were nearing the end of their tour and were keen to do a second tour of 20 trips on 617 Squadron but *Q-Queenie* was one of eight bombers shot down by Oberleutnant Martin 'Tino' Becker on Nuremberg on 30/31 March. Dick Starkey, who was blown out of the aircraft and badly injured when it exploded, left a 2½ft depression on the ground when he landed on his back, and F/Sgt Wally Paris, who escaped from the aircraft before it exploded, were both taken prisoner. Sgt Colin Roberts, Sgt George Walker, Sgt 'Jock' Jamieson and Sgt Joe Ellick were killed. Sgt Johnny Harris, the flight engineer, who was from Biggleswade, was also blown out of the aircraft but his parachute did not open and he was found in a wood 6km from the wreckage of the aircraft. In 1998 Dick Starkey met 83-year-old 'Tino' Becker at his house in Limbourg. The German was ill and frail and the English pilot left with a sadness because he knew he would not see him again. 'But what a pilot he must have been in his career with the Luftwaffe which he joined in 1936.' *A Lancaster Pilot's Impression on Germany* by Richard 'Dick' Starkey (Compaid Graphics, Preston 1999 & 2004).

12 *Thundering Through The Clear Air: No.61 (Lincoln Imp) Squadron at War* by Derek Brammer (Tucann Books 1997).

13 F/L Allan Pluis Whitford and his crew were all killed in action.

14 *Thundering Through The Clear Air: No.61 (Lincoln Imp) Squadron at War* by Derek Brammer (Tucann Books 1997).

15 *S-Sugar* completed its 137th and final operational bombing sortie on 23 April 1945 to Flensburg. It is now on permanent display at the RAF Museum, Hendon, London.

16 *Maximum Effort: The Big Bombing Raids* by Bernie Wyatt (The Boston Mills Press 1986).

Chapter 6

1   Foggo landed safely and was taken prisoner.

2   Brophy, 'Art' de Breyne, F/O Robert Body RCAF the navigator and W/O W.J. 'Jim' Kelly RCAF all evaded capture. Jack Friday and Sgt R.E. Vigars, the flight engineer, became POWs. Altogether, 23 aircraft – 17 of them Halifaxes and six Lancasters – were lost on the raids on communication targets this night and 17 Lancasters failed to return from the operation by just over 300 aircraft to the Nordstern synthetic-oil plant at Gelsenkirchen.

3   The Nôtre-Dame district near the port was devastated but fortunately the inhabitants had been evacuated long before the raid. Other districts were also hit and 700 houses were destroyed, 76 civilians were killed and 150 injured. The following evening a similar operation was mounted against the E-boat pens in Boulogne harbour and, according to one account, over 130 E-boats were sunk along with 13 other vessels wrecked or badly damaged by a tidal wave. This and the operation to Le Havre forced the remnants of the E-boat flotillas to a new base at IJmuiden in Holland, which was bombed by 617 Squadron on 24 August.

4   All of Jeffrey's crew were killed in action.

5   Guilfoyle and two of his crew were taken into captivity. The other four evaded.

6   *Barnes Wallis' Bombs: Tallboy, Dambuster & Grand Slam* by Stephen Flower (Tempus 2002).

7   *Barnes Wallis' Bombs: Tallboy, Dambuster & Grand Slam* by Stephen Flower (Tempus 2002). Cheshire then resumed his previous rank of group captain and was awarded the VC two months later.

8   Together with 64 Halifaxes and 33 Lancasters – all in 6 Group. Four of the targets were marked by 8 Group Mosquitoes and Lancasters and at the target where Oboe failed, the Master Bomber and other Pathfinder crews used visual methods. 1 and 4 Groups, each with more than 100 aircraft, attacked the area around Sannerville; Mondeville was attacked by 3, 5 and 6 Groups; 1, 3 and 4 Groups attacked Sannerville; 3 and 8 Groups attacked Cagny.

9   Sgt Pialucha was later posted to 1586 (Polish) (Special Duties) Squadron. On the night of 1 September 1944, his Halifax was hit by flak over Sombor and crashed south-east of Belgrade. Only the navigator baled out and survived.

10 Jack West spent ten days having shrapnel removed from his body. The Lancaster had sustained 450 bullet holes and over 100 shrapnel holes from the enemy defences. The crew decided to keep quiet about the bomb aimer, though they indicated that they would rather not fly with him. *My Life In the RAF* by Jack West DFM; private memoir, RAF Museum X001-6422.

11 *Maximum Effort: The Big Bombing Raids* by Bernie Wyatt (The Boston Mills Press 1986).

12 See *Barnes Wallis' Bombs: Tallboy, Dambuster & Grand Slam* by Stephen Flower (Tempus 2002) and *Legend of the Lancasters* by Martin W. Bowman (Pen & Sword 2009).

13 See *Barnes Wallis' Bombs: Tallboy, Dambuster & Grand Slam* by Stephen Flower (Tempus 2002).

14 *Luck and a Lancaster: Chance and Survival in World War II* by Harry Yates DFC (Airlife 1999).

15 See *Barnes Wallis' Bombs: Tallboy, Dambuster & Grand Slam* by Stephen Flower (Tempus 2002).

16 The Lancaster crashed while attempting to land at Aston Down airfield in Gloucestershire. F/L Gordon Court Owens and four crew were killed.

17 *Luck and a Lancaster: Chance and Survival in World War II* by Harry Yates DFC (Airlife 1999).

18 On 15 August, W/O Clayton Moore was promoted to pilot officer.

Chapter 7

1 *Rio Rita* and F/O Don Aitkin's crew ditched in the River Orwell on 6/7 December 1944, returning from Leuna.

2 *Luck and a Lancaster: Chance and Survival in World War II* by Harry Yates DFC (Airlife 1999).

3 *Maximum Effort: One Group At War* by Patrick Otter (Grimsby *Evening Telegraph* 1990).

4 *Maximum Effort: The Story of the North Lincolnshire Bombers* by Patrick Otter (Grimsby *Evening Telegraph* 1990).

5 *Luck and a Lancaster: Chance and Survival in World War II* by Harry Yates DFC (Airlife 1999).

6 *Luck and a Lancaster: Chance and Survival in World War II* by Harry Yates DFC (Airlife 1999).

7 *Luck and a Lancaster: Chance and Survival in World War II* by Harry Yates DFC (Airlife 1999). The raid on Koblenz was successful. Only two Lancasters failed to return.

8 *Luck and a Lancaster: Chance and Survival in World War II* by Harry Yates DFC (Airlife 1999). All three towns were virtually destroyed and four Lancasters were lost, but the American advance was slow and costly.

9 *Into the Silk* by Ian Mackersey (Granada 1978).

10 *Luck and a Lancaster: Chance and Survival in World War II* by Harry Yates DFC (Airlife 1999). All aircraft returned safely from the raid on Neuss.

11 *Luck and a Lancaster: Chance and Survival in World War II* by Harry Yates DFC (Airlife 1999). The vic Harry Yates saw go up in flames claimed the lives of two, not three, crews. They were F/O James Alexander McIntosh RNZAF's crew on NF980 *F-Freddie* and seven

on PD367 on 115 Squadron skippered by F/L Frederick George Holloway DFM. The third Lancaster, skippered by F/O J. McDonald on 75 Squadron, was heavily damaged but landed back at Mepal at 1459hrs. All 60 Lancasters of 3 Group returned safely from Bottrop.

12 *Luck and a Lancaster: Chance and Survival in World War II* by Harry Yates DFC (Airlife 1999).

13 See *Luck and a Lancaster: Chance and Survival in World War II* by Harry Yates DFC (Airlife 1999).

14 *Barnes Wallis' Bombs: Tallboy, Dambuster & Grand Slam* by Stephen Flower (Tempus 2002).

15 P/O Richard Justin Aitchison and his crew were killed in action on their first op, on Vohwinkel, on 1 January 1945 flying *N-Nan*. W/C Ray J. Newton DFC MiD RNZAF, the CO, was also killed. F/L Tim Blewett RNZAF was killed on 16/17 January 1945 when his Lancaster crashed at Wood Ditton, Cambridgeshire, returning from Wanne-Eickel.

Chapter 8

1 *Maximum Effort: The Big Bombing Raids* by Bernie Wyatt (The Boston Mills Press 1986).

2 One Lancaster was lost on Dortmund.

3 *Nachtjagd: The Night Fighter Versus Bomber War Over the Third Reich 1939–45* by Theo Boiten (The Crowood Press 1997).

4 292 Halifaxes, 156 Lancasters and 16 Mosquitoes.

5 In another operation, 177 Lancasters and 11 Mosquitoes of 5 Group attacked a section of the Dortmund-Ems Canal near Ladbergen with delayed-action bombs. Later photographs showed that the banks had not been damaged; the bombs had fallen into nearby fields. Three Lancasters were lost.

6 Over 300 B-17s of the 8th Air Force dropped 771 tons of bombs on Dresden on 14 February.

7 *Nachtjagd: The Night Fighter Versus Bomber War Over the Third Reich 1939–45* by Theo Boiten (The Crowood Press 1997).

8 Crew members who died on Lancaster RF121 QR-J were: F/L Greenfield, P/O W.J.A. Gibb, F/Sgt W.J. Haddon, F/Sgt J.R. King and W/O V.P. Smith. The only crew members to survive the jet fighter's attack were the flight engineer and rear gunner.

9 *Thundering Through The Clear Air: No.61 (Lincoln Imp) Squadron at War* by Derek Brammer (Tucann Books 1997).

10 *Bombs Gone! An Elvington Lad's War* by Dickie Parfitt (Riverdale Publications 2001), edited by Derek Leach.

11 Article on Operation Manna, Nanton Lancaster Society Air Museum.

INDEX

Aachen 118

Abbott, Roy 162, 168, 172, 179

Achères 121–2

Ad Extremum 'Press On Regardless' 173

Admiral Hipper 199

Admiral Scheer 199

Aire-sur-la-Lys 149

Allen, Jack 186

Anderson, General Fred L. 33, 208

Annecy 116

Ansfield, P/O A.S. 'Ted' 67–8

Antheor viaduct 89

Antwerp 118, 165

Ardbegen railway bridge 195

Arnold, General Henry H. 'Hap' 202

Arnsberg viaduct 188, 195

Ashaffenburg 174

Aubigne 114

Augsburg 86, 96–8

Augsburg raid 11–3, 17, 19

Aulnoye 137

A² Aussie 69

Austerbury, F/Sgt Francis 171

Balderton 189

Bardney 30, 32, 38–9, 42–4, 54, 56, 61, 64–5, 72, 76, 80, 182, 209

Barmen 27–8, 194

Barron, W/C John Fraser DSO DFC DFM 111, 117

Baxter, W/C R. E. 99

Bayldon, P/O Richard A. 72

Bayreuth 199

Beck, Pip 9–11, 28

Beeson, P/O Stanley 159–60

Belford, F/L W.N. RAAF 209

Bell, F/O J. 186

Bellicose, Operation 29

Benbow, F/Sgt Doug 'Ben' 184–6

Benjamin, W/C Eric Arthur DFC* 190

Bennett, AVM Donald 18

Bennett, Sgt Robert 167

Bennett, W/C Jimmy 72–3

Berchtesgaden 196–7, 200

Bergkamen 191

Berlin 23–5, 33, 45, 47, 49–51, 53–5, 65–7, 69–72, 76–87, 89, 91–4, 100–5, 107, 112, 148–9, 188–9, 200

Berneval 114

Beswick, Sgt Ted 198–9

Bielefeld viaduct 188

Binbrook 58, 77, 88–9, 99, 110–1, 155–6, 184

Blacker, F/O George 185

Blackham, F/L Thomas 87–8

Blagnac airfield 114

Blainville 132

Blaye 147

Bochum 26, 28–9, 58, 173, 184, 196, 216

Böhlen 190, 193–4, 196

Bolougne 117–8, 210

Bonn 178–9, 188

Bonnett, F/L Dorian Dick 20

Bois de Cassan 147

Boscombe Down 187

Bottesford 26

Bottrop 139, 197

Bottrop-Homberg 140

Bourg-Leopold 114, 116, 118, 174

Breaden, P/O Geoffrey 85

Bremen 16–9, 24, 59, 84, 158, 171, 195–6

Brest 30, 114–5, 147, 151, 158, 161

Brophy, F/O George P. RCAF 126, 210

Brown, Captain Eric 'Winkle' 182

Brüx 184–5

Brunswick 58, 84, 86–7, 113, 118, 152, 154–6, 173, 209

Bruntingthorpe 45

Brydon, F/L Knute 92

Burke, P/O Robert William DFC RAAF 93

Burnett, P/O Walter Henry 108

Caen 117, 121, 126, 132, 134–5, 137, 157

Calder S/L C. C. 188

Campbell, F/L Alex RCAF 117, 143–6

Campbell, F/Sgt S.E. 86

Canday, Sgt C.A. 128

Carbutt, P/O Dennis 100

Carroll, F/O Dickie 123

Castrop-Rauxel 162

Chadwick, Margaret 202

Chadwick, Roy 10, 202–3

Châteauroux 99–100

Châtellerault 126, 149

Chemnitz 188–90, 193

Cheshire, W/C Leonard 89, 99, 111, 113, 126, 131–2, 134, 210

Churchill, Winston 10, 154, 190, 208

Clark, F/L J.H. 77

Clermont-Ferrand 99, 101, 209

Cochrane, AVM Sir Ralph A. 107

Cole, F/O Vic 81–2, 103

Cologne 11, 13, 15–9, 28, 30–2, 63, 76, 108–9, 112, 127, 160, 162–3, 169–72, 178–9, 181, 184, 192

Coningsby 10, 14, 99, 107–9, 111, 121, 127, 143, 151–2, 155, 208

Corry, S/L Noel 'Paddy' 133, 141

Corsica 89, 129

Coulombe, W/O J.A.R. 70–1

Courtrai 116, 139–40

Couzins, Bill 30–1

Cox, F/O A. 201

Cox, F/O Sam 191

Crawford, F/L Jack S.G. 95

Currie, Jack 45

Daniels, W/C Sidney Patrick 'Pat' Daniels DSO DFC* 21

Danzig 110

Darmstadt 58, 94, 159, 163

Datteln 194

Day, F/L Roy 190–1

De Breyne, P/O Arthur 125, 210

De Jong, Arie 201

Dear, Reg 116

Dennis, S/L John Mervyn DSO DFC 117

Denton, F/O Harry RNZAF 182

Dessau 158–9, 185, 193

Deverill, Ernie 13

Dieppe 116

Dierkes, P/O W.J. 13

Dimbleby, Richard 23, 188

Dixon, Sgt Joe 'Dixie' 194

Docking 11

Donges 140–1

Dorhill, Patrick 12–3, 26, 182–3

Dortmund 13, 26–7, 58, 63, 86, 89, 118, 162, 171, 174, 176, 182–3, 187, 190, 192, 197, 211

Dortmund-Ems Canal 58, 164, 173–4, 182, 191–2, 211

Doubleday, S/L Arthur W. 108

Douetil, S/L Barry 94–5

Dresden 94, 188–90, 211

Duisburg 17–8, 22, 25–6, 118, 161–2, 167–8, 171, 176, 178, 184–5, 190

Duisburg-Ruhrort 22, 26

Dunholme Lodge 28, 31, 54, 64, 125, 129, 137

Dunstan, P/O Roberts 77

Düsseldorf 13, 17–8, 23, 27–8, 62–3, 65, 77, 112–3, 141, 147, 171–3, 175, 179, 190

East Kirkby 94, 129, 184

Edith 196

Egan, F/Sgt Alan E. RAAF 29–30

Eindhoven 118

El Alamein 19

Elberfeld 87

Elsham Wolds 29, 49, 53, 56–7, 59, 80, 91, 95, 104, 127, 150, 158–60, 163, 194, 201

Emden 199

Essen 11, 19, 22–3, 25, 27, 106, 168, 177, 194

Etchell, F/L R.M. 158

Exeter 19

Exodus, Operation 201–2

Eye 184

Falaise 157

Faldingworth 131, 137

Falgate, Donald 137

Fallersleben 22

Fauquier, W/C 'Johnny' 47, 196

Fawke, Gerry 131

FIDO 68–9, 80, 178

'Fishpond' 38, 40, 101

Fiskerton 45, 47, 54, 63, 72, 178, 200

Flensburg 202

Flushing 165–6, 168, 170–1

Frankfurt-am-Main 18, 21, 25, 45, 59, 61, 67–8, 71, 74–5, 86, 101–2, 109, 163–4, 196, 209

Frankfurt on Oder 90

Frazier, Sgt Ben 80, 91

Freeman, P/O Denny 109

Freiburg 175

Friday, Sgt Jack 125

Friedrichshafen 29, 113

Friend, F/O Arthur 'Bull' 15–6

Fudge, Sgt Cliff 89

Fulbeck 174, 187, 194

Garner, W/C Don 122

Garvey, F/L 'Rick' 33

Garwell, F/O John 'Ginger' DFM 12

Gaston, F/Sgt J.M. RCAF 140

Gee, John 188

Gelsenkirchen 29, 31, 121, 124–5, 137–9, 164, 171, 184, 188, 191–2, 194–5, 210

Gennevilliers 116

Genoa 19–20, 43

Ghent 116

Gibson, W/C Guy 14, 21, 23, 208

Gill, F/Sgt Robert M.J. DFM 131

Givors 155

Glimmer, Operation 120

Goering, Reichsmarschall Hermann 154

Golub, F/O Matthew M. RCAF 128, 134

Gomm, W/C Cosme 29

Gomorrah, Operation 32–9

Goulevitch, F/Sgt Jack 123

Grant-Dalton, S/L Hugh 105

Graveley 18, 21, 68, 74, 80

Gravenhorst 190

Greenfield, F/L Paul DFC 198

Grieg, Captain Nordahl 71

Grimsby 45, 74, 87, 123, 145, 158, 161, 167–8, 186

Hagen 58, 176

Haine-St-Pierre 114

Ham, F/O H.D. 52

Hamburg 11, 13, 17, 20, 23–4, 28, 30, 32–45, 85, 94, 114, 143, 146, 189–90, 193–4, 196–8, 208

Hamburg, Battle of 32–43, 143, 189

Hanau 195

Hanover 45, 58–9, 82, 94, 104, 183, 195, 197, 209

Harburg 193–4, 197–8

Harris, ACM Sir Arthur 9–10, 15–7, 23–4, 27, 31–3, 93, 102, 164, 190, 202

Harrison, Herbert 181

Hartshorn, First Lt Joseph DFC 156

Haslett, F/Sgt Eric 135

Hasselt 116

Hattingen 194–5

Hay, F/L 'Bob' 89

Healy, P/O F.G. 21

Hemmingstedt 193, 196

Henderson, J.W. 25

Henry, David 163

Henry, F/O Colin 161–2, 167

Henry, Gavin 163

Henry, John 163

Herne 194

Herrmann, Oberst Hans-Joachim 30

Higgs, W/O A.J. 112–3

Higman, Sgt N.D. 52–3

Hiroshima 202

Holford, W/C David 73–4

Holme-on-Spalding Moor 20, 26, 28

Homberg 139–40, 160, 168, 172, 174, 194

Howlett, N.L. 177

Howlett's 'Hooligans' 177

Hulance, F/L Freddy 189–90

Hutcheon, F/O Joe C. 145

Hydra, Operation 44–5

Ingram, S/L George F.H. DFC 131

Jarratt, George 156

J-Joe 28

Johnny Walker/Still Going Strong! 88

Johnson, F/O Harold 'Johnny' 81–3

Johnson, Peter 125

Johnston, Sgt J.S. 'Johnny' 29–30, 103

Jones, Sgt Eric 45–51, 58, 60–1, 63–4, 66, 69, 72–5, 83–4, 95–6

Juvisy 110–1, 122

Kamen 191–2

Kammhuber, Generalmajor Josef 10, 22, 32

Karlsruhe 113, 164, 176, 181, 187

Kassel 58, 77, 194

Kelso, F/Sgt Harry 'Paddy' 186–7

Kelstern 73, 94, 178, 186, 188

Kiel 19, 25, 107, 111, 140, 158–9, 164, 171, 193, 199–200

Kirmington 67, 74, 83, 89, 98–9, 110, 102–3, 122, 181
Kleve 188
Knapsack 160
Koblenz 162, 173, 179–80, 211
Konigsberg 159, 161
Krefeld 29–31, 162, 171, 184, 188

La Chapelle 111
La Délivrance 110
La Pallice 149, 151, 159
La Spezia 23, 25–6, 29
Ladbergen 182, 191–2, 211
Langenreer 195
Langston, Sgt John 94
Laon 110
Le Clipon 117
Le Creusot 19–20, 28
Le Havre 126, 145–6, 149, 161–2, 170, 210
Le Mans 117
Leeming 179
Leipzig 58–9, 61, 71, 77, 85–6, 94–6, 108, 160, 181, 184, 188–90, 194, 196, 198–9
Lens 116, 126, 151
Leopoldsburg 116
Letford, F/L Ken 53
Leuna 176, 198
Leverkusen 66
Lille 110, 116
Linton-on-Ouse 86, 107–8
Lisieux Conde 121
Little Staughton 179
Lloyd, Sgt T.A. 116
Lloyd-Davies, F/L 194
Lorient 23–4
Lossberg, Oberst Viktor von 209
Louvain 116
Lübeck 203
Lucheux 150
Ludford Magna 28, 107, 142
Ludwigshafen 54, 65, 178, 182–3, 186
Lutzkendorf 194, 198
Lutzow 200

Maddern, Sgt Geoff 31
Magdeburg 85, 185
Mahoney, Sgt M.O. 'Spud' 69–70
Mailly-le-Camp 113–4
MAN works 11–3

Manhattan Project 202
Manna, Operation 198
Mannheim 44, 54, 58, 187, 192
Manston 95, 108, 181–2
Mantes-la-Jolie 114
Marco, F/O Wilfred T. 201
Martin, S/L Mick 89
Matthews, Sgt Bill 128, 134
McLennan, F/O Burus A. RCAF 74
McNamara's Band 142
Mepal 140, 147, 150, 169, 174, 179
Meredith, LAC Ray 78–9
Merseburg 176
Merville 117
Metheringham 135, 186
Methwold 161
Metz 132
Milan 24, 29, 43–4
Mildenhall 128, 134, 198
Miller, Percy 114
Miller, Sgt Stanley 71, 74
Minnie the Moocher 133, 135
Misburg 195
Mittelland Canal 173, 190–1
Modane 63
Möhne Dam 21
Mönchengladbach 51
Moore, Sgt Clayton C. 33, 35–44, 54–7, 61–2, 64–5, 75–7, 79–80, 83–5, 90–2, 100–2, 105–7, 109, 111–2, 115–6, 120–1, 123, 127–9, 143, 151–3, 162–5
Morgan, F/O Gomer S. 'Taff' 80, 91
Morrison, W/C H.A. RCAF 161
Morrison, F/L Geoffrey Arnold DSO 86, 89
Morsalines 114
Moss, S/L 'Ted' 108–9
Muirhead, Campbell 118–25, 130–5, 137–43, 147–9, 151–2, 154–6
Mulheim 29, 31
Mullock, Major John 67, 209
Munich 22, 25, 58, 175, 178, 183–4
Munro, S/L Les 99
Münster 197
Murrow, Ed 71
Mycock, W/O Tommy 13
Mynarski, P/O Andrew VC 126

Nagasaki 202
Nairne, F/Sgt Colin George RNZAF 147

Nancy 135
Nantes 118
Naughty Nan! 201
Nettleton, S/L John Dering 10, 12–3, 31
Neuss 183
Nienburg 196
Nippes 178
Noisy-le-Sec 110
North Killingholme 89, 107, 133, 146, 173, 176, 186, 201
Nulli Secundus 57
Nuremberg 24–5, 44, 47, 107–9, 182–9, 199, 208
Nutting, WO2 S.H. DFM RCAF 74

Oakington 18, 67
Ollie's Bus 153
Ollis, F/O Hal RCAF 153
Oomph Gal 124
Orléans 119, 133
Oslo fjord 178, 200
Osnabrück 176
Osterfeld 177

Paderborn 197
Palmer, S/L Robert A.M. VC DFC* 179
Parfitt, Dickie 200–1
Pas-de-Calais 131
Peenemünde 44–5, 52
Penman, F/L David 11–3
Penn, Sgt William Harvey RCAF 22
Penrose, F/Sgt 'Penny' 162, 167–8
Pforzheim 191
Phantom of the Ruhr 145–6
Phyllis Dixey 130
Pialucha, Sgt Jozef 137
Piccadilly Princess 185
Pillau 110
Pilzen 26
Plauen 199
Pleasance, F/O Donald 122
Pleasance, G/C Norman Charles 209
Pölitz 178, 184, 188
Portal, Sir Charles, Marshal of the RAF 10
Princess Patricia 200
Pritchard, F/Sgt Frank 193
Purves, F/O Bob 192

Rackley, P/O Lionel N. 'Blue' RAAF 129–30
Ramsey, Dr Norman 202

Recklinghausen 195
Reid, F/L Bill VC 63, 147
Remscheid 208
Rennes 122
Revigny 137
Rhein-Prussen 160
Rheydt 51
Rhodes, Sgt George 11
Rilly-la-Montagne 146
Rix, Sgt Peter 18
Roantree, F/Sgt Clive 69
Robinson Geoff 189
Rokeby, F/O 156
Rokker, Oberleutnant Heinz 149
Rostock 13–4
Rotterdam 201
Rouen 110–1
Royan 183
Ruhr Express 182–3
Ruhr Rover, The 87
Rüsselsheim 159
Ryan, Jack RAAF 136–7

Sablé-sur-Sarthe 114
Salzgitter 197
Sanders, John 97–8
Santa Azucar 194
Sardinia 89
Sarll, Peter 193, 201
Saumur railway tunnel 122
Saunders, F/L A.J. RAAF 155
Scampton 17, 19–20, 26, 153, 188
Schall, Hauptmann Franz 199
Scholefield, S/L T.N. 114, 117
Scholven-Buer 127, 129, 137, 179, 194
Schräge Musik 44, 82, 102, 105, 107, 109, 131, 181
Schweinfurt 45, 86, 95–6, 107–8, 113
Searby, W/C John 43–4, 51
Shannon, S/L Dave 132
Shaw, F/L David 173
Sherwood, S/L John 11–2
Shield, Operation 202
Shipdham 63
Sholto-Douglas, William 32
Siddle, Sgt William 'Bill' 30, 32–3, 43–4, 54–5, 61–2, 64, 90, 101, 115, 121, 151, 153, 164
Skellingthorpe 22, 71, 78–9, 116, 191, 198
Skipton-on-Swale 185
Slee, W/C Leonard 'Slosher' 19, 29, 67

Slessor, John 10
Spirit of Russia 79
Sri Gajah/Jill 62
Stark, F/O Roussel 'Russ' RAAF 194
Starkey, Dick 59–60, 74–5, 94–7, 99–100, 102–4, 210
Stedman, P/O J.R. 'Mike' 127
Stepian, F/Sgt Z. 137
Sterkrade 174, 197
Sterkrade/Holten 126
Stettin 26, 44, 83–4, 102, 158, 161, 178, 184
St-Leu-d'Esserent 134–5
Stockdale, W/C Mike 148
Stockton, Norman 71
Stones, Ted 107, 201
Storms, F/L J.L. 197
Sturgate 178
Stuttgart 15, 96, 99, 141–3, 146, 164, 168, 185
Sumak, F/O Len DFM RCAF 86
Swales, Captain Edwin VC DFC SAAF 191
Swinemünde 111, 200
Syerston 22, 59

Tait, W/C James 'Willie' 47, 134
Tangmere 161
Taxable, Operation 120
Tergnier 110, 119
Tetley, F/Sgt Norman 18
Thiverny 147
Thomas, W/O J.E. DFC 67
Thompson, F/Sgt George VC 182
Thompson, Gwen 26
Thorn, Sidney 202
Thunderclap, Operation 189, 193
Tirpitz 10, 169–70
Topham, F/Sgt 117
Topliss F/L H. 149
Toronto 202
Toulouse 114
Tours 110–1, 114, 117, 122, 135
Trier 178–80
Trossy-St-Maximin 147
Tuddenham 171, 195
Tudor, Avro 202
Turin 13, 20-23, 31, 43–4
Tutty, F/O E.B. 122

Upwood 112

Vairés 132, 135
Valenciennes 126
Vallance, F/O 187
Vaughan, F/L W.E. 22
Vaughan-Thomas, Wynford 53–4
Vergeltungswaffe 87
Vernieuwe, P/O 'Selmo' 158–60
Vernon, F/O H.A. 118, 124, 130, 133, 139–40
Vicious Virgin 192
Victorious Virgin 186
'Village Inn' 90
Villeneuve 110, 119, 133, 137
Villers-Bocage 146
Vlotho 195
Vohwinkel 162, 179–80, 182, 211
Vulture Strikes!, The 184–5

Waddington 9, 11, 13–4, 18–20, 22, 28, 108, 117, 129, 174, 184
Walcheren 165, 170–1
Walker, David 15
Wallis, Dr Barnes 122, 126, 208
Walmsley, F/L 93
Waltham 30–1, 72–3, 84, 87, 105, 145, 149
Wangerooge 200
Wanne-Eickel 142, 160, 174, 184, 187–8, 211
Warboys 18
Warner, F/L A.G.R. RCAF 183
Waterbeach 117, 143, 147, 151, 169
Watten 126–7
We Dood It Too 85
Weir, G/C C.T. DFC 174–5
Wesel 190, 193, 197
Wesseling 125, 127, 129, 137, 171
West, Sgt Jack 140, 157–8
Whamond, F/L Bill 21
Wharton, F/L Ernie 97
Whitehead, Sgt Len 107
Whitmarsh, F/O Arthur 'Whitty' 182–3
Wickenby 26, 28, 45, 70, 118, 122, 133, 135, 139, 141, 148, 154, 161, 167–8, 178, 181, 187, 189, 209
Wickman, John 107–8
Wiesbaden 187
Wilde Sau 30
Wilhelmshaven 19, 24, 164
Willoughby, LAC 114
Wilson, Sgt J. RCAF 22
'Window' 17, 32, 43

Wing 105
Winthorpe 45
Wismar 19
Witchford 105, 122, 140, 151
Witten 177, 195
Wizernes 127, 131–2, 134, 139
Woodbridge 86, 89, 107, 129, 140, 160, 169, 176, 186
Woodhall Spa 10–3, 19, 22, 89, 132

Woodhead, Sgt George 112–3
Woodley, Sgt Frank 176–7
Woodroffe, W/C John 178
Wooldridge, F/O John 14
Worms 190
Wuppertal 27–9, 194
Würzburg 195
Wyatt, W/C Mike 143
Wyton 18, 90

Yates, F/O Harry 150–1, 157, 161, 169–72, 174–7, 179–81, 211

Zentar, F/Sgt M. 137
Zweibrücken 194
Z-Zombie 108–9